Tissue Engineering: Beyond the Basics

Tissue Engineering: Beyond the Basics

Edited by Heidi Lee

Syrawood
PUBLISHING HOUSE

New York

Published by Syrawood Publishing House,
750 Third Avenue, 9ᵗʰ Floor,
New York, NY 10017, USA
www.syrawoodpublishinghouse.com

Tissue Engineering: Beyond the Basics
Edited by Heidi Lee

International Standard Book Number: 978-1-64740-090-3 (Hardback)

Cataloging-in-Publication Data

Tissue engineering : beyond the basics / edited by Heidi Lee.
    p. cm.
Includes bibliographical references and index.
ISBN 978-1-64740-090-3
1. Tissue engineering. 2. Tissue culture. 3. Regenerative medicine. 4. Biomedical engineering. I. Lee, Heidi.
R857.T55 T57 2022
612.028--dc23

TABLE OF CONTENTS

PREFACE

I am honored to present to you this unique book which encompasses the most up-to-date data in the field. I was extremely pleased to get this opportunity of editing the work of experts from across the globe. I have also written papers in this field and researched the various aspects revolving around the progress of the discipline. I have tried to unify my knowledge along with that of stalwarts from every corner of the world, to produce a text which not only benefits the readers but also facilitates the growth of the field.

Tissue engineering refers to the use of science and technology to improve or replace biological tissues. It is used in medicine for the repair or replacement of bone, cartilage, skin, muscle, etc. It involves the development of new viable tissue with the aid of a tissue scaffold. Innovations in stem cells, their growth and differentiation, and biomaterial and biomimetic environments have allowed the fabrication of tissues in the laboratory using combinations of cells, biologically active molecules and engineered extracellular matrices. The goal of tissue engineering is the development of human replacements parts, with functional and biomechanical stability, and complex functionality. This book is a valuable compilation of topics, ranging from the basic to the most complex advancements in the field of tissue engineering. It presents the complex subject of tissue engineering in the most comprehensible and easy to understand language. It is an essential guide for both academicians and those who wish to pursue this discipline further.

Finally, I would like to thank all the contributing authors for their valuable time and contributions. This book would not have been possible without their efforts. I would also like to thank my friends and family for their constant support.

Editor

Recent Advances in 3D Printing of Aliphatic Polyesters

Ioana Chiulan [1],* , **Adriana Nicoleta Frone** [1],*, **Călin Brandabur** [2] and
Denis Mihaela Panaitescu [1]

[1]   Polymer Department, National Institute for R&D in Chemistry and Petrochemistry ICECHIM,
      202 Splaiul Independentei, 060021 Bucharest, Romania; panaitescu@icechim.ro
[2]   Symme3D and LTHD Corporation SRL, 300425 Timisoara, Romania; calin.brandabur@symme3d.com
*   Correspondence: ioana.chiulan@icechim-rezultate.ro (I.C.); ciucu_adriana@yahoo.com (A.N.F.)

Academic Editor: Gary Chinga Carrasco

Abstract: 3D printing represents a valuable alternative to traditional processing methods, clearly demonstrated by the promising results obtained in the manufacture of various products, such as scaffolds for regenerative medicine, artificial tissues and organs, electronics, components for the automotive industry, art objects and so on. This revolutionary technique showed unique capabilities for fabricating complex structures, with precisely controlled physical characteristics, facile tunable mechanical properties, biological functionality and easily customizable architecture. In this paper, we provide an overview of the main 3D-printing technologies currently employed in the case of poly (lactic acid) (PLA) and polyhydroxyalkanoates (PHA), two of the most important classes of thermoplastic aliphatic polyesters. Moreover, a short presentation of the main 3D-printing methods is briefly discussed. Both PLA and PHA, in the form of filaments or powder, proved to be suitable for the fabrication of artificial tissue or scaffolds for bone regeneration. The processability of PLA and PHB blends and composites fabricated through different 3D-printing techniques, their final characteristics and targeted applications in bioengineering are thoroughly reviewed.

Keywords: 3D printing; aliphatic polyesters; scaffolds; tissue engineering; polylactic acid; polyhydroxyalkanoates

1. Introduction

The diversity and complexity of materials expands continuously with a speed that is beyond of any expectations. Traditional manufacturing cannot meet all the requirements of the new products, especially when they are of small dimension and with high shape complexity. 3D printing, usually called "additive manufacturing", is a useful tool for scalable fabrication of high complexity devices, or materials with multiple functions such as smart materials or customized products. It is very important in the process of prototyping and may also lead to the improvement of manufacturing by increasing the speed of production and lowering product cost. The first 3D printer was invented in 1987 and since then, this technology has grown rapidly because it brings multiple advantages over traditional production methods: (i) very complex structures can be created without added costs; (ii) the pieces are fabricated directly in assembled forms and the number of the components is consistently smaller compared to the same piece obtained by classical methods; and (iii) small series of personalized products can be obtained by this technique [1–3]. The interest for this technology is highlighted by the vibrant growth of the sales reported by 3D printer producers, who claim an increase of 17.4% in worldwide revenues, in 2016, as compared with previous years [4]. A substantial amount of research predicts the proliferation of this industry and a potential increase of the products and services from $6 billion in 2016 to $21 billion worldwide by 2021 [4].

3D-printing technology is attractive for many applications: (i) in the research field for prototyping or for a limited production of prototypes; (ii) in medicine to create 3D biomedical structures using digital models obtained with different medical imaging techniques (computer tomography, magnetic resonance imaging, ultrasound); (iii) in industry for prototyping and manufacture of spare parts for automotive, airplanes, etc. 3D-printing development is speeding up annually due to the reduction of the production cycles, waste, limited use of cutting fluids, and it becomes more accessible for small companies etc. [5]. Thus, 3D printing is often used to develop medical devices [6], flexible electronics [7,8], various pieces for automotive or robotics [9], art objects [2,3], precise replica of archeological objects [10] etc. Dental implants and porous scaffolds for tissue engineering, with increased surface roughness and improved mechanical performance and biocompatibility, used for bone fixation [11,12] are among the most studied medical devices.

The additive manufacturing methods are suitable for multiple types of materials, such as thermoplastics (acrylonitrile-butadiene-styrene (ABS), poly (lactic acid) (PLA), polyamide 6 (PA6), high-impact polystyrene, etc.), resins, metals (Al, steel, Au, Ag, Ti, alloys), gypsum-based powders, ceramics, waxed materials, biomaterials, paper, food. Polymers are by far the most used materials for 3D printing [13]. Likewise, aliphatic polyesters are among the most used biopolymers in the biomedical field due to their non-toxic, biodegradable and biocompatible character [14].

This mini-review deals with the use of aliphatic polyesters in 3D printing for medical applications, with a deeper attention on materials and methods suitable to construct scaffolds for tissue engineering. The industrial applications of 3D printing of aliphatic polyesters are quickly reviewed. The motivation behind this work resides from the recent scientific reports that highlight the ability of additive manufacturing to overcome the limitations of traditional methods such as molding, electrospinning, solvent casting, gas foaming, leaching etc. in the fabrication of medical products. Through 3D-printing techniques it is now possible to obtain superior control of the pore size, to manufacture scaffolds with complex architecture, and to implement biological functions in order to mimic the natural tissue [15,16]. A short presentation of the main 3D-printing methods will be followed by an overview of two most important classes of thermoplastic aliphatic polyesters, poly(lactic acid) and polyhydroxyalkanoates (PHA), their blends and composites that were processed by these methods. Finally, a discussion of the future perspectives and research approaches is included.

2. Short Overview of the Main 3D-Printing Techniques

Broadly, the main 3D-printing techniques commercially available are: (i) selective layer/laser sintering (SLS); (ii) fused filament fabrication (FFF), also known as fused deposition modeling (FDM, trademark of Stratasys) or molten polymer deposition; (iii) stereolithography; (iv) digital light processing; (v) polyjet / inkjet 3D printing and (vi) electronic beam melting [3,16]. Only SLS and FDM have been used for the 3D printing of aliphatic polyesters (Figure 1).

2.1. Selective Laser Sintering

In SLS technique, the 3D-designed model is transferred to the printer, where an infrared laser beam fuses the polymeric powder, especially polyamides and thermoplastic polyurethanes (TPU), as well as metal and ceramic powders, into thin layers, one layer at a time [17]. After the completion of a layer, a new layer of powder is applied to it and then subjected to another round of heating action and sintering. The process is repeated and the completed object is removed from the printer, brushed and sandblasted in order to remove any trace of powder [18]. Depending on the application and material used, the printed object can be further polished and/or dyed. This technique is characterized by a high resolution, is suitable for functional polymers and does not require a support material or structures, so the printed structures can be used without further cleaning steps [19]. Polyamide 12 (PA12) or its powdered blends with PA6 were successfully printed through SLS and represents around 90% of the total industrial consumption [20]. Other materials processed to a much lesser extent through SLS are polyamide 11, PLA and polyether ether ketone.

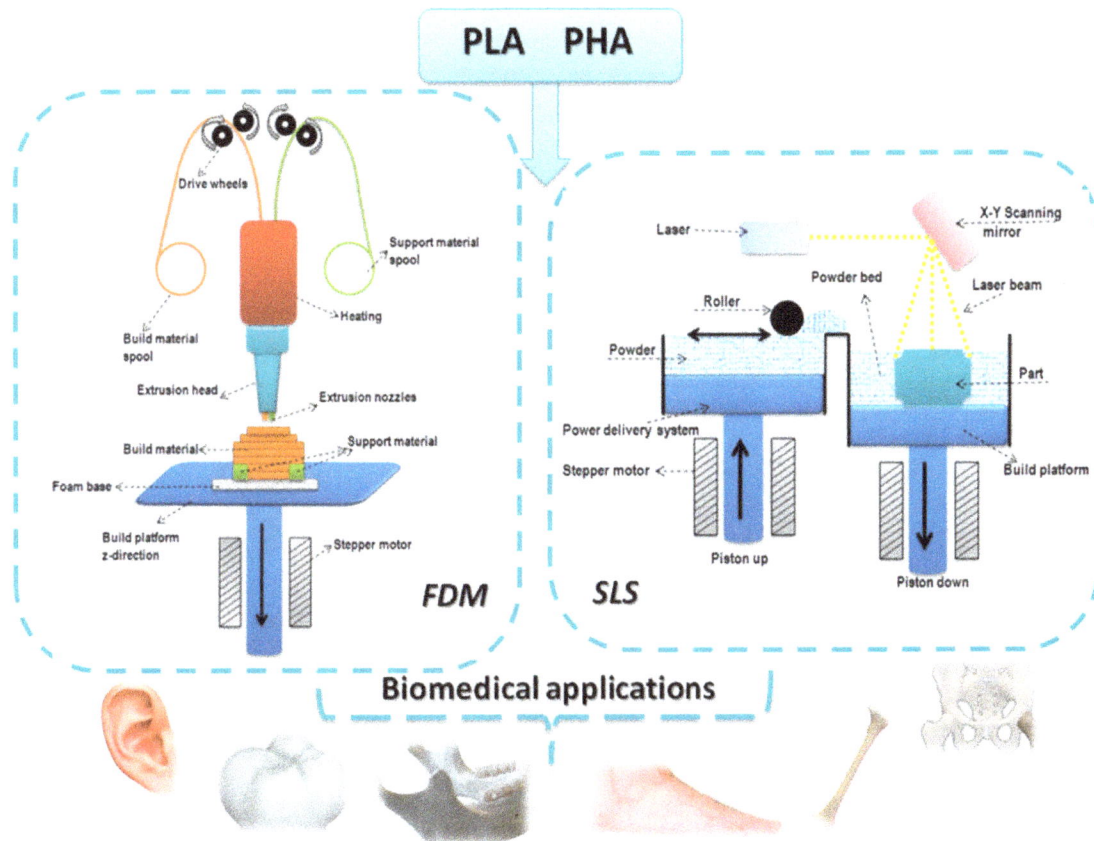

Figure 1. 3D-printing techniques employed for PLA and PHA.

2.2. Fused Deposition Modeling

Through this technique, filaments made of thermoplastic materials are extruded in thin threads and deposited layer by layer in the desired 3D structure and adhere to each other by physical interactions. The layer underneath hardens as it cools and binds with the new layer that is added on the top, remaining a fully solidified structure throughout the process. FDM is already used to produce commercial plastics and, in general, is the most used among all the techniques; this is partially because of the low price of the printer and the facile manipulation, which makes it possible even for home use. Thermoplastic polymers currently processed with FDM are ABS and PLA. Other polymers were also found suitable for this technique: acrylonitrile-styrene-acrylate, PA12, polycarbonate, polyethylene terephtalate, TPU and thermoplastic elastomers. The roughness of the 3D-printed structures is an important issue in the case of FDM, since it affects not only the appearance but also the mechanical resistance of the products. A polishing device connected to the 3D printer [21], the use of the vaporized acetone to melt uniformly the surface of 3D-printed prostheses made of ABS [22] and filling the grooves with the material dissolved by the solvent stored in a pen-style device [23] were among the solutions proposed to remove the layer grooves. FDM technology also allows the printing of cells suspension into a scaffold support. A schematic illustration of a tissue-engineered structure obtained by FDM bioprinter is presented in Figure 2.

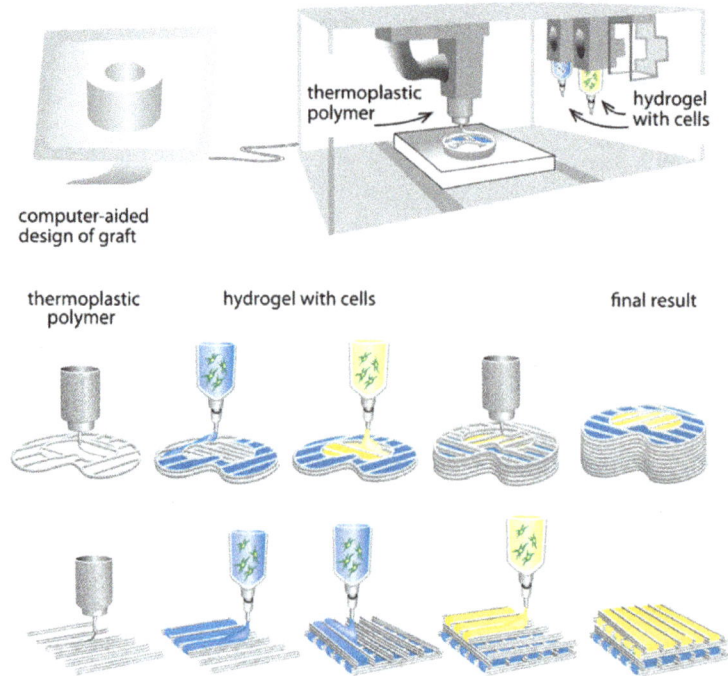

Figure 2. FDM schematic of the bioprinting of tissue and organs.

3. Aliphatic Polyesters for Additive Manufacturing

Well selected and up-to-date information on the additive manufacturing of various polymers were recently reported [13]. Considering the huge importance of aliphatic polyesters for biomedical applications, this review gives thorough information on the use of 3D-printing techniques in the case of PLA and PHA, correlated with the properties of the manufactured products and their applications in bioengineering.

3.1. Poly(Lactic Acid)

PLA is up to now the most used bioplastic for 3D printing by FDM, intended to be used in regenerative medicine, mostly as scaffolds for tissue engineering. PLA is thermoplastic aliphatic polyester (Figure 3a) prepared from fossil fuels or derived from renewable resources such as cornstarch or sugarcanes, rendering it accessible and inexpensive. PLA properties are strongly influenced by even small amounts of enantiomeric impurities. Pure poly(L-lactic acid) (PLLA) or poly(D-lactic acid) are semicrystalline polymers with a glass transition temperature (T_g) around 57 °C and a melting temperature of about 175 °C while PLA with a content of 50–93% L-lactic acid is completely amorphous [24]. Generally, amorphous grades have better processability and wider processing window than the crystalline grades [25] but much lower mechanical properties (Table 1). T_g value is important for amorphous PLA because it determines the maximum usage temperature in most applications while both T_g and T_m vales are important in the case of crystalline PLA applications. Some thermal and mechanical characteristics of PLA are given in Table 1.

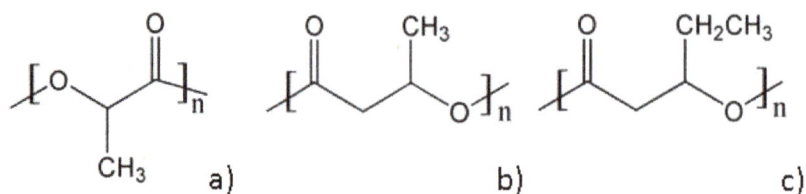

Figure 3. Chemical structures of PLA (**a**), PHB (**b**) and PHV (**c**).

PLA is the most studied aliphatic polyester for biomedical and packaging applications, due to its biocompatibility, biodegradability, clarity, high mechanical strength and modulus, and facile processability through extrusion, injection molding or casting [14]. Moreover, its lower coefficient of thermal expansion and non-adherent properties to the printed surface makes PLA a suitable material for 3D printing. In addition, it is already approved by the Food and Drug Administration (FDA) and European Medicines Agency (EMA), which makes it suitable for rapid transfer from production to clinical trials and fabrication of medical devices, pharmaceutics or various consumer products [26]. This material was intensively studied for applications such as sutures, scaffolds, extracellular matrix, dental implants, drug delivery systems, cell carriers, bioresorbable screws for bones fractures, bioabsorbable meniscus repair and stents, hernia meshes, to name just a few [27].

Table 1. Mechanical and thermal properties of PLA.

Properties	T_g, °C	T_m, °C	Tensile Strength, MPa	Young's Modulus, GPa	References
PLA (Bio-flex®F 6510) solution casting from chloroform	57.5	156.3	15.2	1.17	[28]
PLA (Nature Works™ 4032D) solution casting from DMF	-	-	32.8	2.5	[29]
PLA (Nature Works™ 4031D) extrusion	-	-	40.9	2.9	[30]
PLA film extrusion grade (Nature Works™)	55.3	151.3	40.0	1.4	[31]
PLA (Nature Works™ 4032D) Melt compounding	60.0	167.0	40.0	2.7	[32]

DMF—dimethylformamide

To date, the most common technique for 3D printing of PLA is fused deposition modeling [12,33–51]. Printing parameters such as build orientation, layer thickness, raster angle, raster width, air gap, infill density and pattern, feed rate and others directly influence the quality and the mechanical properties of the FDM printed parts [35]. Considering the importance of mechanical performance for the printed parts, the majority of current studies are focused on the influence of printing parameters on the mechanical properties of the resulted parts [33–35]. Therefore, many recent studies highlighted the mechanical and biocompatibility characteristics of PLA or its composites after 3D printing [36–40].

3.1.1. 3D Printing of PLA through Fused Deposition Modeling

A detailed study comparing the mechanical response of 3D-printed PLA blocks versus that of injection-molded PLA was provided by Song et al. [33]. PLA filament (commercial, diameter 1.75 mm) was deposited in a single direction using FDM method. Specimens cut from the printed blocks were measured along different material directions. 3D printing had a limited influence upon material elasticity; both axial and transverse stiffness being similar to that of injection-molded PLA while the inelastic response of the 3D-printed material was ductile and orthotropic. It was observed that the fracture response of the 3D-printed product was tougher when loaded in the extrusion direction than in the transverse direction. Moreover, the unidirectional 3D-printed material showed an increased toughness as compared to injection-molded PLA, due to its layered and filamentous nature. By controlling the process parameters (extruder temperature, extrusion speed, and deposition speed during 3D printing) the porosity of the material can be controlled.

Other authors used a custom 3D-printing profile for printing the specimen entirely in a single raster orientation in order to evaluate the connection between printing orientation and the material anisotropy [34]. It was found that the 45° raster orientation resulted in a slight improvement of the ultimate tensile strength and fatigue endurance limit as compared to the specimens printed at 0° and 90° raster orientation angles. Still, the mechanical properties of printed specimens were similar to those of PLA filament.

In addition to mechanical properties, 3D-printing process parameters have also great influence on the shape-memory properties of the printed parts, as reported by Wu et al. [26]. Authors used orthogonal experimental design method in order to evaluate the influence of four FDM parameters (layer thickness, raster angle, deformation temperature and recovery temperature) on the shape-recovery ratio and maximum shape-recovery rate of 3D-printed PLA. Authors concluded that the shape-memory effect of 3D-printed PLA parts depended more on recovery temperature and less on the deformation temperature and 3D-printing parameters. These findings could be of great interest for biomedical applications (self-expanding vascular stents, the elimination of thrombus) as well as the selection of parameters for 4D printing.

The possibility to replace conventional processing technique with additive manufacturing is considered by most to be unrealistic and the reasons for this opinion come from some drawbacks of the latter, such as the impossibility of manufacturing very large objects, the limitation to a small range of materials and the cost of high-performance 3D printers. This cost is subsequently reflected by the price of the final product. In order to evaluate the cost of the 3D procedure and the possibility to reduce it, Chacón et al. tried to find a connection between printing parameters and the FDM manufacturing cost [35]. Thus, PLA samples were obtained from a filament with a diameter of 1.75 mm using a low cost desktop 3D printer. Build orientation, layer thickness and feed rate parameters were analyzed and it was found that printing time decreases as layer thickness and feed rate increase. Thus, the manufacturing cost is directly related to the layer thickness and feed rate parameters.

It has been shown previously that it is possible to control the mechanical properties of PLA printed parts using an optimal selection of FDM parameters but other properties are also of great importance when referring, for example, to biomedical applications. In this respect, recent studies focused on the evaluation of PLA printed parts for reconstructive surgery and tissue engineering [36–39]. In a paper by Wurm et al. [39] FDM was successfully employed for the fabrication of PLA discs and the influence of processing technique upon biocompatibility of printed parts was assessed. In vitro tests, using human fetal osteoblasts showed no cytotoxic effects of PLA discs. Since FDM proved no negative influence on the biocompatibility of PLA, this 3D-printing technique could be further used in the reconstructive surgery for the production of individual shaped scaffolds or other implants. The filaments were printed at a nozzle temperature of 225 °C, which led to an enhanced degree of crystallinity of 22% and, finally, to a modulus of elasticity of 3.2 GPa that fits the requirements for maxillofacial implants [39].

PLA membranes, with a thickness of 100 µm and pores diameter of 200 µm, were also fabricated by direct 3D-printing method, using a PLA chloroform solution, of 5%, well dissolved by heating at 45 °C, for 24 h [40]. The PLA membranes were further seeded with human osteoprogenitors and endothelial progenitor cells and then assembled one above the other, to form a layer-by-layer (LBL) structure. After evaluating the properties of LBL constructs in vitro, in 2D – 3D, the authors stated that LBL approach could be suitable for bone tissue engineering in order to promote cells proliferation and a homogenous distribution into the scaffold.

The surface roughness of the 3D structure is very important, since cell attachment and proliferation are mainly influenced by the surface tension, roughness and stiffness of the substrate [41]. In order to enhance the roughness of the surface, Wang et al. used cold atmospheric plasma (CAP) to treat a 3D-printed PLA scaffold fabricated using a FDM printer [12]. They obtained an increase of roughness from 1.20 nm to 27.60 nm upon exposure to CAP for 5 min as compared to the untreated PLA scaffold. A significant increase of the hydrophilicity, revealed by a decrease of the contact angle from 70° to 24°, was obtained after the CAP treatment, which was proven to be a facile route to positively impact the proliferation of the osteoblasts on the PLA scaffold.

Another research study proposed a design process for FDM 3D printing of a prosthetic foot made from PLA which can significantly reduce the prosthetic weight, design and manufacturing cycle [42]. Through this process the initial model was optimized using topology optimization methods. The optimized model was printed directly from a 3D desktop printer. The authors obtained a reduction of the prosthetic feet weight by 62% compared to the initial model and a more accurate 3D-printed product (Figure 4). The proposed method facilitates the manufacturing process and reduces the fabrication time, by skipping the transfer to computer-aided design software. This research can contribute to the improvement of the quality of life of patients who need foot-customized prostheses.

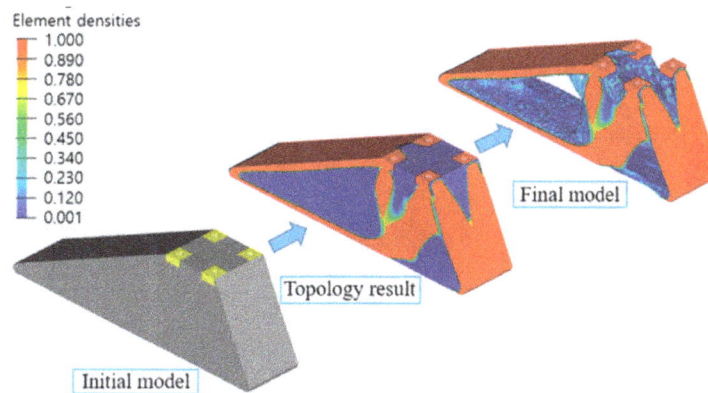

Figure 4. Topology optimization process of designed prosthetic foot. Reproduced with permission from [42].

Flores et al. also emphasized the cost effectiveness, easy manufacturing and high accuracy of the 3D-printing technology. They successfully obtained auricular prosthesis, fully customizable, which replicate in an astonishing degree the skin color and texture of the patient. However, further maintenance and potential replacement of this 3D ear prosthesis may convince the patient to agree with other alternative options [43].

3.1.2. 3D Printing of PLA Composites through Fused Deposition Modeling

PLA has relatively low glass transition temperature (55–60 °C), low toughness and weak heat resistance, which limits its application. Scaffolds made only of PLA do not mimic sufficiently the native bone architecture and they do not ensure properly the cell colonization or mechanical properties. For some uses, PLA needs to be mixed with other polymers or fillers in order to create materials with improved thermal and mechanical properties, or higher biocompatibility for biomedical purposes.

Good improvement of properties was achieved by adding 15 wt.% of nano-hydroxyapatite (HA) to PLA [47,48]. The composite was extruded in filaments and then 3D printed at a nozzle temperature of 220 °C [47]. Long-term creep test revealed a superior hardness of the 3D-printed composite as compared with PLA scaffold and consequently an increase in creep resistance. However, both samples displayed identical delamination destruction, due to limitations of the 3D-printing technique that cannot ensure completely sinterization between the layers. This causes the air to be trapped between layers, which lead to creation of voids. As expected, in vivo tests made on mice showed no inflammatory reaction even after 2 months and a slow biodegradation rate. Corcione et al. used filaments made of PLA and HA in different concentrations to obtain a molar tooth (Figure 5); this was successfully printed using FDM [48].

No noticeable difference was observed for both composites in terms of morphology, thermal behavior and crystallinity. A good dispersion of the filler was observed, but some expectable agglomerations of the nanoparticles took place, both at 5 and 15 wt.% HA. Similar values of the glass transition temperature and crystallization degree were obtained for PLA and PLA/HA samples. The addition of 15% HA influenced the rheological behavior by a significant increase of viscosity and the mechanical properties by the increase with almost 4% of the average compressive modulus as compared with the PLA sample.

Zhuang et al. used 3D printing to obtain plastic items with anisotropic heat and resistance distribution, which allows storing a simple message as color information in the printed objects. These were obtained from conductive graphene doped poly(lactic acid) (G-PLA) [50]. The authors used a method of programmed mixed printing to manufacture PLA composites with anisotropic properties. They stated that the method could be applied to other polymeric materials for a wide range of applications including biomedical ones.

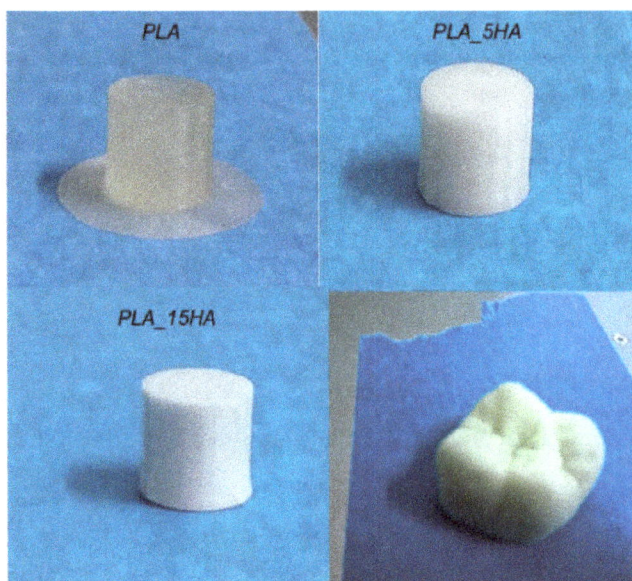

Figure 5. PLA/HA nanocomposites by FDM 3D printer. Reproduced with permission from [48].

For some medical applications, the rigidity and brittleness of the PLA are undesirable and the addition of elastomers is the easiest solution to overcome this drawback. TPU are among the most used polymers in 3D printing. They are also attractive for some biomedical applications due to their biocompatibility, high elongation at break and good abrasion resistance. As shown before, the use of different fillers impart to PLA exceptional mechanical strength, electrical conductivity, and enhanced thermal stability. Among them, composites with carbon fibers and graphene oxide (GO) proved to be also suitable for 3D-printing process. The addition of GO and TPU may have a cumulative effect of increased flexibility and mechanical strength. Chen et al. studied both the influence of the GO concentration and printing orientation on the mechanical properties of a TPU/PLA (7/3) blend [51]. Compression modulus tests have shown an increasing trend with the increase of GO content from 0.5 to 5 wt.% for both printing orientations, but the highest values were found for the specimens having the same printing orientation and height direction. The addition of only 0.5 wt.% GO determined an increase of the tensile modulus by 75% as compared with TPU/PLA sample, further addition of nanofiller determining a reduction of properties. This was explained by the percolation effect, which appeared below 2 wt.% GO content. All TPU/PLA/GO scaffolds supported fibroblast cells growth and proliferation, with the optimum effect at 0.5 wt.%.

3.1.3. 3D Printing of PLA and PLA Composites through SLS

An important requirement for the powders intended for SLS is the sintering behavior, which is greatly influenced by the thermal properties, melt viscosity, melt surface tension, and powder surface energy [13]. Semicrystalline polymers such as PLA exhibit a large change in both viscosity and density within a narrow temperature range upon melting and crystallization, which affects their processing through SLS method. Therefore, the consolidation of semicrystalline powders is conducted by local heating to temperatures slightly above melting temperature [13].

Thus, a porous scaffold was sintered from PLLA using a modified commercial Sinterstation® 2000 system (3D Systems, Valencia, CA, USA), adapted for the use of small amount of raw material [52]. The PLLA was in the form of microsphere of 5–30 μm in diameter, obtained by oil-in-water emulsion solvent evaporation technique. The SLS was conducted at 15 watts, the PLLA powder bed was preheated at 60 °C and the scan spacing was 0.15 mm. The control of the 3D scaffold porosity was difficult, since PLLA microspheres were partially melted and entangled, as revealed by SEM images [52].

The same equipment was further used to manufacture scaffolds made of PLLA and carbonated hydroxyapatite (CHAp) nanospheres, intended for bone tissue reconstruction [52–55]. Both PLLA microspheres and the PLLA/CHAp nanocomposite with 10 wt.% CHAp were prepared by emulsion method. The good dispersion and embedment of the CHAp nanoparticles in the PLLA matrix conducted to the increase of nanocomposite hardness, as revealed by the nanoindentation test. The SLS processing parameters (laser power, scan spacing, part bed temperature, roller speed, scan speed) were optimized in order to obtain adequate porosity, good compression properties, osteoconductivity and biodegradability of the PLLA and PLLA/CHAp scaffolds. The addition of the CHAp was found to influence the thermal behavior, by lowering the glass transition temperature and cold crystallization temperature and increasing to a lesser extent the melting temperature of PLLA. CHAp addition favored the powder deposition but reduced the fusion degree compared with pure PLLA powder. The porosity was mostly influenced by the part bed temperature, being enlarged in the case of nanocomposite [53–55].

Duan et al. reported the fabrication of PLLA/CHAp nanocomposite scaffolds with controllable architecture and pore size for bone tissue engineering starting from PLLA microspheres and PLLA/CHAp nanocomposite microspheres through SLS method [56]. Both raw CHAp microspheres and nanocomposite microspheres were made "in house" using a nanoemulsion method in the first case and double emulsion solvent evaporation method in the second case. More than that, in order to ensure a firm foundation and to facilitate handling of the sintered scaffold a solid base was incorporated into the scaffold design. The sintered PLLA/CHAp nanocomposite scaffolds exhibited a lower porosity value (66.8 ± 2.5%) as compared with the control PLLA scaffolds (69.5 ± 1.3%). The mechanical response (the compressive strength and modulus) of 3D scaffolds under dry conditions was higher than the one obtained under wet conditions (immersion in phosphate-buffered saline at 37 °C). In terms of biological evaluation, the PLLA/CHAp nanocomposite scaffolds exhibited a similar level of cell response compared with control PLLA scaffolds. After 7 days culture, the human osteoblastic cells were found to be well attached and spread over the strut surface and interacted favorably with all scaffolds [56].

3.1.4. Other Directions in 3D Printing of PLA Based Materials

PLA may also fit the requirements for electronic devices and other fields by chemical modification or by the addition of different fillers and polymers [44]. The presence of ionic liquids (IL) in a PLA 3D-printed structure provides unique features to PLA-based electronics; IL were recently added in the process of additive manufacturing of PLA filaments by Dichtl et al. [45]. The mixture was prepared by simply adding IL (5 and 10 wt.%) into a PLA chloroform solution, stirring for 12 h and then casting on a teflon plate. A significant enhancement of the PLA conductivity was noticed after the addition of trihexyl tetradecyl phosphonium decanoate, but further mechanical investigations are required to certify that this mixture is suitable for different applications. Prashantha and Roger studied the mechanical and electrical properties of 3D-printed specimens made from commercially available PLA filaments filled with 10 wt.% graphene [46]. The porosity distribution of the structure and the adhesion between layers were characterized through X-ray computed tomography. The results suggested that a shorter deposition time is favorable to obtain better interactions between the fused filaments and the maximum concentration for a suitable graphene dispersion is 10 wt.%. The increase of the electrical resistivity of the 3D-printed specimens, compared with the same composite before FDM processing, was explained by the alignment of the graphene nanoplatelets in the same direction with the deposited filaments. The reinforcing effect of graphene was highlighted by the increase of the storage modulus with more than 20% and tensile strength with 27%, with respect to PLA, as revealed by the DMA and static mechanical analysis [46].

PLA reinforced with 15 wt.% short carbon fibers (length about 60 mm) was manufactured by 3D printing based on fused filament fabrication and tested for mechanical and morphological properties [49]. The PLA composite showed a higher increase in stiffness in the direction of printing.

This behavior was explained by the morphological results, which revealed that the short carbon fibers were mostly aligned with the length of the 3D-printing filament, and remained aligned with the direction of printing within the PLA composite.

Wood pulp fibers (WPF) are valuable reinforcements for many polymers but the application of FDM technology for 3D printing of biocomposites with WPF is a difficult process [57]. The issues are related to the low thermal degradation temperature of the fibers, the small size of the nozzle used in FDM process and the poor dispersion of the fibers in the hydrophobic matrix, which causes fibers accumulation in the nozzle. A full enzymatic treatment was used to modify the surface of thermomechanical pulp (TMP) fibers [57]; TMP fibers modified via laccase-assisted grafting of octyl gallate (OG) showed improved interfacial adhesion with PLA and a remarkable impact on the mechanical properties of PLA-TMP fibers composites. Moreover, filaments obtained from PLA reinforced with OG-treated fibers showed a good behavior during the 3D printing [57].

3D printing of a recycled PLA composite has proved to be a viable solution to the environmental issues, since the remanufactured 3D structure showed even better mechanical properties than the original one. Tian et al. have managed to recover a PLA/carbon fiber composite in a 100% rate for the carbon fiber and 73% for PLA matrix. They reused the material for the fabrication of new filaments, with a carbon fiber content of 10 wt.%, that were further processed by 3D printing [58]. No increase of the tensile strength was observed for the remanufactured composites as compared with the original composite, but other representative characteristics were improved, such as flexural strength, which increased with around 25%. The aging process of the PLA matrix was impossible to be avoided due to repeated thermal cycles, but the mechanical performances were maintained by the addition of pure PLA in the 3D printing of the recycled composite.

3.2. Polyhydroxyalkanoates

The polyesters of aliphatic hydroxyacids, PHA, are natural polymers with some of their properties similar to those of conventional plastic materials but, in addition, they show biodegradability and biocompatibility. PHA are biosynthesized intracellularly as spherical inclusions by some bacterial strains in unbalanced growing conditions (low concentrations of nitrogen, phosphorus, oxygen or magnesium and an excess of carbon). Depending on the number of carbon atoms in the lateral chain, they may be brittle materials or elastomers. Both types are interesting materials for the biomedical field, especially for scaffolds and implants.

Short-chain-length PHA contain 3–5 carbon atoms and show high stiffness and brittleness in relation to their high crystallinity (50–80%) [24]. Poly(3-hydroxybutyrate) (PHB) (Figure 3b) and poly(3-hydroxybutyrate-co-3-hydroxyvalerate) (PHBV) are by far the most studied of PHA and are commercially available. PHB is biodegradable and biocompatible and can be processed with common plastic manufacturing equipment. However, its brittleness and small processing window limits its applications. PHBV, obtained by copolymerization with hydroxyvalerate (HV), is a more ductile material, with lower melting point and decreased strength and stiffness [24,59]. The properties of PHB or PHBV strongly depend on the processing conditions and composition (Table 2).

Table 2. Mechanical and thermal properties of some PHA.

Properties	T_g, °C	T_m, °C	Tensile Strength, MPa	Young's Modulus, GPa	Reference
PHB (Biocycle)—compression molding		164/174	43	3.5	[60]
PHB—solution casting from chloroform			28	2.1	[61]
PHBV 12 mol% HV (Metabolix Inc.)—solvent casting from DMF		140	17		[59]
PHBV 12 mol% HV (Metabolix Inc.)—solvent casting from DMF	~0	140/154	14	0.8	[62]

Cell attachment and viability tests were performed using various cultures and revealed a good biocompatibility of PHA to these cells. For example, CHL fibroblast cells showed good adhesion and proliferation on PHB scaffolds [60]. Moreover, polyhydroxyalkanoates degrade into non-toxic oligomers being suitable candidates for in vivo use in medical applications.

However, the reconstruction of some parts of the human body and organs using PHA is a very complex and difficult process because of the large differences between patients. The patient specific anatomical data should be considered for reconstruction and 3D printing is a promising technique to produce complex medical devices according to the computer aided design of the damage part or organ. Only few data were reported regarding the application of rapid prototyping techniques (RP) for the fabrication of PHA scaffolds [63–70]. Comparing to PLA, PHA cover a much broader range of properties and, therefore multiple possibilities of 3D printing.

3.2.1. PHA Filaments for Fused Deposition Modeling

PHA filaments can be used to obtain scaffolds by using FDM. Wu et al [63] obtained PHBV/palm fibers (PF) composite for 3D printers by melt mixing PHBV grafted with maleic anhydride (PHBV-g-MA) and silane treated PF. The filaments (diameter 1.75 ± 0.05 mm) were obtained from these composite materials by extrusion at 130–140 °C and 50 rpm [63]. The treatments ensured a better adhesion at polymer–filler interface and avoided the phase separation and fluctuation in the filaments diameter. The treated composites showed enhanced mechanical properties compared to that of PHBV matrix and untreated composites and higher biodegradation rate than that of PHBV when incubated in soil. Increased tensile strength and antibacterial activity were also reported for PHBV-g-MA/wood flower (WF) composites prepared with the same purpose, for 3D-printing filaments [64]. Thus, the tensile strength of PHBV-g-MA/WF composites was 6–18 MPa greater than that of untreated composites and increased with the increase of WF content [63]. Wu and Liao [65] have also prepared 3D-printing filaments from PHBV-g-MA composites with acid oxidized multi-walled carbon nanotubes (MWCNTs) using a similar method. Highly improved thermal stability, Young's modulus and antibacterial activity were obtained for only 1.0 wt.% MWCNTs in PHA-g-MA matrix [65]. However, no study on the behavior of these types of filaments in a real 3D-printing process was reported.

3.2.2. PHA Structures Obtained by Selective Laser Sintering

SLS Applied to Pure PHB

SLS technique is very attractive because porous structures with very controlled pore size may be built up without the need of any additives such as plasticizers. Preliminary RP tests with a polyhydroxyalkanoate were done by Oliveira et al. using SLS technique [66]. They worked with a polyhydroxybutyrate powder in pure form (without additives) and obtained structures of about 2.5 mm in thickness (up to 10 layers) with 1 mm holes by SLS (Figure 6).

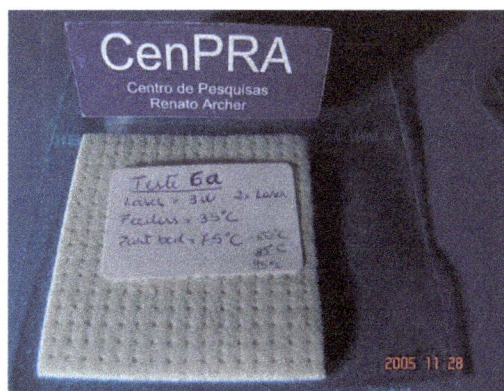

Figure 6. Sintered PHB using SLS containing pores of 1 mm in diameter [66].

They also reported the difficulties encountered with the application of SLS technique to PHB powder, such as excessive dust drag, curbing of the coating or release of vapors and the solutions adopted to solve these problems [66].

Pereira et al synthesized porous 3D cubes with orthogonal channels measuring 0.836 mm in diameter by SLS, starting from a poly(3-hydroxybutyrate) powder from PHB Industrial S/A (Brazil) [67]. A thin layer of powdered PHB was scanned by a CO_2 laser and sintered, the polymer layers being deposited one on the top of each other until the object reached the dimensions of the virtual model. The obtained objects showed geometrical and dimensional features closed to the model [67]. No significant change in crystallinity, glass transition, melting or crystallization temperatures of PHB were detected after SLS process, suggesting no thermal degradation [67]. Moreover, the possibility to recycle PHB through 3 rounds of SLS processes without any sign of degradation was also demonstrated.

SLS Applied to PHA Nanocomposites

One of the most studied applications of PHA based materials is in bone tissue engineering. PHA nanocomposites were designed to obtain 3D scaffolds that mimic the structure and function of an extracellular matrix (ECM) and support cells adhesion and proliferation [56,68–70]. Thus, bionanocomposites microspheres from PHBV and nano-sized osteoconductive inorganic fillers were obtained using a solid-in-oil-in-water emulsion/solvent evaporation method [56]. Nano-sized calcium phosphate (Ca-P) was prepared for this purpose and dispersed in a PHBV-chloroform solution by ultrasonication to form a solid-in-oil nano-suspension which was added to an aqueous solution containing 1% of poly(vinyl alcohol) and maintained at room temperature until total evaporation of the solvent, resulting Ca-P/PHBV nanocomposite microspheres. Tetragonal scaffolds with porosity around 60% were obtained from these nanocomposite microspheres using selective laser sintering. These 3D scaffolds show many advantages related to (i) the nanodimension of the inorganic filler which may provide a better cell response and osteoconductivity, (ii) the nanocomposite microspheres that ensure better dispersion of the nanofiller and (iii) the SLS technique which resulted in a controlled microstructure with totally interconnected pores [56]. Moreover, improved cell proliferation was obtained for Ca-P/PHBV nanocomposite compared to pure PHBV scaffolds.

Porous scaffolds with complex shapes and architecture (Figure 7) were constructed by SLS using Ca–P/PHBV nanocomposite [69]. Moreover, Ca–P/PHBV scaffold representing a human proximal femoral condyle (40% scale-down) was produced by SLS technique. The surface modification of Ca–P/PHBV nanocomposite scaffolds by physically entrapping gelatin and subsequent immobilization of heparin improved the wettability and provided affinity to the growth factor recombinant human bone morphogenetic protein-2 [68]. This osteoconductive nanocomposite with controlled architecture also showed sustained release behavior of osteogenic growth factor and had a great potential for bone tissue engineering [68].

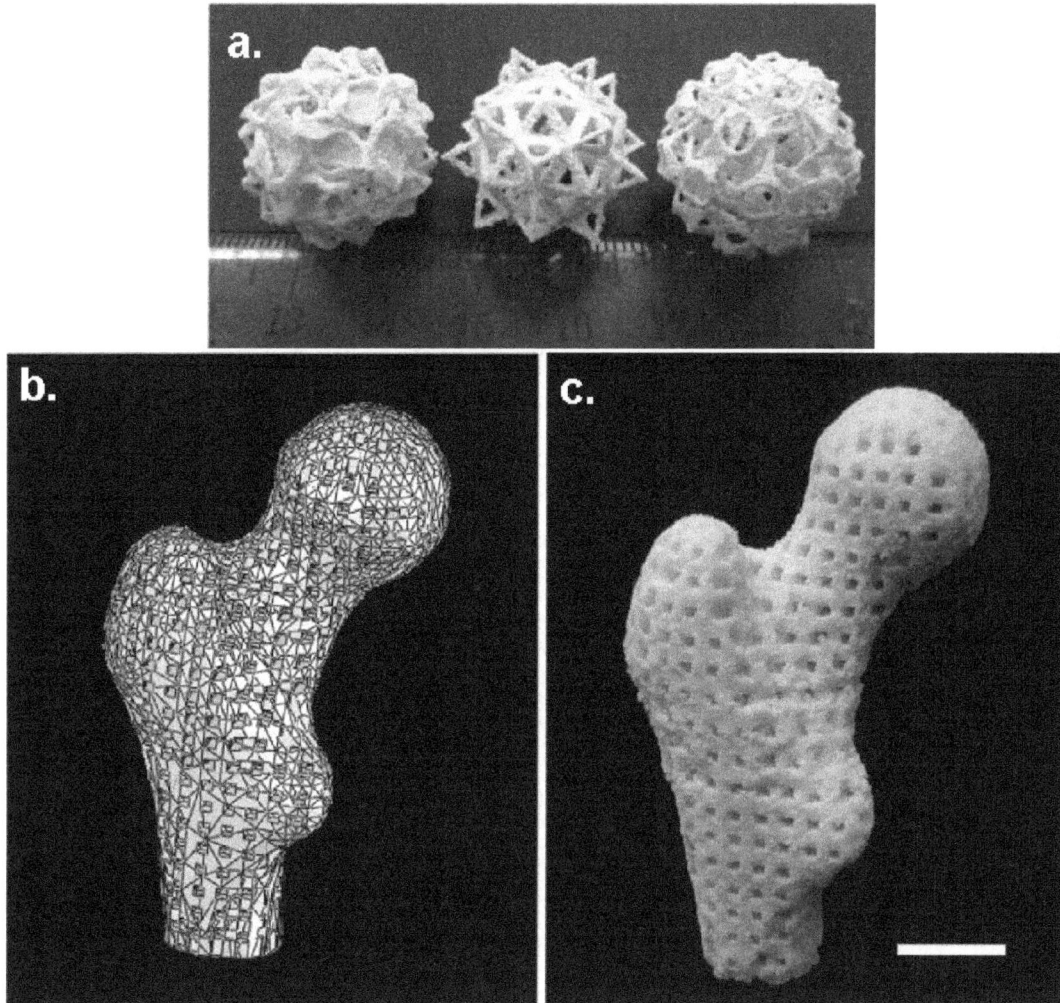

Figure 7. (**a**) Sintered Ca–P/PHBV nanocomposite porous structures based on the following models: salamanders, elevated icosidodecahedron and snarl (from left to right) (**b**) three-dimensional model of a human proximal femoral condyle reconstructed from CT images and then processed into porous scaffold using cubic cells; (**c**) sintered Ca–P/PHBV nanocomposite proximal femoral condyle scaffold. Scale bar, 1 cm. Reproduced with permission from [68].

The technique of preparation of Ca-P/PHBV nanocomposite coupled with SLS also offers the possibility of incorporating biomolecules in the nanocomposite microspheres [69]. The advantage of incorporating biomolecules in nanocomposite microspheres is related to the preservation of their biological activity and controlled release. For this purpose, a model protein, bovine serum albumin (BSA), was encapsulated into Ca-P/PHBV nanocomposite microspheres and Ca-P/PHBV/BSA 3D scaffolds with good dimensional accuracy were produced by SLS [69]. It is worth mentioning that the bioactivity of BSA was maintained during SLS processing. In vitro BSA release test showed an initial high activity followed by a slow release of BSA and a slight degradation of the PHBV matrix after 28 days in vitro test [69].

The influence of the SLS process parameters (laser power, scan spacing, layer thickness) on the quality of Ca–P/PHBV nanocomposite scaffolds was also studied [70]. The quality of the scaffolds was quantified by their structure and handling stability, their dimensional accuracy and their compressive properties and the optimized SLS parameters were determined [70].

The most important results regarding the application of 3D-printing techniques to PLA and PHA-based materials are summarized in Table 3.

Table 3. Summary of 3D-printed PLA-based materials.

Technique	Material	Results	Application	Reference
FDM	PLA	Controllable porosity and pore size by controlling the extrusion and 3D-printing parameters	quantifying anisotropic responses of PLA parts	[33]
FDM	PLA	The 3D-printed samples supports the growth of human fetal osteoblast	Bone reconstruction	[39]
FDM	PLA	The 3D-printed model with optimized design displayed a reduction with 62% of the weight as compared to the initial model	Prosthetic foot	[42]
FDM	PLA	Accurate anatomic aspect, reduced amount of raw material, inexpensive final product	Artificial ear	[43]
FDM	PLA, PLA/ionic liquid (IL)	The addition of IL led to enhanced conductivity	Electronic devices	[45]
FDM	PLA/HA	Good dispersion of the HA in the PLA matrix; increased viscosity and compressive modulus for the composites with 15 wt.% HA	Molar tooth	[48]
FDM	PLA, PLA/graphene	Enhanced electrical resistivity and mechanical strength	Electronics	[46]
FDM	PLA	The increased surface roughness and hydrophilicity conducted to cells attachment and proliferation	Bone regeneration	[12]
FDM	TPU/PLA/GO	0.5 wt.% GO led to the highest tensile modulus and cell proliferation	Tissue engineering scaffolds	[51]
FDM	PHA, PHA-g-MA, PHA/palm fibers, PHA-g-MA/ wood flower	Silane treatment of the palm fibers enhanced the adhesion with the polymer matrix; increased mechanical properties and higher degradation rate of the treated composites as compared to pure PHA and untreated composites; Increased tensile strength and antibacterial activity for PHA-g-MA/ wood flower		[63,64]
SLS	PHB	Fidel replication of the 3D-printed structure with the design model; no thermal degradation of the PHB observed after 3D printing	Tissue engineering	[66,67]
SLS	PHBV/Ca-P	The addition of the inorganic filler led to improved cell proliferation; the SLS process didn't influenced the bioactivity of the incorporated model protein	Bone tissue	[56,68,69]

4. Future Perspectives

The use of additive manufacturing methods for the production of artificial organs, tissues or bone implants is an effervescent research area with a promising future. The new era of artificial tissues and organs started twenty years ago with the production of the first 3D FDM printer and since then significant advancements have been made. However, only a few materials have been transferred to mass production and explored with 3D printing and even less of them were found suitable for medical applications. Aliphatic polyesters and, especially, PLA and PHA are suitable materials for in vivo applications due to their biocompatibility, biodegradability, good mechanical strength and processability. The continuous development of new or more specialized biomaterials is often correlated with the progress in the 3D-printing technology enhancing its potential and forcing its rapid development. There are still some challenges in the introduction of 3D-printing technologies as industrial manufacturing tools competing with injection molding and other well-established techniques. They are related to both material and equipment limits, such as reaching high accuracy of the porosity and morphology of the 3D-printed structure according to design specifications, improving the adhesion between layers, fitting the properties and their spatial distribution are some of these challenges. However, the implementation of 3D printing in biomedicine for building prosthetics, tissue grafts and other surgical implants is much more rapid than in other fields. The actual bioprinting technology is suited for the production of artificial organs or implants containing living cells, which requires a sterile environment, but avoiding contamination while handling and keeping the cells alive until they are placed into the patient are still challenges. Likewise, tuning the mechanical and biological properties of artificial tissues and organs is still a challenge and new biocompatible materials are needed to replicate parts of the human body. In addition, it is important for these future materials to be easily combined and manufactured in order to obtain adjustable properties (strength, elasticity, color) for each individual, in respect to its age, gender or race. Besides the aliphatic polyesters presented in this mini-review, some elastomers such as TPU or silicones, which can be processed through different 3D-printing technologies, deserve more attention.

Acknowledgments: This work was partially supported by a grant of Ministry of Research and Innovation, CNCS–UEFISCDI, project number PN-III-P4-ID-PCE-2016-0431, within PNCDI III, Contract no. 148/2017, (CELL-3D) and partially by a grant of the Romanian National Authority for Scientific Research and Innovation, CCCDI-UEFISCDI, project number 107 BM.

Author Contributions: Ioana Chiulan, Adriana Nicoleta Frone and Denis Mihaela Panaitescu have equally revised current literature. Calin Brandabur provided the schematic illustration of the FDM bioprinting technique and the technical information about 3D printing.

Conflicts of Interest: The authors declare no conflict of interest.

References

1. Wüst, S.; Müller, R.; Hofmann, S. Controlled positioning of cells in biomaterials—Approaches towards 3D tissue printing. *J. Funct. Biomater.* **2011**, *2*, 119–154. [CrossRef] [PubMed]
2. Balletti, C.; Ballarin, M.; Guerra, F. 3D printing: State of the art and future perspectives. *J. Cult. Herit.* **2017**, *26*, 172–182. [CrossRef]
3. Lee, J.-Y.; An, J.; Chua, C.K. Fundamentals and applications of 3D printing for novel materials. *Appl. Mater. Today* **2017**, *7*, 120–133. [CrossRef]
4. Wohlers, T. *Wohlers Report 2016*; WOHLERS Associates: Fort Collins, CO, USA, 2016.
5. Kamei, K.-I.; Mashimo, Y.; Koyama, Y.; Fockenberg, C.; Nakashima, M.; Nakajima, M.; Li, J.; Chen, Y. 3D printing of soft lithography mold for rapid production of polydimethylsiloxane-based microfluidic devices for cell stimulation with concentration gradients. *Biomed. Microdevices* **2015**, *17*, 36. [CrossRef] [PubMed]
6. Shahali, H.; Jaggessar, A.; Yarlagadda, P.K.D.V. Recent advances in manufacturing and surface modification of titanium orthopaedic applications. *Procedia Eng.* **2017**, *174*, 1067–1076. [CrossRef]
7. Patel, D.K.; Sakhaei, A.H.; Layani, M.; Zhang, B.; Ge, Q.; Magdassi, S. Highly stretchable and UV curable elastomers for digital light processing based 3D printing. *Adv. Mater.* **2017**, *29*, 1606000. [CrossRef] [PubMed]
8. Muth, J.T.; Vogt, D.M.; Truby, R.L.; Mengüç, Y.; Kolesky, D.B.; Wood, R.J.; Lewis, J.A. Embedded 3D printing of strain sensors within highly stretchable elastomers. *Adv. Mater.* **2014**, *26*, 6307–6312. [CrossRef] [PubMed]
9. Bâlc, N.; Vilău, C. Design for additive manufacturing, to produce assembled products, by SLS. *MATEC Web Conf.* **2017**, *121*, 04002. [CrossRef]
10. Additive Manufacturing in Archeology. Available online: http://3dprintingcenter.net/2017/06/16/additive-manufacturing-in-archeology/ (accessed on 3 November 2017).
11. David, O.T.; Szuhanek, C.; Tuce, R.A.; David, A.P.; Leretter, M. Polylactic acid 3D printed drill guide for dental implants using CBCT. *Rev. Chim.-Bucharest* **2017**, *68*, 341–342.
12. Wang, M.; Favi, P.; Cheng, X.; Golshan, N.H.; Ziemer, K.S.; Keidar, M.; Webster, T.J. Cold atmospheric plasma (CAP) surface nanomodified 3D printed polylactic acid (PLA) scaffolds for bone regeneration. *Acta Biomater.* **2016**, *46*, 256–265. [CrossRef] [PubMed]
13. Ligon, S.C.; Liska, R.; Stampfl, J.; Gurr, M.; Mülhaupt, R. Polymers for 3D Printing and Customized Additive Manufacturing. *Chem. Rev.* **2017**, *117*, 10212–10290. [CrossRef] [PubMed]
14. Panaitescu, D.M.; Frone, A.N.; Chiulan, I. Nanostructured biocomposites from aliphatic polyesters and bacterial cellulose. *Ind. Crops Prod.* **2016**, *93*, 251–266. [CrossRef]
15. Mondschein, R.J.; Kanitkar, A.; Williams, C.B.; Verbridge, S.S.; Long, T.E. Polymer structure-property requirements for stereolithographic 3D printing of soft tissue engineering scaffolds. *Biomaterials* **2017**, *140*, 170–188. [CrossRef] [PubMed]
16. Wu, G.-H.; Hsu, S. Polymeric-based 3D printing for tissue engineering. *J. Med. Biol. Eng.* **2015**, *35*, 285–292. [CrossRef] [PubMed]
17. Kruth, J.P.; Wang, X.; Laoui, T.; Froyen, L. Lasers and materials in selective laser sintering. *Assem. Autom.* **2003**, *23*, 357–371. [CrossRef]
18. 3D Printing Material: Alumide. Available online: https://www.sculpteo.com/en/materials/alumide-material/ (accessed on 3 November 2017).

19. Türk, D.-A.; Kussmaul, R.; Zogg, M.; Klahn, C.; Leutenecker-Twelsiek, B.; Meboldt, M. Composites part production with additive manufacturing technologies. *Procedia CIRP* **2017**, *66*, 306–311. [CrossRef]

20. Schmid, M.; Amado, A.; Wegener, K. Polymer powders for selective laser sintering (sls). *AIP Conf. Proc.* **2015**, *1664*, 160009.

21. Dieste, J.A.; Fernández, A.; Roba, D.; Gonzalvo, B.; Lucas, P. Automatic grinding and polishing using spherical robot. *Procedia Eng.* **2013**, *63*, 938–946. [CrossRef]

22. He, Y.; Xue, G.; Fu, J. Fabrication of low cost soft tissue prostheses with the desktop 3D printer. *Sci. Rep.-UK* **2014**, *4*, 6973. [CrossRef] [PubMed]

23. Takagishi, K.; Umezu, S. Development of the improving process for the 3D printed structure. *Sci. Rep.-UK* **2017**, *7*, 39852. [CrossRef] [PubMed]

24. Panaitescu, D.; Frone, A.N.; Chiulan, I. Green Composites with Cellulose Nanoreinforcements. In *Handbook of Composites from Renewable Materials*; Thakur, V.K., Thakur, M.K., Kessler, M.R., Eds.; Scrivener Publishing LLC: Beverly, MA, USA, 2017; Volume 7, pp. 299–338.

25. Farah, S.; Anderson, D.G.; Langer, R. Physical and mechanical properties of PLA, and their functions in widespread applications—A comprehensive review. *Adv. Drug Deliv. Rev.* **2016**, *107*, 367–392. [CrossRef] [PubMed]

26. Wu, W.; Ye, W.; Wu, Z.; Geng, P.; Wang, Y.; Zhao, J. Influence of layer thickness, raster angle, deformation temperature and recovery temperature on the shape-memory effect of 3D-printed polylactic acid samples. *Materials* **2017**, *10*, 970. [CrossRef] [PubMed]

27. Tyler, B.; Gullotti, D.; Mangraviti, A.; Utsuki, T.; Brem, H. Polylactic acid (pla) controlled delivery carriers for biomedical applications. *Adv. Drug Deliv. Rev.* **2016**, *107*, 163–175. [CrossRef] [PubMed]

28. Abdulkhani, A.; Hosseinzadeh, J.; Ashori, A.; Dadashi, S.; Takzare, Z. Preparation and characterization of modified cellulose nanofibers reinforced polylactic acid nano composite. *Polym. Test.* **2014**, *35*, 73–79. [CrossRef]

29. Gu, J.; Catchmark, J.M. Polylactic acid composites incorporating casein functionalized cellulose nanowhiskers. *J. Biol. Eng.* **2013**, *7*, 31. [CrossRef] [PubMed]

30. Oksman, K.; Mathew, A.P.; Bondeson, D.; Kvien, I. Manufacturing process of cellulose whiskers/polylactic acid nanocomposites. *Compos. Sci. Technol.* **2006**, *66*, 2776–2784. [CrossRef]

31. Ambrosio-Martın, J.; Fabra, M.J.; Lopez-Rubio, A.; Lagaron, J.M. Melt polycondensation to improve the dispersion of bacterial cellulose into polylactide via melt compounding: Enhanced barrier and mechanical properties. *Cellulose* **2015**, *22*, 1201–1226. [CrossRef]

32. Frone, A.N.; Panaitescu, D.; Chiulan, I.; Nicolae, C.A.; Vuluga, Z.; Vitelaru, C.; Damian, C.M. The effect of cellulose nanofibers on the crystallinity and nanostructure of poly(lactic acid) composites. *J. Mater. Sci.* **2016**, *51*, 9771–9791. [CrossRef]

33. Song, Y.; Li, Y.; Song, W.; Yee, K.; Lee, K.Y.; Tagarielli, V.L. Measurements of the mechanical response of unidirectional 3D-printed pla. *Mater. Des.* **2017**, *123*, 154–164. [CrossRef]

34. Letcher, T.; Waytashek, M. Material property testing of 3D-printed specimen in PLA on an entry-level 3D printer. *Adv. Manuf.* **2014**, *2A*, IMECE2014-39379.

35. Chacón, J.M.; Caminero, M.A.; García-Plaza, E.; Núñez, P.J. Additive manufacturing of pla structures using fused deposition modelling: Effect of process parameters on mechanical properties and their optimal selection. *Mater. Des.* **2017**, *124*, 143–157. [CrossRef]

36. Guo, R.; Lu, S.; Page, J.M.; Merkel, A.R.; Basu, S.; Sterling, J.A.; Guelcher, S.A. Fabrication of 3D scaffolds with precisely controlled substrate modulus and pore size by templated-fused deposition modeling to direct osteogenic differentiation. *Adv. Healthc. Mater.* **2015**, *4*, 1826–1832. [CrossRef] [PubMed]

37. Pedro, F.C.; Cédryck, V.; Jeremy, B.; Mohit, C.; Manuela, E.G.; Rui, L.R.; Christina, T.; Dietmar, W.H. Biofabrication of customized bone grafts by combination of additive manufacturing and bioreactor knowhow. *Biofabrication* **2014**, *6*, 035006.

38. Almeida, C.R.; Serra, T.; Oliveira, M.I.; Planell, J.A.; Barbosa, M.A.; Navarro, M. Impact of 3-d printed pla- and chitosan-based scaffolds on human monocyte/macrophage responses: Unraveling the effect of 3-d structures on inflammation. *Acta Biomater.* **2014**, *10*, 613–622. [CrossRef] [PubMed]

39. Wurm, M.C.; Möst, T.; Bergauer, B.; Rietzel, D.; Neukam, F.W.; Cifuentes, S.C.; Wilmowsky, C.V. In-vitro evaluation of polylactic acid (PLA) manufactured by fused deposition modeling. *J. Biol. Eng.* **2017**, *11*, 29. [CrossRef] [PubMed]

40. Guduric, V.; Metz, C.; Siadous, R.; Bareille, R.; Levato, R.; Engel, E.; Fricain, J.-C.; Devillard, R.; Luzanin, O.; Catros, S. Layer-by-layer bioassembly of cellularized polylactic acid porous membranes for bone tissue engineering. *J. Mater. Sci. Mater. Med.* **2017**, *28*, 78. [CrossRef] [PubMed]

41. Chiulan, I.; Mihaela Panaitescu, D.; Nicoleta Frone, A.; Teodorescu, M.; Andi Nicolae, C.; Căşărică, A.; Tofan, V.; Sălăgeanu, A. Biocompatible polyhydroxyalkanoates/bacterial cellulose composites: Preparation, characterization, and in vitro evaluation. *J. Biomed. Mater. Res. A* **2016**, *104*, 2576–2584. [CrossRef] [PubMed]

42. Tao, Z.; Ahn, H.-J.; Lian, C.; Lee, K.-H.; Lee, C.-H. Design and optimization of prosthetic foot by using polylactic acid 3D printing. *J. Mech. Sci. Technol.* **2017**, *31*, 2393–2398. [CrossRef]

43. Flores, R.L.; Liss, H.; Raffaelli, S.; Humayun, A.; Khouri, K.S.; Coelho, P.G.; Witek, L. The technique for 3D printing patient-specific models for auricular reconstruction. *J. Cranio Maxill. Surg.* **2017**, *45*, 937–943. [CrossRef] [PubMed]

44. Nakatsuka, T. Polylactic acid-coated cable. *Fujikura Tech. Rev.* **2011**, *40*, 39–45.

45. Dichtl, C.; Sippel, P.; Krohns, S. Dielectric properties of 3D printed polylactic acid. *Adv. Mater. Sci. Eng.* **2017**, *2017*, 10. [CrossRef]

46. Prashantha, K.; Roger, F. Multifunctional properties of 3D printed poly(lactic acid)/graphene nanocomposites by fused deposition modeling. *J. Macromol. Sci. A* **2017**, *54*, 24–29. [CrossRef]

47. Niaza, K.V.; Senatov, F.S.; Stepashkin, A.; Anisimova, N.Y.; Kiselevsky, M.V. Long-term creep and impact strength of biocompatible 3D-printed PLA-based scaffolds. *Nano Hybrids Compos.* **2017**, *13*, 15–20. [CrossRef]

48. Esposito Corcione, C.; Gervaso, F.; Scalera, F.; Montagna, F.; Sannino, A.; Maffezzoli, A. The feasibility of printing polylactic acid–nanohydroxyapatite composites using a low-cost fused deposition modeling 3D printer. *J. Appl. Polym. Sci.* **2017**, *134*. [CrossRef]

49. Ferreira, R.T.L.; Amatte, I.C.; Dutra, T.A.; Bürger, D. Experimental characterization and micrography of 3D printed PLA and PLA reinforced with short carbon fibers. *Compos. B-Eng.* **2017**, *124*, 88–100. [CrossRef]

50. Zhuang, Y.; Song, W.; Ning, G.; Sun, X.; Sun, Z.; Xu, G.; Zhang, B.; Chen, Y.; Tao, S. 3D–printing of materials with anisotropic heat distribution using conductive polylactic acid composites. *Mater. Des.* **2017**, *126*, 135–140. [CrossRef]

51. Chen, Q.; Mangadlao, J.D.; Wallat, J.; De Leon, A.; Pokorski, J.K.; Advincula, R.C. 3D printing biocompatible polyurethane/poly(lactic acid)/graphene oxide nanocomposites: Anisotropic properties. *ACS Appl. Mater. Interfaces* **2017**, *9*, 4015–4023. [CrossRef] [PubMed]

52. Zhou, W.Y.; Lee, S.H.; Wang, M.; Cheung, W.L. Selective Laser Sintering of Tissue Engineering Scaffolds Using Poly(L-Lactide) Microspheres. *Key Eng. Mater.* **2007**, *334–335*, 1225–1228. [CrossRef]

53. Zhou, W.Y.; Lee, S.H.; Wang, M.; Cheung, W.L.; Ip, W.Y. Selective laser sintering of porous tissue engineering scaffolds from poly(L-lactide)/carbonated hydroxyapatite nanocomposite microspheres. *J. Mater. Sci.-Mater. Med.* **2008**, *19*, 2535–2540. [CrossRef] [PubMed]

54. Zhou, W.Y.; Duan, B.; Wang, M.; Cheung, W.L. Crystallization Kinetics of Poly(L-Lactide)/Carbonated Hydroxyapatite Nanocomposite Microspheres. *J. Appl. Polym. Sci.* **2009**, *113*, 4100–4115. [CrossRef]

55. Zhou, W.Y.; Wang, M.; Cheung, W.L.; Ip, W.Y. Selective Laser Sintering of Poly(L-Lactide)/Carbonated Hydroxyapatite Nanocomposite Porous Scaffolds for Bone Tissue Engineering. In *Tissue Engineering*; Eberli, D., Ed.; InTech: Vukovar, Croatia, 2010; pp. 179–204.

56. Duan, B.; Wang, M.; Zhou, W.Y.; Cheung, W.L.; Li, Z.Y.; Lu, W.W. Three-dimensional nanocomposite scaffolds fabricated via selective laser sintering for bone tissue engineering. *Acta Biomater.* **2010**, *6*, 4495–4505. [CrossRef] [PubMed]

57. Filgueira, D.; Holmen, S.; Melbø, J.K.; Moldes, D.; Echtermeyer, A.T.; Chinga-Carrasco, G. Enzymatic-assisted modification of TMP fibers for improving the interfacial adhesion with PLA for 3D printing. *ACS Sustain. Chem. Eng.* **2017**, *5*, 9338–9346. [CrossRef]

58. Tian, X.; Liu, T.; Wang, Q.; Dilmurat, A.; Li, D.; Ziegmann, G. Recycling and remanufacturing of 3D printed continuous carbon fiber reinforced PLA composites. *J Clean. Prod.* **2017**, *142*, 1609–1618. [CrossRef]

59. Ten, E.; Turtle, J.; Bahr, D.; Jiang, L.; Wolcott, M. Thermal and mechanical properties of poly(3-hydroxybutyrate-co-3-hydroxyvalerate)/cellulose nanowhiskers composites. *Polymer* **2010**, *51*, 2652–2660. [CrossRef]

60. Pinto, C.E.D.S.; Arizaga, G.G.C.; Wypych, F.; Ramos, L.P.; Satyanarayana, K.G. Studies of the effect of molding pressure and incorporation of sugarcane bagasse fibers on the structure andproperties of poly (hydroxy butyrate). *Compos. A-Appl. Sci. Manuf.* **2009**, *40*, 573–582. [CrossRef]

61. Cai, Z.; Ynag, G.; Kim, J. Biocompatible nanocomposites prepared by impregnating bacterial cellulose nanofibrils into poly(3-hydroxybutyrate). *Curr. Appl. Phys.* **2011**, *11*, 247–249.

62. Jiang, L.; Morelius, E.; Zhang, J.; Wolcott, M. Study of the poly(3-hydroxybutyrate-co-3-hydroxyvalerate)/cellulose nanowhisker composites prepared by solution casting and melt processing. *J. Compos. Mater.* **2008**, *42*, 2629–2645. [CrossRef]

63. Wu, C.-S.; Liao, H.; Cai, Y.-X. Characterisation, biodegradability and application of palm fibre-reinforced polyhydroxyalkanoate composites. *Polym. Degrad. Stabil.* **2017**, *140*, 55–63. [CrossRef]

64. Wu, C.-S.; Liao, H. Fabrication, characterization, and application of polyester/wood flour composites. *J. Polym. Eng.* **2017**, *37*, 689–698. [CrossRef]

65. Wu, C.-S.; Liao, H. Interface design of environmentally friendly carbon nanotube-filled polyester composites: Fabrication, characterisation, functionality and application. *Express Polym. Lett.* **2017**, *11*, 187–198. [CrossRef]

66. Oliveira, M.F.; Maia, I.A.; Noritomi, P.Y.; Nargi, G.C.; Silva, J.V.L.; Ferreira, B.M.P.; Duek, E.A.R. Construção de Scaffolds para engenharia tecidual utilizando prototipagem rápida. *Matéria* **2007**, *12*, 373–382. [CrossRef]

67. Pereira, T.F.; Oliveira, M.F.; Maia, I.A.; Silva, J.V.L.; Costa, M.F.; Thire, R.M.S.M. 3D Printing of Poly(3-hydroxybutyrate) Porous Structures Using Selective Laser Sintering. *Macromol. Symp.* **2012**, *319*, 64–73. [CrossRef]

68. Duan, B.; Wang, M. Customized Ca–P/PHBV nanocomposite scaffolds for bone tissue engineering: design, fabrication, surface modification and sustained release of growth factor. *J. R. Soc. Interface* **2010**, *7*, S615–S629. [CrossRef] [PubMed]

69. Duan, B.; Wang, M. Encapsulation and release of biomolecules from Ca-P/PHBV nanocomposite microspheres and three-dimensional scaffolds fabricated by selective laser sintering. *Polym. Degrad. Stabil.* **2010**, *95*, 1655–1664. [CrossRef]

70. Duan, B.; Cheung, W.L.; Wang, M. Optimized fabrication of Ca–P/PHBV nanocomposite scaffolds via selective laser sintering for bone tissue engineering. *Biofabrication* **2011**, *3*, 015001. [CrossRef] [PubMed]

2

Photobiomodulation Therapy (PBMT) in Peripheral Nerve Regeneration

Marcelie Priscila de Oliveira Rosso [1], Daniela Vieira Buchaim [2,3], Natália Kawano [2], Gabriela Furlanette [2], Karina Torres Pomini [1] and Rogério Leone Buchaim [1,2,*]

[1]   Department of Biological Sciences (Anatomy), Bauru School of Dentistry, University of São Paulo (USP), Alameda Dr. Octávio Pinheiro Brisola 9-75, Vila Nova Cidade Universitária, Bauru, São Paulo CEP 17012-901, Brazil; marcelierosso@usp.br (M.P.d.O.R.); karinatorrespomini@gmail.com (K.T.P.)

[2]   Medical School, Discipline of Human Morphophysiology, University of Marilia (UNIMAR), Av. Higino Muzi Filho, 1001 Campus Universitário, Jardim Araxa, Marília, São Paulo CEP 17525-902, Brazil; danibuchaim@usp.br (D.V.B.); natalia.kawano@hotmail.com (N.K.); gafurla@hotmail.com (G.F.)

[3]   Medical School, Discipline of Neuroanatomy, University Center of Adamantina (UNIFAI), Rua Nove de Julho, 730, Centro, Adamantina, São Paulo CEP 17800-000, Brazil

*   Correspondence: rogerio@fob.usp.br

Abstract: Photobiomodulation therapy (PBMT) has been investigated because of its intimate relationship with tissue recovery processes, such as on peripheral nerve damage. Based on the wide range of benefits that the PBMT has shown and its clinical relevance, the aim of this research was to carry out a systematic review of the last 10 years, ascertaining the influence of the PBMT in the regeneration of injured peripheral nerves. The search was performed in the PubMed/MEDLINE database with the combination of the keywords: low-level laser therapy AND nerve regeneration. Initially, 54 articles were obtained, 26 articles of which were chosen for the study according to the inclusion criteria. In the qualitative aspect, it was observed that PBMT was able to accelerate the process of nerve regeneration, presenting an increase in the number of myelinated fibers and a better lamellar organization of myelin sheath, besides improvement of electrophysiological function, immunoreactivity, high functionality rate, decrease of inflammation, pain, and the facilitation of neural regeneration, release of growth factors, increase of vascular network and collagen. It was concluded that PBMT has beneficial effects on the recovery of nerve lesions, especially when related to a faster regeneration and functional improvement, despite the variety of parameters.

Keywords: low-level laser therapy; nerve regeneration; peripheral nerve repair; photobiomodulation therapy; tissue regeneration

1. Introduction

Low-level laser therapy (LLLT), now commonly referred to as photobiomodulation therapy (PBMT), using low-level infrared light spectrum lasers is considered a therapeutic advance. Its effects are related to tissue biostimulation, presenting therapeutic responses from photoelectric, photoenergetic, and photochemical reactions [1]. Scientific research has shown the application of PBMT in bone tissue and peripheral nerves with good results whether or not it is associated with other supporting methods in tissue repair [2–7].

Laser photobiomodulation presents itself as an electromagnetic technology that is being inserted into clinical practice due to its characteristics that differ from other conventional thermal sources [8,9]. It was observed that there are several features of PBMT that are related to the reduction of tissue repair time and its capacity to increase cell proliferation [10].

In rehabilitative health, PBMT was inserted to promote the repair and recovery of tissues. For example, in physical therapy, the use of PBMT is applied in postoperative phases as an aid in the muscular, nervous, joint, and other functional recovery processes, and in dentistry it is applied in the processes of dental extraction, grafting, osteonecrosis, and periodontal lesions [11–13].

The wavelength of infrared irradiation is easily absorbed by tissues and the loss of intensity is minimal, affecting metabolic modifications, DNA activity, adenosine triphosphate (ATP) formation, and the mitochondrial chain. The effect of photobiomodulation is due to the absorption of the photons by cytochrome C oxidase in the mitochondrial respiratory chain, consequently increasing the cytochrome C oxidase activity and therefore ATP formation. ATP from injured or regions of impaired blood perfusion can reactivate injured cells and metabolic disorders [10]. PBMT is also related to pain and inflammation relief and prevention of tissue death to avoid neurological degeneration [14,15].

The wavelength is the key point that regulates the depth and penetration of the laser irradiance in the tissue, noting that the absorption and dispersion coefficients are larger at the lower wavelengths. Regarding the type of wave, whether continuous or pulsed, there are still divergences in which is the best and for which factors are the pulse parameters to be chosen [16]. PBMT presents difficulties in selecting the most suitable parameters for its application due to the lack of standardization, since wavelength, power density, irradiation time, and light polarization have repercussions on the biological effects [9].

Due to the photochemical and photobiological effects of PBMT at the cellular level, there is a relationship between the improvement of trophic conditions and the reduction of inflammatory processes, closely related to a more efficient nervous regeneration and, also, promoting the secretion of neural factors [16,17]. Thus, photobiomodulation therapy in the neurological area acts as an adjuvant in the treatment of traumatic brain degeneration/injury, spinal cord trauma, and in the process of peripheral nerve regeneration.

Peripheral nerve lesions are a reality today, but there is a deficit in relating effective treatments for recovery of the nerves, resulting in considerable functional changes in the daily life of the individual. When injured, the nerve can lose its function, causing motor or sensitive deficits. There is retrograde axonal degeneration to the area of the lesion, so regeneration occurs slowly and sometimes incompletely [18,19].

At the end of the 80's, the scientific interest in the therapeutic approach of rehabilitation for neural lesions was initiated [20], due to the good results with the use of PBMT in the recovery of injured peripheral nerves but, until the present day, there are still difficulties related to the application parameters [19,20]. Its beneficial effects are independent of the repair technique, neurorrhaphy techniques, and the use of fibrin sealants [3,6,7,21].

PBMT leads to changes in important vascular levels such as elevation of the secretion of antiapoptotic factors in ischemic organs, providing a better wound healing [22,23]; the presence of angiogenesis when ischemic organs were injured [24,25]; a decrease in the site of infarction in rats; as well as elevation in neurological scores following embolic stroke in rats [26,27].

Due to the high range of benefits that PBMT has shown and its clinical relevance of application, the aim of this research was to carry out a systematic review of the scientific papers published in the last 10 years verifying the relation of PBMT with the regeneration of injured peripheral nerves.

2. Materials and Methods

A search was performed in the PubMed/MEDLINE database, combining low-level laser therapy AND nerve regeneration keywords, over the last 10 years and restricted to the English language. The next step was to restrict the verification and consultation of articles that used animals as a study object (non-human species).

We verified those articles that presented titles and summaries that approached the subject of this research, as well as methodology, results, and relevance for its practical application.

The articles included should necessarily be presented with full access to the text. The acquired texts were analyzed and synthesized in a reflexive way in order to obtain consistent information on the subject.

3. Results

Initially, 54 articles were obtained from the PubMed/MEDLINE database, of which 28 were excluded because they were not included in the search criteria (in English, study in animals, and full access to content). At the end, 26 articles related to the subject were included. Figure 1 schematizes the search system, according to PRISMA Flow Diagram [28].

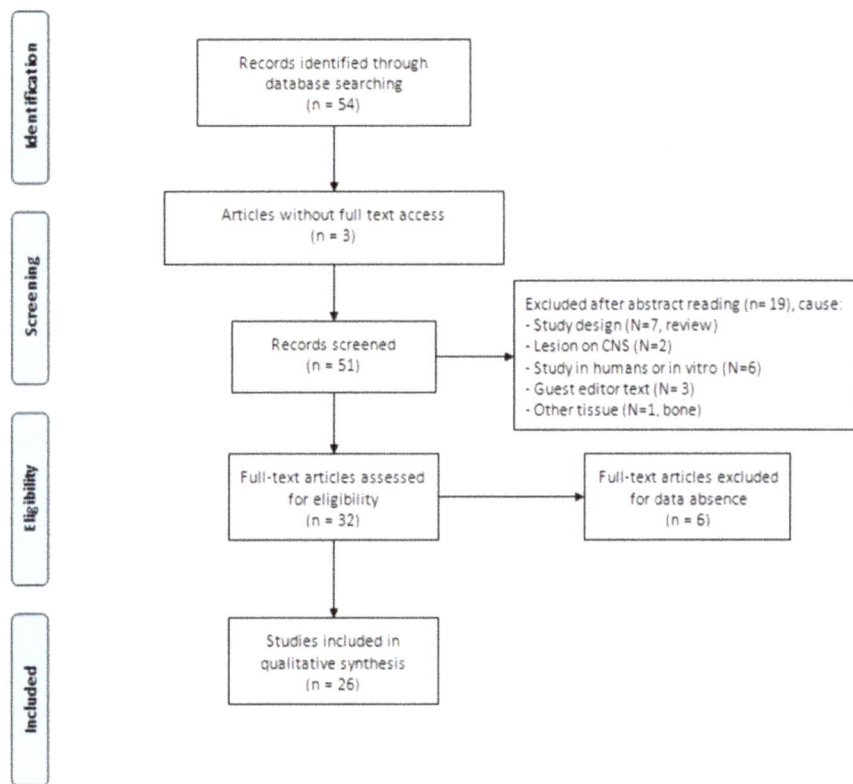

Figure 1. Design used to select the articles.

Table 1 summarizes the data presented in the 26 articles selected for this research.

Table 1. Data of selected articles.

Authors	Type of Laser (Manufacturer)	Wavelength (nm)/Spot Beam	Energy (mW)	Energy Density (J/cm²)	Radiation Amount	Variables	Irradiation Site	Evaluation Time	Main Results
Buchaim et al. [2]	GaAlAs (Laserpulse IBRAMED, Brazil)	660/0.116	30	4	16 s per point; 3 points	Sural nerve graft was coapted to the vagus nerve using the fibrin glue.	Right side of the neck.	Application on the 1st day post-operatory, 5 weeks, 3 times a week. Evaluation 30 days after irradiation.	LLLT improved the nerve regeneration.
Buchaim et al. [3]	GaAlAs (Laserpulse IBRAMED, Brazil)	830/0.116	30	6	24 s per point; 3 points	Neurotmeses of buccal branch of facial nerve, followed by end-to-end suture or coaptation with heterologous fibrin sealant derived from snake venom	On the surgical site, on both sides of the face	Application 1st day post-operatory, 3 times/week for 5 weeks. Evaluation 5 and 10 weeks after the surgery.	LLLT showed satisfactory results on facial nerve regeneration.
Buchaim et al. [6]	GaAlAs (Laserpulse IBRAMED®, Brazil)	830/0.116	30	6.2	24 s per point; 3 points	Neurotmeses of buccal branch of facial nerve, end-to-end anastomosis. Use of epineural suture or coaptation with heterologous fibrin sealant derived from snake venom.	On the surgical site, on both sides of the face	Application on the 1st day post-operatory, 3 times a week, for 5 weeks.	Laser stimulated axonal regeneration accelerated the process of functional recovery of whisker, and the two techniques used allowed the growth of axons.
Rosso et al. [7]	GaAlAs (Laserpulse IBRAMED®, Brazil)	830/0.116	30	6.2	24 s per point; 3 points	Neurotmeses in buccal branch of facial nerve, end-lateral anastomosis in the zygomatic branch of the facial nerve with epineural suture or heterologous sealant of fibrin derived from snake venom.	On the surgical site, on both sides of the face	Application on the 1st day post-operatory, 3 times a week, for 10 weeks.	Laser groups presented faster functional recovery, similar results to the control group. It was observed that PBMT provided accelerated morphological and functional repair in the two techniques used.
Ziago et al. [19]	GaAlAs (Twin Laser, MMO, São Carlos, SP, Brazil)	780/0.04	40	4 10 50	4, 10 e 50 s per point; 3 points	Crushing of the left sciatic nerve.	On the surgical site	Application during 6 sessions on alternate days.	Best morphological quantitative and morphometric results on L10 group after 15 days of nerve lesion.
Alessi Pissulin et al. [29]	GaAs (Endophoton, KLD Biosystems, Amparo, Brazil)	904/0.035	50	69	48 s per point	0.5% bupivacaine injection to the right and 0.9% sodium chloride injection to the left on sternocleidomastoid muscle and accessory nerve exposed in surgery.	Ventral side of the neck	Application 1st day post-operatory, during 5 successive days.	LLLT reduced the aggressive effects of bupivacaine on the nerve and the muscle, of muscular degeneration, of myonecrosis and fibrosis, kept the morphology of the axon and the myelin sheath.
Takhtfooladi; Sharifi [30]	GaAlAs (pulsed) LED (red and blue) (—)	680/0.04 650/1.5 red 450/1.5 blue	10	10	200 s per point; 3 points	Neurotmeses of right sciatic nerve followed by epineural neurorrhaphy.	On the surgical site, sciatic nerve	Application 1st day post-operatory, during 14 successive days	LLLT increased Schwann cells on the great myelinic axons and on neurons, sped up and potentialized nerve regeneration.

Table 1. *Cont.*

Authors	Type of Laser (Manufacturer)	Wavelength (nm)/Spot Beam	Energy (mW)	Energy Density (J/cm²)	Radiation Amount	Variables	Irradiation Site	Evaluation Time	Main Results
Takhtfooladi et al. [31]	InGaAlP (Teralaser; DMC®São Carlos, SP, Brazil)	685/0.028	15	3	10 s per point	Crushing of the left sciatic nerve.	On the surgery site on sciatic nerve.	Application on the 1st day post-operatory, during 21 successive days.	LLLT accelerated and improved the nerve function after crushing lesion.
Wang et al. [32]	GaAlAs (Transverse IND. CO., LTD., Taipei, Taiwan)	808/3.8	170	3 8 15	67.2 s 179 s 335.6 s	Crushing of the right sciatic nerve.	On lesion on sciatic nerve.	Application during 20 successive days.	LLLT (3 and 8 J/cm²) accelerated functional and morphologic recovery of the nerve, increased the expression of the marker GAP43.
Shen; Yang; Liu [33]	AlGaInP (Megalas1-AM-800, Konftec Co, Taipei, Taiwan, ROC)	660/——	0.0032	3.84	5 min per day	Neurotmeses of the left sciatic nerve, 10 mm gap and use of biodegradable tube containing genipin-cross-linked gelatin annexed with β-tricalcium phosphate ceramic particles (genipin-gelatin-tricalcium phosphate, GGT)	Applied to the surgical site.	Application 1st day post-operatory, during 20 successive days. Euthanasia after 8 weeks.	LLLT obtained better functional, electrophysiological and histomorphometric results and assisted on neural repair.
Shen; Yang; Liu [34]	AlGaInP (MegalasVR-AM-800; Konftec, Taipei, Taiwan)	660/——	50	Immediate post-surgery (5.76) 9 following days (0.96)	Immediate post-surgery (30 min) 9 successive consecutive (5 min)	Neurotmeses of the left sciatic nerve, 15 mm gap and the use of 1-ethyl-3-(3-dimethylaminopro-pyl) carbodiimide (EDC) cross-linked gelatin, annexed with β-tricalcium phosphate (TCP) ceramic particles (EDC-Gelatin-TCP, EGT).	On the surgery site.	Application immediately after the lesion, during 9 successive days. Euthanasia after 12 weeks.	LLLT showed better results on the functional index, on development, on electrophysiology, on nerve regeneration, larger neural tissue area, larger axon, and myelin sheath diameter.
Medalha et al. [35]	GaAlAs (Teralaser, DMC São Carlos, São Paulo, Brazil)	660/0.028 808/0.028	30 30	10 e 50 10 e 50	9 s and 47 s; 3 points 9 s and 47 s; 3 points	Neurotmeses of the sciatic nerve, approximately 3 mm distal to the tendon of the internal obturator. Anastomosis with 3 sutures using nylon monofilament 10-0.	Applied to the surgical site.	Application 1st day post-operatory during 5 successive days and 2 days interval until completing 15 days.	LLLT 808 nm on 50 J/cm² obtained higher fiber density. LLLT 660 nm on 50 J/cm² presented larger diameters of axons and of fibers of gait functional recovery.
Shen et al. [36]	GaAlAsP (Aculas-AM-100A, Konftec Co, Taipei, Taiwan)	660/0.1	50	2	2 min per day; 2 points at the same time	A biodegradable nerve conduit containing genipin-cross-linked gelatin was annexed using beta-tricalcium phosphate (TCP) ceramic particles (genipin-gelatin-TCP, GGT) with a 15 nm sciatic nerve transection gap.	On the sciatic nerve.	Application 1st day post-operatory during 10 successive days.	LLLT accelerated the nerve regeneration due to the larger neural tissue, larger diameter and thicker myelin sheath, motor function, electrophysiology and muscular innervation.

Table 1. *Cont.*

Authors	Type of Laser (Manufacturer)	Wavelength (nm)/Spot Beam	Energy (mW)	Energy Density (J/cm²)	Radiation Amount	Variables	Irradiation Site	Evaluation Time	Main Results
Chen et al. [37]	GaAlAs (Transverse IND. CO., LTD., Taipei, Taiwan)	808 ± 5/≤0.5	190	8	207 s	Chronic compression on dorsal root ganglion. A thin L shaped needle (0.6 mm of diameter) was inserted 4 mm in the L4 and L5 intervertebral foramen.	On the dorsal root of L4 and L5.	Application 1st day post-operatory, per 8 successive days. Euthanasia 4 e 8 days.	LLLT decreased the levels of inflammatory cytokines and of pain, facilitating the nerve regeneration, demonstrated by levels of TNF-a, IL-1b e GAP-43.
Belchior et al. [38]	GaAlAs (KLD® Endophoton model)	660/0.63	26.3	4	96.7 s; 3 points	Crushing of the right sciatic nerve.	On the surgical site.	Application 1st day post-operatory, during 20 successive days.	LLLT was positive on the functional index after the 21st day.
Barbosa et al. [39]	GaAlAs (Ibramed® Equipamentos Médicos)	660/0.06 830/0.116	30	10 10	20 s 38.66 s	Crushing of the right sciatic nerve.	On the surgical site.	Application 1st day post-operatory, during 20 successive days.	LLLT 660 nm promoted functional recovery in a faster manner.
Marcolino et al. [40]	AlGaAs (Laser Diode, Ibramed)	830/0.116	30	10 40 80	38.66 s 154.66 s 309.33 s	Crushing of the right fibular nerve.	On the right sciatic nerve.	Application immediately after surgery and during the 21 successive days.	40 J/cm² and 80 J/cm² LLLT influenced the functional recovery of the nerve.
Akgül; Gulsoy; Gulcur [41]	Laser diode (model: DH650-24-3(5), Huanic, China)	650/≈0.14	25	10	57 s on 3 points	Crushing of the sciatic nerve.	On the sciatic nerve.	Early group: Application after surgery, up to the 14th day. Delayed group: Application on the 7th day post-operatory and up to the 21st day.	LLLT accelerated nervous recovery. The group with delayed application showed better functional results.
Gigo-Benato et al. [42]	GaAlAs (TWIN LASER; MM Optics, São Carlos, SP, Brazil)	660/0.04 780/0.04	40 40	10, 60 and 120 10, 60 and 120	0.3 s, 1 min and 2 min 0.3 s, 1 min and 2 min; 2 points	Crushing of the left sciatic nerve.	Applied to the surgical site.	Application 1st day post-operatory, during 10 successive days.	LLLT (660 nm, 10 J/cm² or 60 J/cm²) accelerated the neuromuscular recuperation.
dos Reis et al. [43]	AlGaAs (KLD®; Endophoton model)	660/0.63	26.3	4	96.7 s per point; 3 points	Neurotmeses and epineural anastomosis on the right sciatic nerve.	On the surgical site.	Application 1st day post-operatory, 20 successive days.	LLLT significantly changed the morphometry (myelin sheath), but did not interfere on functionality.
Yang et al. [44]	GaAlAs (Aculas-Am series, Multi-channel LLLT System, Konftec Corp., Taipei, Taiwan)	660/≈0.2	30	9	60 s per point; 4 points	Use of Mesenchymal stem cells (MSC) on the lesion by crushing of sciatic nerve.	On the sciatic nerve	7 successive days.	LLLT+MSC improved the electrophysiologic function, S100 immunoreactivity, less inflammatory cells and less vacuole formation.
de Oliveira Martins et al. [45]	GaAs (Laserpulse-Laser, Ibramed Brazil) pulsado	904/0.1	70 Wpk	6	18 s on 5 points	Pulsed LLLT. Lesion on alveolar nerve, by a hemostatic Crile clamp.	On the sciatic nerve.	10 sessions every 10 days.	LLLT obtained better nociception, higher expression of neural growth factor (NGF) 53% and of expression of neurotrophic factor (BDNF) 40%.

Table 1. *Cont.*

Authors	Type of Laser (Manufacturer)	Wavelength (nm)/Spot Beam	Energy (mW)	Energy Density (J/cm^2)	Radiation Amount	Variables	Irradiation Site	Evaluation Time	Main Results
Gomes; Dalmarco; André [46]	HeNe (—)	632.8/0.1	5	10	20 s on 10 points	Crushing of the right sciatic nerve.	On the sciatic nerve.	1st Application 24 h after surgery; 7, 14 and 21 successive days.	LLLT increased the expression of mRNA and the factors BDNF and NGF after 14 days and maximum expression was observed on the 21st day.
Hsieh et al. [47]	GaAlAs (Aculas-Am series, Multi-channel laser system; Konftec, Taipei, Taiwan)	660/≈0.2	30	9	60 s per point; 4 points	Lesion on the sciatic nerve with 4 ligatures, using chromic suture 4-0.	On the surgery site.	Application 7th post-operatory, during 7 successive days.	LLLT improved functional index, decreased HIF-1a, TNF-a, and IL-1b, increased VEGF, NGF, and S100, reduced tissue ischemia and inflammation, helped the nerve recovery.
Sene et al. [48]	GaAsAl (Physiolux Dual, BIOSET, Rio Claro, Brazil)	830/0.02	30	5 10 20	Maximum time of application was 40 s	Crushing of the right fibular nerve.	Application fibular nerve region.	Application immediately after the lesion, during 21 successive days.	LLLT simulation group obtained a larger nerve transverse area; group 10 J/cm^2 obtained higher density of the fiber. LLLT did not speed up nerve recovery.
Dias et al. [49]	GaAlAs (Mm Twin Laser Optics, São Carlos, Brazil)	780/0.4	30	15	20 s per point; 3 points	Latex protein (F1) on lesion per crushing of sciatic nerve.	On the surgery site, sciatic nerve.	Application per 6 sessions on alternate days.	LLLT associated to the F1 protein did not present positive results and did not potentialize the effects of this protein.

4. Discussion

With the evolution of the technology in the health field and the evolution of the adjunct methods for rehabilitation and functional restoration of injured nerves [3,6–9], the PBMT has shown a wide range of benefits with clinical relevance. Thus, the aim of this research was to carry out a review of the scientific papers published in the last 10 years in order to verify the relation of PBMT in the regeneration of injured peripheral nerves. Regarding the varied benefits of PBMT, the highlight is the reduction of regeneration time and the aid in nerve function.

Among the effects of PBMT on nerve injury, it was verified that the laser minimized the side effects of bupivacaine on the nerve and on the muscle [29], potentiated the process of nerve regeneration observed by morpho-quantitative analysis of the axons and of the nerve fibers [2,3,19,30–35], in addition to assisting muscular reinnervation [36].

Photobiomodulation in the nerve injury was also related to a decrease in inflammatory cytokine levels, in pain, and to the facilitation of neural regeneration, demonstrated by the levels of TNF-a, IL-1b, and GAP-43 [32,37].

The functional analysis evidenced the evolution of functional recovery associated with PBMT [6,7,34,38,39]. Marcolino et al. [40] found a functional recovery with both 40 J/cm^2 and 80 J/cm^2 (830 nm), Akgul; Gulsoy; Gulcur [41] also scored improvement in functionality with late application PBMT (650 nm) (7 days after injury), as well as Medalha et al. [35] at 660 nm at 50 J/cm^2. PBMT 660 nm, 10 J/cm^2, or 60 J/cm^2 accelerated neuromuscular recovery when compared to 780 nm and 830 nm PBMT [42]. Differently, dos Reis et al. [43] observed that PBMT significantly altered morphometry (myelin sheath thickness values) but did not interfere with the functionality.

Yang et al. [44], when associating PBMT with MSC, demonstrated a better electrophysiological function, immunoreactivity of S100, and fewer inflammatory cells. de Oliveira Martins et al. [45] demonstrated that PBMT (904 nm) had better nociception, greater expression of neural growth factor (NGF) 53% and neurotrophic factor expression (BDNF) 40%. As seen, Gomes; Dalmarco; André [46] evidenced that PBMT (632.8 nm) increased mRNA expression, BDNF and NGF factors after 14 days and maximum expression was observed on day 21. PBMT (660 nm) improved functional index, reduced HIF-1a, TNF-a, and IL-1b, elevated VEGF, NGF, and S100, and decreased tissue ischemia and inflammation [47]. Sene et al. [48] (830 nm) observed that PBMT did not accelerate nerve recovery and the study by Dias et al. [49] when associating PBMT (780 nm) with latex protein also did not find positive results.

The effects of PBMT on nerve damage were verified in the sciatic nerve in 17 articles [19,30–36,38,39,41–44,46,47,49], facial nerve in 3 [3,6,7], fibular nerve in 2 articles [40,48], and vagus nerve [2], accessory nerve [29], alveolar nerve [45], and dorsal root [37] in one article each. Of the 26 articles inserted in this review, it was observed that 14 [19,31,32,37–42,44–46,48,49] presented compression as nerve damage (crushing) and 11 [2,3,6,7,30,33–36,43,47] articles evaluated the effects of PBMT on neurotmeses, which is the worst type of nerve injury.

It has been observed that the diversity of PBMT application protocols in nerve lesions is large, with the wavelength varying from 632.8 to 904 nm, a varied range of energy and energy density, in addition to the time of application, despite the similarity in the type of lesion targeted in each experiment. As shown, the infrared spectrum has good experimental results. The red spectrum (600 to 700 nm) [50] was seen in 15 studies with satisfactory morphological and electrophysiological results, immunological factors, and tissue markers [2,30,31,33–36,38,39,41–44,46,47]. It was also possible to verify the lack of standardization in relation to the application protocols, noting that 6 studies were discarded due to lack of data information regarding energy density and time of application of PBMT.

In a general critical analysis of the articles for the detailed study, a consensus was observed on the effectiveness of PBMT, with the use of low-level laser therapy on the improvement of the morphological and morphometric aspects of the regenerated peripheral nerve, as well as on the reduction of events inflammatory and painful sensitivity, providing faster and higher quality functional recovery [51,52].

In the perspective of new fronts of study, in the last decade, optogenetic and chemogenetic techniques have been used more frequently in the investigation of neuronal circuits, as well as in the study of non-neuronal cells in the brain and peripheral nerves. Optogenetics is effective in generating patterns that mimic neuron responses using a pulse generator that produces lights with different frequencies and pulse durations. Photostimulation can be performed in different subcellular regions, being useful for the study of neuronal circuits in the brain. Chemogenetics are less invasive in animal experiments and do not require the installation of a fiber optic cable into the brain or the connection of the cable to a light source, such as a laser or a light emitting diode (LED).

5. Conclusions

At the end of the present study, it can be seen that the data presented in the current articles helped us to understand the beneficial and helpful effects of photobiomodulation on regeneration and functionality after nerve injury. In spite of the great variety of parameters presented, great results were observed, mainly when related to the faster nervous regeneration process.

Conflicts of Interest: The authors declare that they have no conflicts of interest.

References

1. Angeletti, P.; Pereira, M.D.; Gomes, H.C.; Hino, C.T.; Ferreira, L.M. Effect of low-level laser therapy (GaAlAs) on bone regeneration in midpalatal anterior suture after surgically assisted rapid maxillary expansion. *Oral Surg. Oral Med. Oral Pathol. Oral Radiol. Endod.* **2010**, *109*, e38–e46. [CrossRef] [PubMed]
2. Buchaim, R.L.; Andreo, J.C.; Barraviera, B.; Ferreira Junior, R.S.; Buchaim, D.V.; Rosa Junior, G.M.; de Oliveira, A.L.; de Castro Rodrigues, A. Effect of low-level laser therapy (LLLT) on peripheral nerve regeneration using fibrin glue derived from snake venom. *Injury* **2015**, *46*, 655–660. [CrossRef] [PubMed]
3. Buchaim, D.V.; de Castro Rodrigues, A.; Buchaim, R.L.; Barraviera, B.; Junior, R.S.; Junior, G.M.; Bueno, C.R.; Roque, D.D.; Dias, D.V.; Dare, L.R.; et al. The new heterologous fibrina sealant in combination with low-level laser therapy (LLLT) in the repair of the buccal branch of the facial nerve. *Lasers Med. Sci.* **2016**, *31*, 965–972. [CrossRef] [PubMed]
4. De Oliveira Gonçalves, J.B.; Buchaim, D.V.; de Souza Bueno, C.R.; Pomini, K.T.; Barraviera, B.; Júnior, R.S.F.; Andreo, J.C.; de Castro Rodrigues, A.; Cestari, T.M.; Buchaim, R.L. Effects of low-level laser therapy on autogenous bone graft stabilized with a new heterologous fibrin sealant. *J. Photochem. Photobiol. B* **2016**, *162*, 663–668. [CrossRef] [PubMed]
5. De Vasconcellos, L.M.; Barbara, M.A.; Rovai, E.S.; de Oliveira França, M.; Ebrahim, Z.F.; de Vasconcellos, L.G.; Porto, C.D.; Cairo, C.A. Titanium scaffold osteogenesis in healthy and osteoporotic rats is improved by the use of low-level laser therapy (GaAlAs). *Lasers Med. Sci.* **2016**, *31*, 899–905. [CrossRef] [PubMed]
6. Buchaim, D.V.; Andreo, J.C.; Ferreira Junior, R.S.; Barraviera, B.; Rodrigues, A.C.; Macedo, M.C.; Rosa Junior, G.M.; Shinohara, A.L.; Santos German, I.J.; Pomini, K.T.; et al. Efficacy of Laser Photobiomodulation on Morphological and Functional Repair of the Facial Nerve. *Photomed. Laser Surg.* **2017**, *35*, 442–449. [CrossRef] [PubMed]
7. Rosso, M.P.O.; Rosa Júnior, G.M.; Buchaim, D.V.; German, I.J.S.; Pomini, K.T.; de Souza, R.G.; Pereira, M.; Favaretto Júnior, I.A.; Bueno, C.R.S.; Gonçalves, J.B.O.; et al. Stimulation of morphofunctional repair of the facial nerve with photobiomodulation, using the end-to-side technique or a new heterologous fibrin sealant. *J. Photochem. Photobiol. B* **2017**, *175*, 20–28. [CrossRef] [PubMed]
8. Sulewski, J.G. Historical survey of laser dentistry. *Dent. Clin. N. Am.* **2000**, *44*, 717–752. [PubMed]
9. Ginani, F.; Soares, D.M.; Barreto, M.P.; Barboza, C.A. Effect of low-level laser therapy on mesenchymal stem cell proliferation: A systematic review. *Lasers Med. Sci.* **2015**, *30*, 2189–2194. [CrossRef] [PubMed]
10. Morries, L.D.; Cassano, P.; Henderson, T.A. Treatments for traumatic brain injury with emphasis on transcranial near-infrared laser phototherapy. *Neuropsychiatr. Dis. Treat.* **2015**, *20*, 2159–2175. [CrossRef]
11. Chang, W.D.; Wu, J.H.; Wang, H.J.; Jiang, J.A. Therapeutic outcomes of low-level laser therapy for closed bone fracture in the human wrist and hand. *Photomed. Laser Surg.* **2014**, *32*, 212–218. [CrossRef] [PubMed]
12. Gladsjo, J.A.; Jiang, S.I. Treatment of surgical scars using a 595-nm pulsed dye laser using purpuric and nonpurpuric parameters: A comparative study. *Dermatol. Surg.* **2014**, *40*, 118–126. [CrossRef] [PubMed]

13. Keaney, T.C.; Tanzi, E.; Alster, T. Comparison of 532 nm potassium titanyl phosphate laser and 595 nm pulsed dye laser in the treatment of erythematous surgical scars: A randomized, controlled, Open-label study. *Dermatol. Surg.* **2016**, *42*, 70–76. [CrossRef] [PubMed]
14. Li, W.T.; Chen, H.L.; Wang, C.T. Effect of light emitting diode irradiation on proliferation of human bone marrow mesenchymal stem cells. *J. Med. Biol. Eng.* **2006**, *26*, 35–42.
15. Wu, Y.H.; Wang, J.; Gong, D.X.; Gu, H.Y.; Hu, S.S.; Zhang, H. Effects of low-level laser irradiation on mesenchymal stem cell proliferation: A microarray analysis. *Lasers Med. Sci.* **2012**, *27*, 509–519. [CrossRef] [PubMed]
16. Hashmi, J.T.; Huang, Y.Y.; Osmani, B.Z.; Sharma, S.K.; Naeser, M.A.; Hamblin, M.R. Role of low-level laser therapy in neurorehabilitation. *PM&R* **2010**, *2*, 292–305. [CrossRef]
17. Yazdani, S.O.; Golestaneh, A.F.; Shafiee, A.; Hafizi, M.; Omrani, H.A.; Soleimani, M. Effects of low level laser therapy on proliferation and neurotrophic factor gene expression of human schwann cells in vitro. *J. Photochem. Photobiol. B* **2012**, *107*, 9–13. [CrossRef] [PubMed]
18. Martínez de Albornoz, P.; Delgado, P.J.; Forriol, F.; Maffulli, N. Non-surgical therapies for peripheral nerve injury. *Br. Med. Bull.* **2011**, *100*, 73–100. [CrossRef] [PubMed]
19. Ziago, E.K.; Fazan, V.P.; Iyomasa, M.M.; Sousa, L.G.; Yamauchi, P.Y.; da Silva, E.A.; Borie, E.; Fuentes, R.; Dias, F.J. Analysis of the variation in low-level laser energy density on the crushed sciatic nerves of rats: A morphological, quantitative, and morphometric study. *Lasers Med. Sci.* **2017**, *32*, 369–378. [CrossRef] [PubMed]
20. Gigo-Benato, D.; Geuna, S.; Rochkind, S. Phototherapy for enhancing peripheral nerve repair: A review of the literature. *Muscle Nerve* **2005**, *31*, 694–701. [CrossRef] [PubMed]
21. Rosso, M.P.O.; Buchaim, D.V.; Rosa Junior, G.M.; Andreo, J.C.; Pomini, K.T.; Buchaim, R.L. Low-Level Laser Therapy (LLLT) Improves the Repair Process of Peripheral Nerve Injuries: A Mini Review. *Int. J. Neurorehabilit.* **2017**, *4*, 260. [CrossRef]
22. Gigo-Benato, D.; Geuna, S.; de Castro Rodrigues, A.; Tos, P.; Fornaro, M.; Boux, E.; Battiston, B.; Giacobini-Robecchi, M.G. Low-power laser biostimulation enhances nerve repair after end-to-side neurorrhaphy: A double-blind randomized study in the rat median nerve model. *Lasers Med. Sci.* **2004**, *19*, 57–65. [CrossRef] [PubMed]
23. Rochkind, S.; Drory, V.; Alon, M.; Nissan, M.; Ouaknine, G.E. Laser phototherapy (780 nm), a new modality in treatment of long-term incomplete peripheral nerve injury: A randomized double-blind placebo-controlled study. *Photomed. Laser Surg.* **2007**, *25*, 436–442. [CrossRef] [PubMed]
24. Khullar, S.M.; Brodin, P.; Messelt, E.B.; Haanaes, H.R. The effects of low level laser treatment on recovery of nerve conduction and motor function after compression injury in the rat sciatic nerve. *Eur. J. Oral Sci.* **1995**, *103*, 299–305. [CrossRef] [PubMed]
25. Stainki, D.R.; Raiser, A.G.; Graça, D.L.; Becker, C.; Fernandez, G.M.S. Gallium arsenide (GaAs) laser radiation in the radial nerve regeneration submitted secondary to surgical repair. *Braz. J. Vet. Res. Anim. Sci.* **1999**, *35*, 37–40.
26. De Medinaceli, L.; Freed, W.J.; Wyatt, R.J. An index of the functional condition of rat sciatic nerve based on measurements made from walking tracks. *Exp. Neurol.* **1982**, *7*, 634–643. [CrossRef]
27. Gasparini, A.L.P.; Barbieri, C.H.; Mazzer, N. Correlation between different methods of gait functional evaluation in rats with ischiatic nerve crushing injuries. *Acta Ortop. Bras.* **2007**, *15*, 285–289. [CrossRef]
28. Moher, D.; Liberati, A.; Tetzlaff, J.; Altman, D.G.; The PRISMA Group. referred Reporting Items for Systematic Reviews and Meta-Analyses: The PRISMA Statement. *PLoS Med.* **2009**, *6*, e1000097. [CrossRef] [PubMed]
29. Alessi Pissulin, C.N.; Henrique Fernandes, A.A.; Sanchez Orellana, A.M.; Rossi, E.; Silva, R.C.; Michelin Matheus, S.M. Low-level laser therapy (LLLT) accelerates the sternomastoid muscle regeneration process after myonecrosis due to bupivacaine. *J. Photochem. Photobiol. B* **2017**, *168*, 30–39. [CrossRef] [PubMed]
30. Takhtfooladi, M.A.; Sharifi, D. A comparative study of red and blue light-emitting diodes and low-level laser in regeneration of the transected sciatic nerve after an end to end neurorrhaphy in rabbits. *Lasers Med. Sci.* **2015**, *30*, 2319–2324. [CrossRef] [PubMed]
31. Takhtfooladi, M.A.; Jahanbakhsh, F.; Takhtfooladi, H.A.; Yousefi, K.; Allahverdi, A. Effect of low-level laser therapy (685 nm, 3 J/cm^2) on functional recovery of the sciatic nerve in rats following crushing lesion. *Lasers Med. Sci.* **2015**, *30*, 1047–1052. [CrossRef] [PubMed]

32. Wang, C.Z.; Chen, Y.J.; Wang, Y.H.; Yeh, M.L.; Huang, M.H.; Ho, M.L.; Liang, J.I.; Chen, C.H. Low-level laser irradiation improves functional recovery and nerve regeneration in sciatic nerve crush rat injury model. *PLoS ONE* **2014**, *13*, e103348. [CrossRef] [PubMed]

33. Shen, C.C.; Yang, Y.C.; Liu, B.S. Large-area irradiated low-level laser effect in a biodegradable nerve guide conduit on neural regeneration of peripheral nerve injury in rats. *Injury* **2011**, *42*, 803–813. [CrossRef] [PubMed]

34. Shen, C.C.; Yang, Y.C.; Liu, B.S. Effects of large-area irradiated laser phototherapy on peripheral nerve regeneration across a large gap in a biomaterial conduit. *J. Biomed. Mater. Res.* **2013**, *101*, 239–252. [CrossRef] [PubMed]

35. Medalha, C.C.; Di Gangi, G.C.; Barbosa, C.B.; Fernandes, M.; Aguiar, O.; Faloppa, F.; Leite, V.M.; Renno, A.C. Low-level laser therapy improves repair following complete resection of the sciatic nerve in rats. *Lasers Med. Sci.* **2012**, *27*, 629–635. [CrossRef] [PubMed]

36. Shen, C.C.; Yang, Y.C.; Huang, T.B.; Chan, S.C.; Liu, B.S. Low-Level Laser-Accelerated Peripheral Nerve Regeneration within a Reinforced Nerve Conduit across a Large Gap of the Transected Sciatic Nerve in Rats. *Evid. Based Complement. Altern. Med.* **2013**, *2013*, 175629. [CrossRef] [PubMed]

37. Chen, Y.J.; Wang, Y.H.; Wang, C.Z.; Ho, M.L.; Kuo, P.L.; Huang, M.H.; Chen, C.H. Effect of low level laser therapy on chronic compression of the dorsal root ganglion. *PLoS ONE* **2014**, *9*, e89894. [CrossRef] [PubMed]

38. Belchior, A.C.; dos Reis, F.A.; Nicolau, R.A.; Silva, I.S.; Perreira, D.M.; de Carvalho, P.D.T.C. Influence of laser (660 nm) on functional recovery of the sciatic nerve in rats following crushing lesion. *Lasers Med. Sci.* **2009**, *24*, 893–899. [CrossRef] [PubMed]

39. Barbosa, R.I.; Marcolino, A.M.; de Jesus Guirro, R.R.; Mazzer, N.; Barbieri, C.H.; de Cássia Registro Fonseca, M. Comparative effects of wavelengths of low-power laser in regeneration of sciatic nerve in rats following crushing lesion. *Lasers Med. Sci.* **2010**, *25*, 423–430. [CrossRef] [PubMed]

40. Marcolino, A.M.; Barbosa, R.I.; das Neves, L.M.; Mazzer, N.; de Jesus Guirro, R.R.; de Cássia Registro Fonseca, M. Assessment of functional recovery of sciatic nerve in rats submitted to low-level laser therapy with different fluences. An experimental study: Laser in functional recovery in rats. *J. Hand Microsurg.* **2013**, *5*, 49–53. [CrossRef] [PubMed]

41. Akgul, T.; Gulsoy, M.; Gulcur, H.O. Effects of early and delayed laser application on nerve regeneration. *Lasers Med. Sci.* **2014**, *29*, 351–357. [CrossRef] [PubMed]

42. Gigo-Benato, D.; Russo, T.L.; Tanaka, E.H.; Assis, L.; Salvini, T.F.; Parizotto, N.A. Effects of 660 and 780 nm low-level laser therapy on neuromuscular recovery after crush injury in rat sciatic nerve. *Lasers Surg. Med.* **2010**, *42*, 673–682. [CrossRef] [PubMed]

43. Dos Reis, F.A.; Belchior, A.C.G.; de Carvalho, P.T.C.; da Silva, B.A.; Pereira, D.M.; Silva, I.S.; Nicolau, R.A. Effect of laser therapy (660 nm) on recovery of the sciatic nerve in rats after injury through neurotmesis followed by epineural anastomosis. *Lasers Med. Sci.* **2009**, *24*, 741–747. [CrossRef] [PubMed]

44. Yang, C.C.; Wang, J.; Chen, S.C.; Hsieh, Y.L. Synergistic effects of low-level laser and mesenchymal stem cells on functional recovery in rats with crushed sciatic nerves. *J. Tissue Eng. Regener. Med.* **2016**, *10*, 120–131. [CrossRef] [PubMed]

45. De Oliveira Martins, D.; Martinez dos Santos, F.; Evany de Oliveira, M.; de Britto, L.R.; Benedito Dias Lemos, J.; Chacur, M. Laser therapy and pain-related behavior after injury of the inferior alveolar nerve: Possible involvement of neurotrophins. *J. Neurotrauma* **2013**, *30*, 480–486. [CrossRef] [PubMed]

46. Gomes, L.E.; Dalmarco, E.M.; André, E.S. The brain-derived neurotrophic factor, nerve growth factor, neurotrophin-3, and induced nitric oxide synthase expressions after low-level laser therapy in an axonotmesis experimental model. *Photomed. Laser Surg.* **2012**, *30*, 642–647. [CrossRef] [PubMed]

47. Hsieh, Y.L.; Chou, L.W.; Chang, P.L.; Yang, C.C.; Kao, M.J.; Hong, C.Z. Low-level laser therapy alleviates neuropathic pain and promotes function recovery in rats with chronic constriction injury: Possible involvements in hypoxia-inducible factor 1α (HIF-1α). *J. Comp. Neurol.* **2012**, *520*, 2903–2916. [CrossRef] [PubMed]

48. Sene, G.A.; Sousa, F.F.; Fazan, V.S.; Barbieri, C.H. Effects of laser therapy in peripheral nerve regeneration. *Acta Ortop. Bras.* **2013**, *21*, 266–270. [CrossRef] [PubMed]

49. Dias, F.J.; Issa, J.P.; Coutinho-Netto, J.; Fazan, V.P.; Sousa, L.G.; Iyomasa, M.M.; Papa, P.C.; Watanabe, I.S. Morphometric and high-resolution scanning electron microscopy analysis of low-level laser therapy and latex protein (Hevea brasiliensis) administration following a crush injury of the sciatic nerve in rats. *J. Neurol. Sci.* **2015**, *349*, 129–137. [CrossRef] [PubMed]

50. Moore, P.; Ridgway, T.D.; Higbee, R.G.; Howard, E.W.; Lucroy, M.D. Effect of wavelength on low-intensity laser irradiation-stimulated cell proliferation in vitro. *Lasers Surg. Med.* **2005**, *36*, 8–12. [CrossRef] [PubMed]

51. Bang, J.; Kim, H.Y.; Lee, H. Optogenetic and Chemogenetic Approaches for Studying Astrocytes and Gliotransmitters. *Exp. Neurobiol.* **2016**, *25*, 205–221. [CrossRef] [PubMed]

52. Iyer, S.M.; Vesuna, S.; Ramakrishnan, C.; Huynh, K.; Young, S.; Berndt, A.; Lee, S.Y.; Gorini, C.J.; Deisseroth, K.; Delp, S.L. Optogenetic and chemogenetic strategies for sustained inhibition of pain. *Sci. Rep.* **2016**, *6*, 30570. [CrossRef] [PubMed]

3

Electrospun Fibers as a Dressing Material for Drug and Biological Agent Delivery in Wound Healing Applications

Mulugeta Gizaw [1], Jeffrey Thompson [1], Addison Faglie [1], Shih-Yu Lee [2], Pierre Neuenschwander [3] and Shih-Feng Chou [1,*]

[1] Department of Mechanical Engineering, College of Engineering, The University of Texas at Tyler, Tyler, TX 75799, USA; mgizaw@patriots.uttyler.edu (M.G.); jthompson42@patriots.uttyler.edu (J.T.); AFaglie@patriots.uttyler.edu (A.F.)

[2] School of Nursing, College of Nursing and Health Sciences, The University of Texas at Tyler, Tyler, TX 75799, USA; ShihYuLee@uttyler.edu

[3] Department of Cellular and Molecular Biology, The University of Texas Health Science Center at Tyler, Tyler, TX 75708, USA; Pierre.Neuenschwander@uthct.edu

* Correspondence: schou@uttyler.edu

Abstract: Wound healing is a complex tissue regeneration process that promotes the growth of new tissue to provide the body with the necessary barrier from the outside environment. In the class of non-healing wounds, diabetic wounds, and ulcers, dressing materials that are available clinically (e.g., gels and creams) have demonstrated only a slow improvement with current available technologies. Among all available current technologies, electrospun fibers exhibit several characteristics that may provide novel replacement dressing materials for the above-mentioned wounds. Therefore, in this review, we focus on recent achievements in electrospun drug-eluting fibers for wound healing applications. In particular, we review drug release, including small molecule drugs, proteins and peptides, and gene vectors from electrospun fibers with respect to wound healing. Furthermore, we provide an overview on multifunctional dressing materials based on electrospun fibers, including those that are capable of achieving wound debridement and wound healing simultaneously as well as multi-drugs loading/types suitable for various stages of the healing process. Our review provides important and sufficient information to inform the field in development of fiber-based dressing materials for clinical treatment of non-healing wounds.

Keywords: electrospun fibers; drug release; small molecule drugs; proteins and peptides; gene vectors; composites

---

1. Introduction

Wound healing is a complex tissue regeneration process that the body undergoes as a response to wound openings or missing cellular structures as a result of various types of traumatic injury. To facilitate effective wound healing, a wound site is typically covered with a sterile dressing material to avoid infection and to promote the healing process. Gels and creams with a typical drug load of only 5% at the highest are typically applied to the wound site as a treatment method in the clinical setting [1]. These dressings require frequent changing and monitoring/cleaning of the wound site. In patients with non-healing wounds such as diabetic wounds and venous ulcers, the insufficient therapeutic efficacy of the creams and gels coupled with the frequent wound manipulation can often be painful for the patient and cost- and labor-intensive for the healthcare system. With improvements in advanced medical fabrics, a new generation of the wound dressing materials is expected to be able to carry a higher level of drugs and thus provide sustained release properties that will enhance the wound healing process and alleviate much of the painful repetitive procedures of frequent changes of dressing materials. In addition, these new medical fabrics can potentially be incorporated into a multifunctional wound bandage, providing treatment strategies for various types of wounds, locations of the wounds, and conditions of the wounds. The goal is to improve the quality and rate of wound healing while still being able to customize the therapeutic procedures during wound healing.

Different methods have been studied for the fabrication of polymeric fibers, such as melt blowing, phase separation, self-assembly, and temple synthesis. Among these available technologies, drug-eluting fibers made by electrospinning are potential candidates for the formulation of medical fabrics for wound healing. Electrospinning is a simple, robust, and cost-effective method to produce drug-containing fibers with diameters ranging from tens of nanometers to several micrometers [2]. The result is a layer of non-woven fiber mesh that exhibits the texture of typical textiles with a porous structure allowing drainage of the wound exudates and permeation of atmospheric oxygen to the wound. In addition to these advantages, electrospinning is compatible with the incorporation of various types of drugs and/or other drug delivery systems to facilitate generation of various composite dressing materials suitable for controlled release of biological agents at different stages of the wound healing process.

In this review, we discuss different types of wounds and the stages in the wound healing process with respect to the potential use of electrospun fibers as the dressing material in wound healing. In particular, we focus our review on the types of drugs that can be incorporated in electrospun fibers and their drug release behaviors in the context of wound healing. These drugs include small molecule drugs, growth factors, peptides, and non-viral gene vectors. In addition, we provide a brief overview of fiber-based composites that include nanoparticles and/or micelles for generation of multifunctional wound dressings. This review provides a comprehensive discussion on the current status of electrospun fibers for wound healing applications.

2. Wound Healing

A wound can be defined as the disruption or loss of the cellular and/or tissue structure that prevents local tissues from performing their normal biological functions [3]. Since wounds vary in sizes, shapes, and conditions, several methods are used to classify wounds (Table 1).

Table 1. Wound classifications and methods [4].

Classification Method	Subcategory	Characteristics	Examples
Time frame of healing	Acute	Faster healing (5–10 days)	Traumatic wounds, surgical wounds
	Chronic	Takes long time to heal	Leg ulceration
Wound closing method	Primary intention	Treated by closing the surface around the wound	Traumatic lacerations or surgical
	Secondary intentions	Treated by filling the gaps with granulating tissue	Leg ulcers, pressure damage, and lacerations
	Tertiary intention	Open intentionally to allow for drainage to take place	Abdominal wound
Wound tissue types	Black coloration	Shows black discoloration	Necrotic tissue
	Green	Shows green discoloration	Infected tissue
	Yellow	Shows yellow discoloration	Sloughy tissue
	Red	Shows red discoloration	Granulating tissue
	Pink	Shows pink discoloration	Epithelial tissue
Depth of wound	Superficial	Affect the epidermis	Abrasions
	Partial thickness	Affect both the epidermis and the inner dermal layer	Pressure sores and severe scale exits

2.1. Classification of Wounds

The most common way to classify wounds is based on the nature of the wound healing process involved (i.e., acute versus chronic wounds). Specifically, acute wounds are the result of mechanical injuries from external factors resulting in tissue abrasions or tears to the skin and/or flesh. In contrast, chronic wounds are defined as occurring when the normal healing mechanisms of the body are somehow inhibited or disrupted or when tissues are constitutively exposed to damaging environmental factors [5,6]. Particularly problematic is that the healing process of chronic wounds can be readily disrupted by the necessary presence of inflammatory factors, which stimulate the immune system to recruit more macrophages and neutrophils to the wound bed. This causes the additional release of inflammatory cytokines whose presence results in mass production of metalloproteinase that can subsequently disrupt the healing cycle, all of which is characterized by a prolonged inflammatory phase [5]. This is one of the contributing factors that chronic wounds typically take longer to heal (>12 weeks) in comparison to acute wounds (8–12 weeks) that follow a more normal healing and remodeling process [7]. The most common types of chronic wounds that account for 70% of the cases are venous, pressure, or neuropathic ulcers [8]. These non-healing wounds lead to growth of bacteria and other pathogens at the wound sites that elicit increased inflammation, subsequently inhibiting the healing process and resulting in greater chances of complications.

Wounds can also be classified based on the types of closure methods used during the healing process and include Primary Intention, Secondary Intention and Tertiary Intention wound categories. Primary Intention wounds are those that can be treated by closing the surface surrounding the wound using stiches, staples, skin glue, or tape. Secondary Intention wounds are those that involve a wider area of damaged tissue that cannot be closed or stitched together. Examples of Secondary Intention wounds include leg ulcers, pressure damage wounds, and lacerations. Tertiary Intention wounds can also be referred to as delayed Primary Intention wounds and are wounds that are left open intentionally to allow for drainage to occur. These wounds are typically closed once drainage has completed satisfactorily [3,4,6].

A third wound classification method used is based on the color displayed by the damaged or injured tissue. This category includes wounds exhibiting black (necrotic tissue), green (infected tissue), yellow (sloughy tissue), red (granulating tissue), or pink (epithelial tissue) colorations. These categorizations are often used for selecting the appropriate dressing to facilitate a speedy healing process [4]. Additionally, wounds can be classified by the depth of the skin layers that have been affected. Injuries affecting only the epidermis are referred to as superficial wounds, while those affecting both the epidermis and the inner dermal layer (which contains blood vessels, sweat glands and hair follicles) are referred to as partial thickness wounds [9]. These are used largely to describe the severity of pressure sores and severe scale exits [10].

2.2. Wound Healing Cycles

Wound healing is a complex biological process that involves crosstalk between different biological systems for the proper regeneration of cells and tissues and to restore homeostasis and normal biological function [3,11,12]. The typical healing process for a wound includes four overlapping and interdependent stages: hemostasis and coagulation [13]; inflammation; proliferation; and remodeling (Figure 1) [6,14].

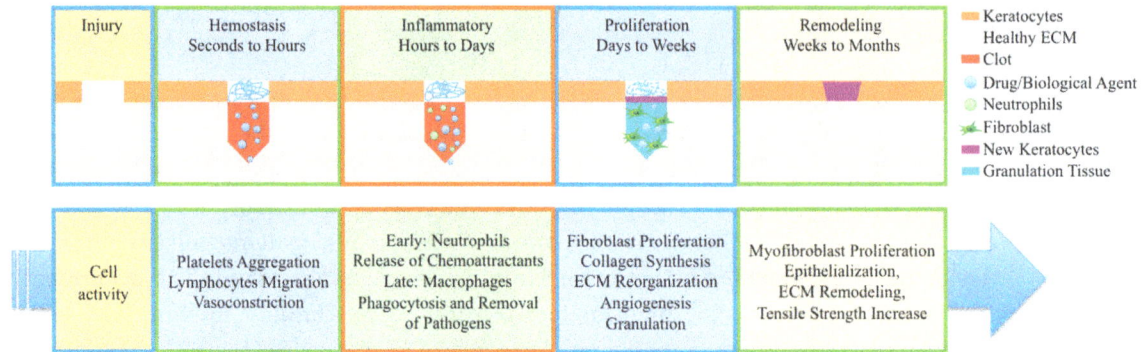

Figure 1. Schematics of typical wound healing cycles and the corresponding cellular activities in each stage.

Hemostasis occurs immediately after tissue injury with the purpose of stopping the loss of blood (exsanguination). Tissue damage results in leakage of blood out of the vessels into the adventitia, where high levels of Tissue Factor protein are present. The contact of plasma factors with Tissue Factor triggers the blood coagulation cascade in the area of damage to generate the fibrin clot. In conjunction with this activity, blood platelets that come into contact with extracellular matrix collagen are activated to form an interim platelet plug that seals the breach until the fibrin clot can form and consolidate the wound. Protein and peptide byproducts of the enzymatic reactions involved in coagulation serve as signals to surrounding cells to recruit immune cells to the clot, which is ultimately composed of fibrin molecules, fibronectin, vitronectin, and thrombospondins [3,6,15]. Although initially involved in hemostasis, the clot subsequently serves as a matrix for immune cells in the subsequent stages of tissue repair and wound healing [14,16].

The inflammatory stage is activated during the hemostasis and coagulation phase and can last up to three days [9,17,18]. Recruiting of neutrophils is triggered at early stages due to the presence of degranulated platelets and byproducts of bacterial degranulation followed by the appearance and transformation of monocytes at later stages [19]. Neutrophils help in the early phase of the inflammatory stage in cleaning bacteria by engulfing them and degrading necrotic tissue. Macrophage enters the injured site within three days and promotes phagocytosis of pathogens and cell debris. They also up-regulate the secretion of growth factors, chemokines, cytokines, and activating the next phase of the wound healing process. Macrophages provide benefits in the wound healing process by boosting defense, promoting and resolving inflammation, removing apoptotic cells, and supporting cell proliferation and tissue restoration following injury [3,6].

The proliferative stage lasts from the third day till the 24th day. When current injury has stopped, hemostasis has been attained and an immune response effectively set in place, the acute wound moves toward tissue repair [18]. The fibrin/fibronectin matrix is replaced with newly formed granulation tissue [14]. At this stage the tissue is very delicate and does not have the normal organization of the surrounding tissues. At the macroscopic level, this phase of wound healing can be perceived as an ample construction of granulation tissue. A fibrous network is formed by the multiplication of fibroblasts forming collagen fibrils. Inflammation also starts to subside at this stage, but in open abnormal wounds the proliferative phase can be prolonged due to more fiber networks needing to be formed [3,6,20].

The remodeling stage occurs from the 24th day and extends to one year [18]. As the concluding phase of wound healing, the remodeling phase is accountable for the growth of new epithelium and final scar tissue formation by achieving a balance between synthesis, deposition, and degradation of the tissues [14]. The wound contracts and becomes smaller due to being pulled by collagen fibers in response to the wound being filled by granulation tissue. While the preliminary deposition of collagen bundles is vastly disorganized, the new collagen matrix becomes more oriented and cross-linked over the phase [3,6]. The healed tissue archives 80% of original tensile strength since some cellular components and their organization cannot be fully recovered during healing [16,17].

2.3. Non-Healing Wounds

A widely accepted mechanism in forming chronic wounds is the interruption or deregulation of one or more of the wound healing cycles that lead to development of non-healing chronic wounds from acute wounds [17]. Furthermore, chronic wounds show improper deregulation of protease and their inhibitors [21]. Matrix components, including fibronectin, as well as various key growth factors are degraded by serine proteases. Non-healing wounds have also been observed with a depletion or small number of chemokine that are needed to recruit bone marrow and endothelial progenitors to the site of injury [22]. The presence and growth of blood vessels surrounding the injured tissue are paramount to the wound healing process, since they supply nutrition and oxygen. Angiogenesis and vasculogenesis contribute to the formation of blood vessels during injury tissue repair [23]. A shortage of any of these factors will lead to the development of chronic wounds.

Management of chronic wounds includes the application of an antimicrobial dressing that help in reducing inflammation and regulation of other pathogens [20]. Wound dressings have been used since ancient times to stop bleeding, and they consist of honey, animal oils or fat, cobwebs, mud, leaves, sphagnum moss, and animal dung. Current commonly used dressing is cotton gauze, and some of its shortcomings include damaging newly formed tissue during removal and causing rapid dehydration of the wound bed, leakage of exudate the might result in infection, and reaction of cotton fibers with foreign body [24]. Ideal wound dressings must meet one or several of the following functions: (1) spot bleeding and protect wound from pathogens, (2) restoration of normal bacterial balance in wound, (3) reduction of inflammation due to unregulated matrix metalloproteinase, and (4) a suitable environment for the control of odor and promotion of autolysis [24]. Based on the types of treatment, developments of wound dressings are commonly categorized in to four groups: passive, interactive, advanced, and bioactive wound dressing. Passive dressings are used to protect mechanical trauma and limit the entrance of pathogens to the wound. Interactive dressings are made from polymeric films, and they facilitate the flow of moisture and air from the environment while providing a barrier from bacteria or other environmental contaminants. Advanced dressings are able to provide and retain a moist environment for the wound and facilitate the healing process. Bioactive dressing works by including drug delivery systems and/or biological agents to stimulate cellular responses in the healing process. The advance of materials science and biomedical engineering has enabled the development of bioactive dressings using natural/synthetic polymers in the form of fibers as a carrier to deliver drugs and biological agents. Fiber based dressing are advantageous to traditional dressing since they provide wide range of advantages including creation of moist and warm environment, remove excess exudates, allow gaseous exchange, and do not release fiber materials to the wound hence minimizing risk of infection [25]. One of the most popular and promising methods in preparation of fibers is electrospinning.

3. Electrospun Fibers

Fibers obtained from electrospinning have gained popularity in the field of drug delivery and are considered as ideal dressing materials for non-healing wounds since the method is versatile and can deliver various biological agents long-term to local tissues at the wound site [26–28]. Not only do they provide physical protection to the wound, but they also have the capacity to be incorporated with a high amount of drugs (up to 40% loading), where the release of which can be adjusted by changing the types and compositions of the materials in the fibers [29]. A large variety of materials can be used to produce electrospun fibers in the pursuit of medical fabrics in wound dressing [26], and these materials can be categorized into natural and synthetic polymers [27,30]. In addition, hydrophilic polymers are ideal for encapsulation of small molecules, proteins, peptides, and gene vectors. The release rate of hydrophilic systems is fast and hence limiting its long-term applications. In contrast, the use of hydrophobic polymers is ideal for controlled release purpose. However, the process requires the use of harmful organic solvents that might affect the stability of the biological agents resulting in the decrease of pharmacological efficacy. In this section, we summarize the most common polymers employed in electrospinning process for wound dressing materials.

3.1. Natural Polymers

Natural polymers have several benefits including being fairly abundant and accessible, as well as being biocompatible, biodegradable, and nontoxic in most cases [31]. In addition, their structural similarity to the ECM promotes and stimulates wound healing process. Others have shown the benefits of using natural polymers for the repair of damaged tissues and consequently in skin regeneration [32]. However, natural polymers are sometimes extremely difficult to electrospin alone due to their molecular structure. This issue can be overcome with the introduction of a synthetic polymer as a carrier to pair with the natural polymer for electrospinning (Table 2). For example, chitosan, a natural occurring polysaccharide, has become a fairly popular material for wound dressing since it is hemostatic, antimicrobial, nontoxic/biocompatible and biodegradable, and capable of sustaining drug release to facilitate wound healing [33,34]. Chitosan is insoluble in water, so the use of organic solvents is necessary during electrospinning. It also exhibits high viscosity at low concentrations when dissolved in organic solvents, which makes electrospinning very difficult. Pakravan et al. showed how these problems can be overcome in electrospinning by paring chitosan with polyethylene oxide (PEO) using 4wt.% stock solution of each at various ratios of 50/50, 70/30, 80/20, and 90/10 chitosan/PEO blends [35]. The paring of these two polymers improved the electrospinnability through strong hydrogen bonding between chitosan and PEO chains. Other popular natural polymers commonly used in blends for electrospinning in application of wound healing include alginate [36], gelatin [37,38], cellulose [39], collagen [40,41], hyaluronic acid [42–44], keratin [45], silk fibroin [46], and zein [47,48].

Table 2. Representative natural and synthetic polymers used in electrospun fibers for wound healing and their corresponding electrospinning parameters.

Polymer(s) [†]	Solvent(s)	Voltage (kV)	Distance (cm)	Flow Rate (mL/h)	Ref.
Natural					
Chitosan/PEO	50% Acetic Acid	15–35	15	0.1–2	[35]
Alginate/Soy Protein/PEO	Deionized Water	15	15	0.5	[49]
Gelatin	20% Acetic Acid	28–35	10	0.1–1	[50]
Cellulose	Acetic Acid	30–40	15	1	[39]
Collagen	PBS/Ethanol	18	15	0.3	[40]
Hyaluronic Acid/PCL	Formic Acid/Acetic Acid (75/25)	13	13	1	[51]
Keratin/PEO	88% Formic Acid	14	15	0.5	[52]
Silk Fibroin	Lithium Bromide	15	18	-	[53]
Synthetic					
PCL	Acedic Acid	9.5–22	15	0.15–1.2	[54]
PLGA/GT	1,1,1,3,3,3 hexafluoro-2-propanol	15	15	1	[55]
PU	N,N-dimethylformamide	35–45	10–15	0.5–1.5	[56]
PVDF	Dimethylformamide and Acetone	25	15	0.75	[57]
PVA/Silk Sericin	Deionized Water	8–12	20	3	[58]
PEO	Ethanol, Chloroform, and Deionized Water	13	10	3	[59]
PVP	Ethanol	15	10	1	[60]

[†] PEO: Poly(ethylene oxide); PCL: Poly(ε-caprolactone); PLGA: Poly(lactic-co-glycolic acid); GT: Gum tragacanth; PU: Polyurethane; PVA: Polyvinyl alcohol; and PVP: Polyvinylpyrrolidone.

3.2. Synthetic Polymers

Synthetic polymers that are biodegradable and biocompatible have been widely used in electrospinning for wound healing purposes. They can be blended with other synthetic or natural polymers to provide sustained release of drugs [2,61]. These polymers can be separated into water soluble and insoluble polymers (Table 2). This characteristic strongly affects the ability of polymers to degrade over time and therefore determine the mechanism in drug release for wound healing applications.

Poly(ε-caprolactone) (PCL) belongs to aliphatic polyester that is extremely popular in the biomedical research sector due to its ease in processing of biomaterials. In addition, its properties, such as being nontoxic, biodegradable, biocompatible to many drugs, and easily accessible, have lent it to be a prime candidate as a long-term drug-delivery carrier. PCL is highly hydrophobic and degrades over several months, and therefore, it is possible to adjust its degradability by blending with a hydrophilic polymer. Ponjavic et al. demonstrated blending water soluble PEO to PCL and showed that the surface properties were greatly improved due to the hydrophilic nature of PEO [62]. Studies have also been done to determine the effect of hydrophilic drugs on the fiber formation and release profile of PCL fibers. For example, Luong-Van et al. showed that increasing loading of the hydrophilic heparin resulted in the decrease of fiber diameter and a sustained release behavior for up to 14 days [63]. Similar to PCL, PLGA is another synthetic polyester that is biodegradable and biocompatible, where its strong mechanical properties make it an ideal candidate as a drug release vehicle. PLGA and PCL exhibit prolonged degradation times, where $t_{1/2} = 30$ days for PLGA and $t_{1/2}$ >18 months for PCL [64–66]. Ranjbar-Mohammadi et al. explored the potential of tuning of hydrophilic tetracycline hydrochloride from blend and core shell fibers using PLGA and gum tragacanth [55]. Results suggested that blended fibers exhibited a much higher cumulative release and initial burst of tetracycline hydrochloride than pure PLGA fibers over 75 days, whereas the core shell structure displayed an intermediate cumulative release profile between blend fibers and pure PLGA fibers. PU and PVDF are synthetic water insoluble polymers that exhibit similar hydrophobic characteristics as PCL and PLGA with various degradation rates that range from several weeks to months [67–69].

In contrast to water insoluble polymers, poly(vinyl alcohol) (PVA) is a synthetic polymer that is soluble in water, biocompatible, and nontoxic. PVA fibers rapidly disintegrate in aqueous solutions resulting in the fast release of drug due to dissolution of the carrier materials [70]. PVA is quite compatible with chitosan as suggested by a study using blend fibers of PVA/chitosan for dressing materials on diabetic wounds [28]. Results showed that the rats treated with PVA/chitosan fiber dressings had an increased wound healing rate as compared to those untreated rats. Similar to PVA, the hydrophilic nature of PEO causes it to disintegrate faster than the hydrophobic polymers. Kim et al. showed the ability to tune release characteristics of protein lysozyme through blending hydrophobic polymers (PCL, PLLA, and PLGA) with hydrophilic PEO [71]. In vitro drug release showed that the PCL/PEO blend received the smallest burst and the most prolonged release profile. Polyethylene glycol (PEG) is also a hydrophilic polymer that has been studied for drug release using blends with a hydrophobic counterpart. Studies have shown that PEG can aid in the healing process of in vivo wounds. Bui et al. explored the effects of curcumin loaded PCL/PEG fibers on wound closure rates in rats, and found that the curcumin loaded blended fibers achieved a 99% wound closure as compared to 90% with just curcumin loaded PCL fibers at 10 days [72]. By controlling the hydrophobicity of these synthetic polymers, a blend polymeric fiber platform is ideal for controlled release of drugs in wound healing applications [60,73,74].

3.3. Electrospinning Parameters

There are several parameters in electrospinning that must be determined in order to produce homogeneous and uniform fibers. These parameters, including solution properties, applied voltage, distance from the tip of the needle to the collector plate, and feeding rate of the polymeric solution, vary with the types of polymers and solvents used during the process. Solvents with moderate boiling points are usually ideal as they are volatile enough to evaporate between the needle tip and collector plate without evaporating too fast and clogging the needle tip [75]. The solvent is also very important when determining the drug that will be incorporated for wound healing purposes. The drug must be able to dissolve in the solvent for complete encapsulation in the fibers. The parameter that dictates the determination of subsequent parameters is the polymer solution viscosity. Megelski et al. showed the influence of solution viscosity on the formation of polystyrene (PS) fibers as fiber diameter increased with increasing solution viscosity [76]. The next parameter that affects electrospun fibers is the applied voltage, which varies with types of polymer solutions used. Each polymer has a critical voltage, or the electric field, that will produce fibers as studies showed that using applied voltage past the critical voltage while holding all other parameters steady created beads during electrospinning of PVA [77]. The solution flow rate has a critical value to produce smooth fibers, and is closely related to the size of the Taylor cone when combined with applied voltage. The increase in flow rate increased fiber diameter and allowed less time for the evaporation of solvent during fiber formation, which can cause beading [76]. The distance from the tip of the needle to the collector has shown an effect on the creation of fibers during electrospinning. Matabola and Moutloali varied the distance when electrospinning poly(vinyledene fluoride) (PVDF) solutions, and found that fiber diameter decreased when increasing the distance, suggesting a complete evaporation of the solvent [78]. While each of these parameters is considered to have its own unique effect on fiber formation during electrospinning, usually these parameters are adjusted together to find the most efficient electrospinning condition. Table 2 provides commonly used natural and synthetic polymers and their electrospinning parameters.

Producing fibers with good structure are important for prolonged delivery of wound healing drugs. Another factor that affects the drug release in wound healing is the type of encapsulation technique used when creating the fibers. There are two main types of methods to create drug-loaded fibers (i.e., uniaxial blends and coaxial core-shell electrospinning). When creating blend fibers, the drug is dissolved in the blend polymer solution and allowed for mixing. For this method, the choice of solvent is important since the particular solvent must be able to dissolve polymers and drugs to create fibers. Uniaxial blends allow the use of hydrophilic and hydrophobic polymers and drugs, if miscible; however, drug partitioning in hydrophilic/hydrophobic phase of the fibers may lead to unexpected release characteristics. Usually this means the drug and polymer must have the same hydrophobicity. Coaxial core-shell electrospinning is a method that allows the formation of layered structure in the radial direction of the fibers. It accommodates much more polymer combinations as compared to the blend fibers since the core and the shell can be from different polymer-solvent system. Core-shell fibers provide the ability to intentionally incorporate hydrophilic and/or hydrophobic drugs at the core of the fibers while the shell serves as a protective layer to prevent burst of the surface drugs [30]. This method not only allows more possible formations of drugs and polymers, but also can prolong the release of the drug to the wound site. One of the drawbacks in core-shell fibers is the swelling of the core polymer, which can cause rupture of the shell layer and expose the drug-containing core to outside environment [79].

4. Release of Small Molecule Drugs

Small molecules represent the majority of the drugs used in treatment of non-healing wounds. A wide variety of small molecule drugs have been incorporated into electrospun fibers for various drug release applications [80]. In this section, we review some of the popular hydrophilic and hydrophobic small molecule drugs incorporated in electrospun fibers and their release behaviors for the potential application in wound healing.

4.1. Hydrophilic Drugs

The use of hydrophilic small molecule drugs to achieve sustained release can be a challenge due to their high solubility in physiological solutions and compatibility with polymers during electrospinning leading to poor or preferential partitioning in the polymer matrix. In addition, the compatibility of the drug in the polymer matrix is indicative of encapsulation efficiency and ability to achieve sustained release behavior. For example, hydrophilic small molecule drugs receive low solubility with nonpolar solvent-polymer systems, and therefore, drugs are likely to partition at the fiber surface, which contributes to burst release. Nonetheless, several hydrophilic drugs for wound healing applications have been incorporated in electrospun fibers and extensively studied for their release characteristics (Table 3). The physicochemical properties of small molecule drugs (i.e., aqueous solubility and Log P) and their loading in the polymer matrix are important factors to consider for drug release behaviors. For example, ciprofloxacin, a hydrophilic antibiotic drug for wound healing, showed burst release behavior in 2 min when incorporated into a water-soluble polymer [81] whereas sustained release of the same drug was reported up to 10 days (80% cumulative) when using a hydrophobic polymer [82]. Similar release behaviors with respect to the polymer used for small molecule hydrophilic antibiotics such as ampicillin [83–85], metronidazole [86–88], and cefazolin [89,90] were reported by others. In addition, drug loading in the polymer matrix plays an important role in the release mechanism of small molecule drugs from fibers. Current trends to achieve sustained release form uniaxial fibers have been utilizing hydrophobic polymers to provide large surface tension to the release media leading to slow release. However, by increasing the loading of hydrophilic drug in the fibers, the overall hydrophobicity changes allowing a better penetration of the release media within fiber mesh. In addition, high drug loading increases surface drug content in fibers, which promotes the burst release behavior. Therefore, a better understanding of the physicochemical properties of hydrophilic small molecule drugs, such as their aqueous solubility and LogP (Table 3), in addition to the polymer used for electrospinning may benefit the development of sustained release wound dressing materials from electrospun fibers.

Table 3. Characteristics of small molecule drugs used in wound healing and their release behaviors from electrospun fibers.

Small Molecules Drugs	Agent		Fiber		Release (Units)			Ref.
	Aq. Sol. † (mg/mL)	Log P †	Polymer(s) ‡	Loading (% w/w)	1 h	2 h	Others	
Hydrophilic								
Ciprofloxacin	1.35	−0.57	PVP	0.4	-	-	60% (1 min)	[81]
			PLCL/PDEGMA	10	12%	20%	80% (220 h)	[82]
			PVA/Alginate	-	30%	40%	85% (6 h)	[91]
Ampicillin	0.605	0.88	AL-BSA	5	23%	37%	99% (96 h)	[83]
				10	17%	25%	81% (96 h)	
				20	7%	10%	40% (96 h)	
			PMMA/Nylon6	1–20	-	-	30% (6 h) 50% (12 days)	[84]
			PCL	16.7	75%	80%	98% (24 h)	[85]
Captopril	4.52	1.02	PLLA	10	-	-	98% (48 h)	[92]
			PLGA	10	-	-	100% (48 h)	
			PLCL	10	-	-	78% (48 h)	
Metronidazole	5.92	-0.15	PCL	1–40	-	-	45% (1 day) 85% (5 days)	[86]
			PCL	4.8–14.4	20%	40%	90% (24 h)	[87]
			Chitosan/PEO	1	52%	75%	-	[88]
				5	70%	80%	-	
				15	70%	100%	-	
Cefazolin	0.487	−0.4	Chitosan/PEO	1	-	26%	65% (24 h)	[89]
			Gelatin	10	10%	30	95% (17 h)	[90]

Table 3. *Cont.*

Small Molecules Drugs	Agent		Fiber		Release (Units)			Ref.
	Aq. Sol. [†] (mg/mL)	Log P [†]	Polymer(s) [‡]	Loading (% *w/w*)	1 h	2 h	Others	
Hydrophobic								
Asiaticoside	[93]	[93]	Alginate/PVA/Chitosan	0.5	20%	23%	83%(12 h)	[94]
				1	20%	40%	45% (5 h)	
Curcumin	0.006	3.62	PHBV	3	55%	65%	70% (5 h)	[95]
				4.7	65%	67%	78% (5 h)	
			PCL/GT	3	-	-	65% (20 days)	[96]
Ketoprofen	0.0213	3.29	PCL/Gelatin	5	-	-	40% (20 h) 80% (45 h)	[97]
			PVA	5	50%	-	62% (48 h)	[98]
			PNVCL-*co*-MAA	20	5%	-	35% (24 h)	[99]
			Cellulose Acetate	15	10%	-	60% (48 h)	[100]
Nifedipine	0.0177	2.49	Eudragit®	10	40%	50%	70% (8 h)	[101]
			PU	4.2	15%	-	75% (72 h)	[102]
			PNIPAAm/PU	12	8%	10%	23% (30 h)	[103]
			PVA	2	27%	29%	88% (48 h)	
Phenytoin	0.0711	2.26	PCL	2	5%	8%	16% (48 h)	[104]
			PVA/PCL	2	11%	15%	47% (48 h)	
Vancomycin	0.255	1.11	Alginate	10	10%	-	60% (48 h)	[105]
Methylene Blue	0.0296	3.61	PHB/PEG	-	32%	-	90% (7 days)	[106]

[†] *DrugBank v5.0.10*: Calculated using ALOGPS v2.1; (accessed 30 November 2017); [‡] PVP: Polyvinylpyrrolidone; PLCL: Poly(lactic-*co*-ε-caprolactone); PDEGMA: Poly(di(ethylene glycol) methyl ether methacrylate); PVA: Polyvinyl alcohol; AL-BSA: Amyloid-like bovine serum albumin; PMMA: Poly(methyl methacrylate); PCL: Poly(ε-caprolactone); PLLA: Poly(L-lactic acid); PLGA: Poly(lactic-*co*-glycolic acid); PEO: Poly(ethylene oxide); PHBV: Poly(3-hydroxybutyric acid-co-3-hydroxyvaleric acid); GT: Gum tragacanth; PNVCL-*co*-MAA: Poly(*N*-vinylcaprolactam-*co*-methacrylic acid); PU: Polyurethane; PNIPAAm: Poly(*N*-isopropylacrylamide); PHB: Poly(R-3-hydroxybutyrate); and PEG: Polyethylene glycol.

4.2. Hydrophobic Drugs

Hydrophobic drugs are generally able to provide a sustained release profile for an extended period of time as comparing to hydrophilic drugs due to their poor solubility in physiological conditions and preferred partitioning in the insoluble polymer matrix than diffusion into the release media (Table 3). In this section, we provide information on the most widely used hydrophobic small molecule drugs in wound healing as well as the commonly used polymeric fibers to achieve the sustained release behavior. One of the most widely studied topical wound healing medications is phenytoin where its mechanism of action includes up-regulating collagen deposition that promotes the formation of fibroblasts, granulation, and other connective tissues with the ability to down-regulate collagenase activity, bacterial colonization, and the formation of wound exudate [104,107,108]. In particular, Zahedi et al. demonstrated the encapsulation of phenytoin in PVA (20% *w/w*) and PCL (17.4% *w/w*) electrospun fibers [104]. In vitro release of phenytoin suggested a cumulative release of 90% and 15% in PVA and PCL fibers, respectively. The differences in release profiles were attributed to the hydrophobicity of PVA and PCL polymer. In addition, in vivo wound closure study showed a 50% reduction of wound area after 6 days using phenytoin incorporated PVA fibers as compared to control phenytoin containing ointment. Another hydrophobic drug that has been extensively studied for wound healing is nifedipine, which serves as calcium antagonist to facilitate blood flow to the wound [109]. Using a water-soluble polymer (i.e., Eudragit), cumulative release of nifedipine in vitro showed 40% and 70% release at 1 h and 8 h, respectively [101]. In another study, nifedipine was encapsulated with PCL-based polyurethane fibers at loadings of 2.7% and 4.2% [102]. In vitro drug concentration in the release media showed positive correlations with drug loading; however, no significant change was found in cumulative drug release with 50% nifedipine release at 24 h followed by 70% release at the end of the study (i.e., 72 h). The slow release behaviors were also observed for curcumin [95,96], ketoprofen [97–100], vancomycin [105], and methylene blue [106] providing the use of proper insoluble drug carriers.

In general, drug types and loading combined with the use of soluble and/or insoluble polymer matrix in electrospinning play a significant role in the release behavior of small molecule drugs. Given that the development in advanced dressing materials using electrospun medical fabric often requires the use of multiple drugs at different stage of the healing process, a better understanding of drug–polymer interactions will benefit the design process of the fiber-based dressing materials.

5. Release of Macromolecules

Wound healing involves complex cellular activities controlled by signaling networks from various growth factors, cytokines, and chemokines [110]. The delivery of macromolecules to the wound site becomes of particular interests for non-healing wounds since it provides a method to remodel normal wound healing cycles. In this section, we review the release of macromolecules (i.e., growth factors and peptides) using electrospun fibers for wound healing.

5.1. Growth Factors

Growth factors are cellular protein secretions that are crucial for tissue remodeling due to their influences on cell cycles and cell fate. Methods to incorporate growth factors in/on electrospun fibers include blend polymers, emulsion, coaxial fibers, and immobilization. The release rates are dictated by the types of technique used. For example, a fast release (>90%) of epidermal growth factor (EGF) and basic fibroblast growth factor (bFGF) were observed in 5 days and 17 days, respectively, whereas a sustained release of platelet-derived growth factor (PDGF-BB) and vascular endothelial growth factor (VEGF) over 1 month was found using blend fibers made from dual electrospinning of collagen and hyaluronic acid [111]. The sustained release is due to the encapsulation of the growth factors in gelatin nanoparticles within the fibers. Others demonstrated the release of bFGF and EGF from coaxial fibers made from block copolymer of PCL-PEG where bFGF solution was encapsulated in the core surrounded by EGF immobilized PCL-PEG shell [112]. The conjugation of EGF to PCL-PEG shell showed a slow release of 2% in a week whereas 30% of bFGF was released in 12 h due to diffusion. In another study, PDGF-BB was absorbed into PCL/collagen/nanohydroxyapatite blend fibers and pure PCL fibers where the blend fibers showed a higher absorption of the growth factor followed by a higher concentration of the PDGF-BB in the release media from the blend fibers as compared to those from pure PCL fibers over 8 weeks of investigation (Figure 2) [113]. Others incorporated transforming growth factor (TGF-β3) into hyaluronic acid (HA) fibers and PCL fibers, and in vitro release study suggested 78% of TGF-β3 was released from HA fibers as compared to 18% from PCL fibers at 2 days followed by 95% release from both fibers at 21 days [114]. Furthermore, others have shown effectiveness of fibroblast growth factor (FGF2) [115], insulin-like growth factor (IGF) [116], granulocyte macrophage colony stimulating factor (GM-CSF) [117], and connective tissue growth factor (CTGF) [118] in promoting cell proliferation, which benefits wound healing (Table 4). In general, with proper incorporation techniques of the growth factors in electrospun fibers, sustained release of the macromolecule agents can be achieved for days to months, which is beneficial for non-healing wounds.

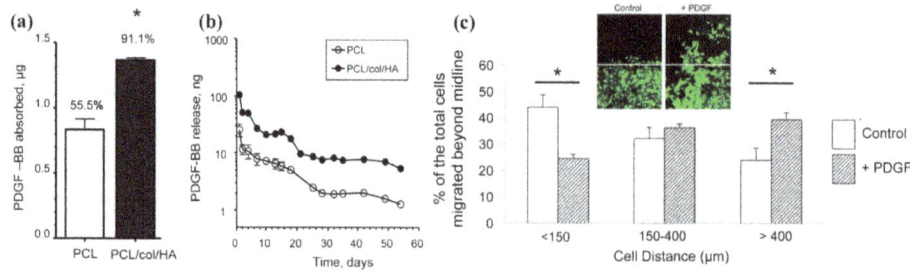

Figure 2. (**a**) PDGF-BB (1.5 µg) was passively absorbed by PCL and PCL/col/HA fibers at 4 °C for 24 h using a PBS bath (300 µL); (**b**) In vitro release profiles of PDGF-BB from fibers over 56 days; (**c**) Release of PDGF-BB promoted MSCs migration using a stringent migration assay (inset shows fluorescent image of the MSC migration) [113]. An asterisk denotes $p < 0.01$.

Table 4. The use of growth factors in electrospun fibers for wound healing applications with respect to solvent used during electrospinning, types of cells studies, and methods to incorporate growth factors in/on fibers.

Growth Factor	Polymer	Solvent	Cell	Method	Ref.
EGF	PCL and PCL–PEG/PCL	Methanol/Chloroform	Human Primary Keratinocyte	Immobilization	[119]
	PLGA and Gelatin	Chloroform/Acetone and Acetic Acid	Human Fibroblasts	Emulsion	[120]
	Silk Fibroin	Lithium Bromide	Human Dermal Fibroblasts	Blend	[121]
	PCL and PCL/Collagen	DMF/DCM and HFIP	Human Dermal Keratinocyte	Immobilization	[122]
	Gelatin/PLA-*co*-PCL	HFIP	Human Dermal Fibroblasts	Coaxial	[123] [124]
	Silk/PEO	Lithium Bromide	-	Blend/Coating	[125]
bFGF/EGF	PCL-PEG	Methanol and Chloroform	Keratinocyte and fibroblast	Coaxial/Immobilization	[112]
	PLGA/PEO	Chloroform and DMF/Water	Human Skin Fibroblasts	Fiber containing GFs encapsulated microspheres.	[126]
bFGF, EGF, VEGF, PDGF	Collagen-Hyaluronic Acid/Gelatin Nanoparticle	Hyaluronic Acid: NaOH/DMF Collagen: Acetic Acid	HUVEC	Blend: bEGF/EGF In nanoparticle: VEGF/PDGF	[111]
PDGF	PCL/Collagen/Hyaluronic Acid	HFP, PBS	MSC	Blend	[113]
FGF2	PHBV, PEO	2, 2, 2-trifluoroethanol	MSC	FGF2-miR-218 induction on aligned PHBV fibers	[115]
KGF	PLA/PCL	Chloroform, Acetone	Fibroblasts	Seeded scaffolds with mouse fibroblast in DMEM with FBS	[127]
TGF-β	MeHA, HH, PCL, HA	DI Water	Cartilage	Composite scaffolds of HA and PCL with TGFβ3	[114]
VEGF	PLGA	Water-in-oil emulsions, Dichloromethane, PBS, BSA	HUVEC, Endothelials	PVEES, and NVEES Scaffolds containing VEGF	[128]
GM-CSF	Chitosan	HCl	In vivo mouse model	Hydrogels containing ovalbumin and GM-CSF	[117]
CTGF	PCL	Chloroform	MSC	Aligned fibers as a guide	[118]

5.2. Peptides

Peptides, such as human cathelicidin peptide LL37, have shown their ability in controlling wound infections through the antibiotic efficacy. In a study, an antimicrobial peptide motif (Cys-KR12) originated from LL37 was immobilized onto silk fibroin fibers [129]. Results showed immobilization processes at various Cys-KR12 concentrations achieved more than 90% yield, and Cysk-KR12 immobilized fibers were able to maintain antibacterial properties for 3 weeks. Furthermore, the study suggested the important role of Cysk-KR12 in wound healing by activating biological activities of keratinocytes, fibroblasts, and monocytes. In another study, a proline-rich peptide (Chex1-Arg20) was electrospun with PVA into fibers for treatment of *Acinetobacter baumannii* infected wounds in mice [130]. Results showed a significant decrease in wound size after 3 days when using Chex1-Arg20 incorporated PVA fibers, whereas the antimicrobial activity of the peptide-loaded fibers was significantly improved. Similarly, Lee et al. incorporated bone forming peptide1 (BFP1) into electrospun PLGA fibers coated in polydopamine (PD) for use of bone regeneration in vivo. Results showed increased bone growth in mice treated with PLGA, PLGA/DP, and PLGA/DP/BFP1 with PLGA/DP/BFP1 having the greatest increase in bone growth [131]. Shao et al. conjugated peptide sequence E7 on electrospun PCL fibers and studied the effects of E7 on the formation of mesenchymal stem cells (MSCs) [132]. After implantation of the E7/PCL fiber meshes into cartilage defective rat knees for 7 days, immunofluorescent staining suggested that the cell growth on the PCL/E7 fibers had a higher percentage of MSC surface markers than the Arg-Gly-Asp peptide (RGD) control group. They also found that the PCL/E7 fibers absorbed less inflammatory cells than the PCL/RGD fibers.

6. Release of Gene Vectors

Gene therapy, as its name implies, is a medical approach that utilizes the delivery of genes to the target cells and/or the use of biological agents such as growth factors to trigger genetic events to further modulate cell behaviors. Similar to other disease states, the success of gene therapy in wound healing is closely associated with the development of delivery systems for gene vectors, which will determine the encapsulation efficiency and release characteristic of the gene. Specifically, therapeutic efficacy and pharmacological results in wound healing depends on gene release rate, which is mediated by cellular uptake during endocytosis followed by biological events of transcription and translation in target cells resulting in the production of proteins. In this section, we review current successes in gene delivery using electrospun fibers for wound healing.

6.1. Non-Viral Genes Vectors

Even though viral vectors (i.e., retrovirus and adenoviruses) possess a higher effectiveness and a better efficacy as compared to non-viral vectors in gene therapy, the use of viral vectors provides a greater chance to trigger immune response [133]. As a result of regulatory, non-viral gene delivery remains the primary method in gene therapy of non-healing wounds. For example, a recent clinical study (NCT01657045) was conducted using a non-viral gene vector (i.e., stromal cell-derived factor-1: SDF-1) for sternal wound edges after open heart surgery, and the results showed significance decreases in scar width (placebo: 35.9 mm and SDF-1: 18.5 mm) and defect volume (placebo: 13.9 mL and SDF-1: 1.4 mL) after 6 months of follow-up on 26 patients [134]. In addition, others observed an increase in diabetic skin wound healing rate after 12 days of follow-up using a mouse model on the delivery of minicircle-VEGF (20 µL) and pβ-EGF (20 µL) cDNA, suggesting that gene therapy can improve wound healing processes (Figure 3) [135]. In parallel to this study, histology observation on the skin tissue at caudal zone of the mice dorsal showed an increase thickness of epithelial tissue after topical administration of keratinocyte growth factor-1 (KGF-1) DNA (control: 16 ± 4 and KGF-1: 26 ± 2 µm) after 48 h while dermal thickness increased in the KGF group (255 ± 36 µm) as compared to the control group with transfected skin (162 ± 16 µm) after 120 h of follow-up [136]. Overall, these examples show that the delivery of non-viral gene vectors is a promising treatment strategy for non-healing wounds.

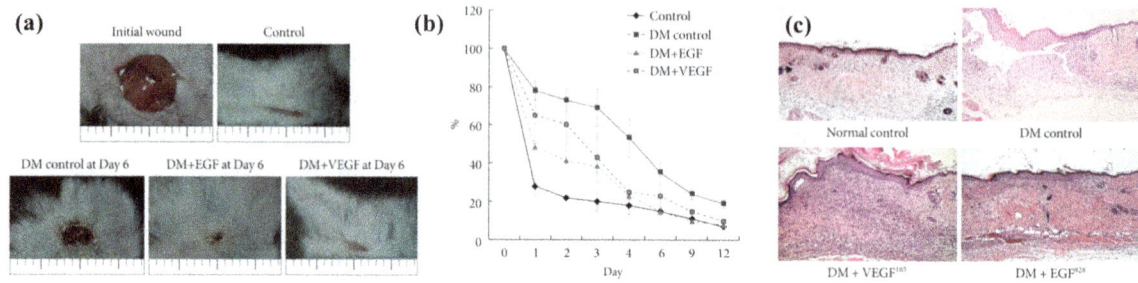

Figure 3. (**a**) Wound healing of a diabetic mouse (DM) model for comparison of effects on delivery of EGF cDNA and VEGF cDNA after 6 days; (**b**) Percent wound closure after receiving gene therapy from the mouse model; (**c**) Histology of the wound tissues from the animal model where tissues receiving EGF and VEGF showed restoration of the tissue structure [135].

6.2. Non-Viral Genes Vectors Delivered by Fiber Platform

The incorporation of gene vectors in electrospun fibers includes encapsulation of the vectors during electrospinning process and immobilization of the vectors after the formation of the fibers. Lee et al. reviewed various gene vectors that were encapsulated in or immobilized on fibers and their release characteristics [137]. In addition, plasmid DNA was attached to linear poly(ethylene imine) segments that were further immobilized with MMP-cleavable peptides onto the amine groups of PCL-PEG blend fibers [138]. Results from the in vitro release study suggested a strong dependence of release characteristics on DNA loading in the fibers. At a plasmid DNA loading of 6.4 ± 0.5 µg per $4~\text{cm}^2$ of fibers, the release curve showed an initial burst of 60% at 12 h followed by 82% release of plasmid DNA at 72 h of incubation time in the presence of MMP-2. Using the same immobilization technique, the researchers attached MMP-responsive siRNA onto PCL-PEG fibers for treatment of diabetic ulcers [139]. The incorporation efficiency was around 77% and the release of siRNA ranged from 30% to 46% depending on the amount of siRNA incorporated onto the fibers at 72 h in the presence of MMP-2. Wound recovery from diabetic ulcer using an in vivo mouse model suggested a 65% recover rate at 7 days when using siRNA immobilized PCL-PEG fibers. Similar to these works, the immobilization of plasmid human epidermal growth factor (phEGF) onto PCL-PEG fibers showed a 2-fold increase in wound recovery rate at 7 days when comparing to the control (no treatment) on an in vivo animal model with diabetic ulcer [140]. The amount of phEGF in the re-epithelized tissue at 14 days was $11~\text{pg}/\text{cm}^2$ when using fibers as compared to $6~\text{pg}/\text{cm}^2$ from the phEGF solution. These examples show that gene delivery from fibers can promote the reconstruction of the tissue in wound and therefore improve the healing process.

The incorporation of plasmid DNA in electrospun fibers is challenging due to the harsh chemicals used for insoluable polymers, which have the potential to provide diffusion-based release mechanism rather than dissolution. Luu et al. demonstrated the ability to electrospin PLGA copolymer and PLA-PEG block copolymer using N,N-dimethyl formamide as the solvent for electrospinning where pCMVβ plasmid was added to the solution [141]. Results suggested a burst release of DNA at around 18% and 36% of cumulative release, depending on block copolymer content, in the first 15 min, perhaps due to the presence of DNA on the surface of the fibers instead of encapsulated inside the fibers followed by 68% to 80% of cumulative release of pCMVβ plasmid at 20 days. Others used a layer-by-layer technique to encapsulate plasmids encoding keratinocyte growth factor (KGF) in blends of PCL and PLA fibers [127]. In vitro release showed an initial 14% burst release of KGF followed by 16% release at 7 days from the fibers. In addition, core-shell fibers show a promising potential to encapsulate DNA inside the fiber structure. In a study, plasmid DNA was encapsulated in the core with the shell composed of poly(ethylenimine)-hyaluronic acid (PEI-HA) at various core-shell compositions, and the results suggested a sustained release behavior up to 60 days [142]. Furthermore, studies showed that electrospinning polyplexes of basic fibroblast growth factor-encoding plasmid (pbFGF) with poly(ethylene imine) in the core and poly(ethylene glycol) as the shell provided a sustained release up to 25 days with 12–19% of initial burst at 12 h (Figure 4) [143,144]. Overall, electrospun fibers demonstrate the potential to encapsulate DNA in the fibers and the sustained release of DNA can provide a much improved therapeutic efficacy in non-healing wounds.

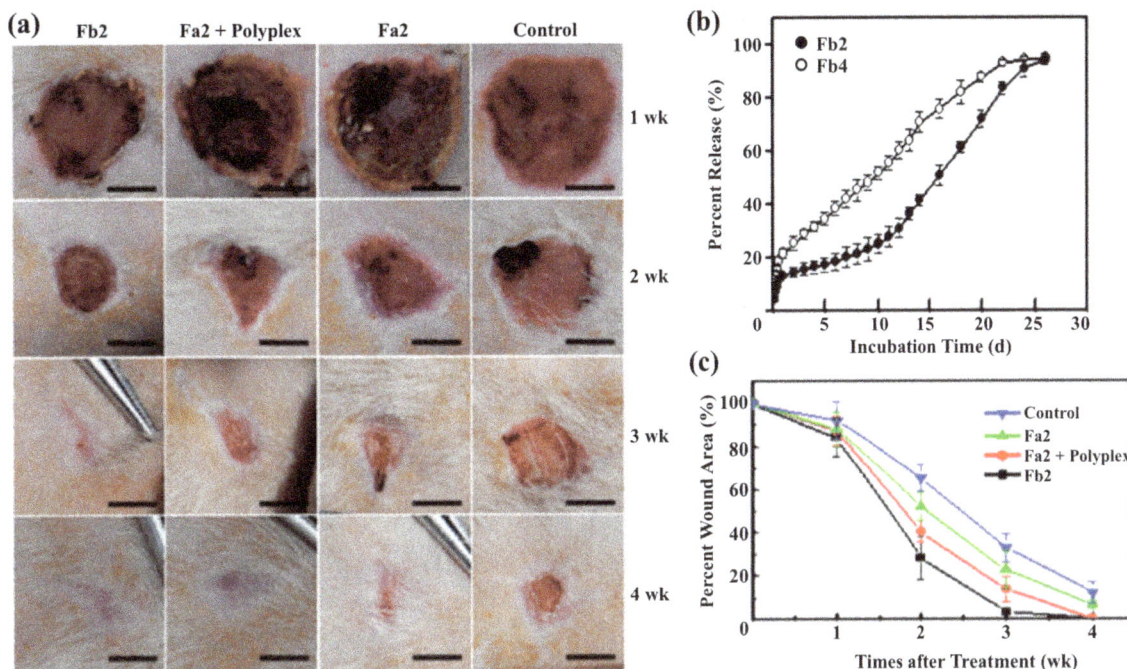

Figure 4. (**a**) Diabetic skin wound using a rat model for comparison of control and those subjected to delivery of pbFGF polyplexes from electrospun poly(ethylene imine)/PEG (2 kDa) core/shell fibers (Fa2: blank fibers and Fb2: fibers with pbFGF polyplexes in the core); (**b**) In vitro release profiles of pbFGF from electrospun poly(ethylene imine)/PEG core/shell fibers (Fb2: 2 kDa PEG and Fb4: 4 kDa PEG); (**c**) Percentage of wound area from the diabetic rat model [143]. Reprinted with permission from American Chemical Society. Copyright (2017) American Chemical Society.

7. Fiber Composites

Nanoparticles and micelles have been widely used in wound healing due to the ability to achieve controlled drug release. In addition, they can be easily incorporated into the traditional cream and gel formulation. In this section, we review the incorporation of nanoparticles and micelles into electrospun fibers for the use of multifunctional wound dressing materials.

7.1. Fiber-Micelle Composites

Micelles have a structure of hydrophobic core surrounded by hydrophilic segments making them an ideal candidate to encapsulate drugs and release them upon contact with physiological fluids. This core shell structure creates a barrier that protects the drug which in turn allows the drug to circulate longer for a prolonged release [145,146]. In work by Redhead et al., Poloxamer 407 and 908 coatings were used with PLGA nanoparticles loaded with Rose Bengal to show the protective effect that polymeric micelles provide [147]. In vivo studies using rats showed that 30% of the dye was still present in the bloodstream 1 h after injection when loaded into the poloxamer coated PLGA nanoparticles. This was a stark contrast to the 8% left in the blood stream after only 5 min when the dye was injected singularly. Others reported the ability of using polymeric micelles from chitosan and palmitic acid to encapsulate and protect an anti-cancer drug (i.e., tamoxifen) [148]. Tamoxifen release profiles showed a much more linear release when encapsulated in chitosan/palmitic acid micelles than with the free drug. Furthermore, polymeric micelles synthesized from phenylboronic acid-functionalized polycarbonate/PEG (PEG-PBC), urea-functionalized polycarbonate/PEG (PEG-PUC), and their diblock copolymers have been reported as drug delivery vehicles for an anti-fungal drug (i.e., amphotericin B) [149]. Result suggested that the PEG-PBC and diblock copolymers of PEG-PBC/PEG-PUC sustained the release of amphotericin B while PEG-PUC showed a burst release profile (Figure 5). Therefore, the use of polymeric micelles has demonstrated the ability to be a potential biomaterial for drug delivery and wound healing processes.

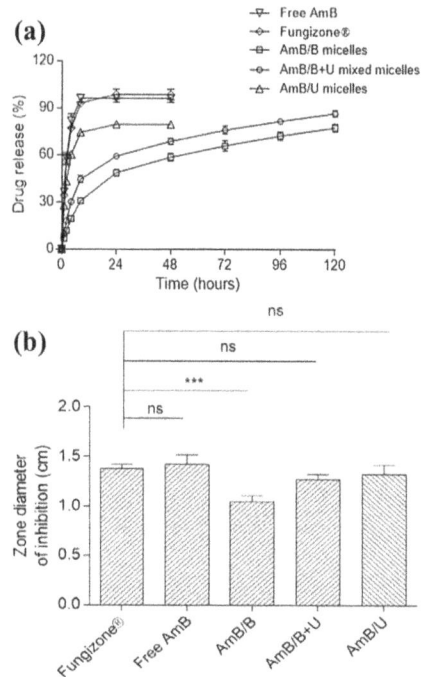

Figure 5. (a) Amphotericin B release profiles from micelles of phenylboronic acid-functionalized polycarbonate/PEG (denotes as B) and urea-functionalized polycarbonate/PEG (denotes as U) in comparison of free drug and Fungizone® using dialysis; (b) Comparison of zone inhibition from C. *albicans* growth after applying amphotericin B containing micelles [149]. "***" denotes $p < 0.001$ and "ns" denotes no significant difference.

Studies have been conducted to explore the ability to incorporate polymeric micelles and/or nanoparticles in electrospun fibers for purposes of drug release and wound healing. Pan et al. demonstrated the biocompatibility of a bilayer scaffold electrospun from PLCL/poloxamer fibers and dextran/gelatin hydrogel (Figure 6) [150]. They found that the fiber scaffolds maintained cell viability and supported cell proliferation of adipose derived stem cells. Similarly, polymeric micelles have also been combined with other delivery systems such as hydrogels. Gong et al. compared the effects of drug release and wound healing characteristics on curcumin loaded PEG-PCL micelles with a combined micelle/hydrogel dressing [151]. The curcumin loaded PEG-PCL micelles showed a sustained drug release over 14 days and achieved a higher cumulative release than the micelle/hydrogel combination. In the in vivo model however, rats treated with the micelle/hydrogel combination showed higher tensile strength with a thicker epidermis during wound breaking testing. There was also an enhancement in wound closure rate using the micelle/hydrogel combination. These findings show the importance of selecting delivery vessels for drug-loaded micelles as the carriers can affect drug release rate and wound healing performance (Table 5). The use of polymeric micelles in drug delivery for wound healing is promising as they provide biocompatibility, extended drug release properties, and shorter healing time that make them ideal for future research.

Table 5. Summary of polymeric micelles used in wound healing.

Polymeric Micelles	Drug	Functions	Ref.
Poloxamer 407 and 908/PLGA nanoparticles	Rose Bengal Dye	Showed protective effects of Poloxamer 407 and 908 micelles.	[147]
Chitosan/Palmitic Acid	Tamoxifen	Release profiles showed much more linear release when encapsulated in micelle structures.	[148]
phenylboronic acid-functionalized polycarbonate/PEG (PEG-PBC)/urea-functionalized polycarbonate/PEG (PEG-PUC)/diblock copolymers	Amphotericin B	Used to study delivery of anti-fungal medication. PEG-PBC and diblock copolymers of PEG-PBC and PEG-PUC showed sustained release of drug while PEG-PUC had burst release profile.	[149]
Poly(L-aspartic acid)-b-poly(ethylene glycol)-b-poly(L-aspartic acid) (PLD-PEG-PLD)	Doxorubicin	Showed effect pH of release media has on release profiles of doxorubicin loaded PLD-PEG-PLD micelles. Found more acidic environment correlated to higher release rates.	[152]
PLCL/poloxamer with dextran/gelatin hydrogel	No Drug	Showed fibers supported cell viability and proliferation when tested with stem cells. Mechanical properties increased with addition of of Poloxamer at 9/1 ratio.	[150]
PEG-PCL and PEG-PCL/hydrogel	Curcumin	Micelle structure sustained release 14 days and achieved higher cumulative release rate than micelle/hydrogel. In Vivo model showed micelle. Hydrogel combination produced higher tensile strength and thicker epidermis during wound healing breaking test. Micelle/Hydrogel also showed enhanced wound closure rate.	[151]

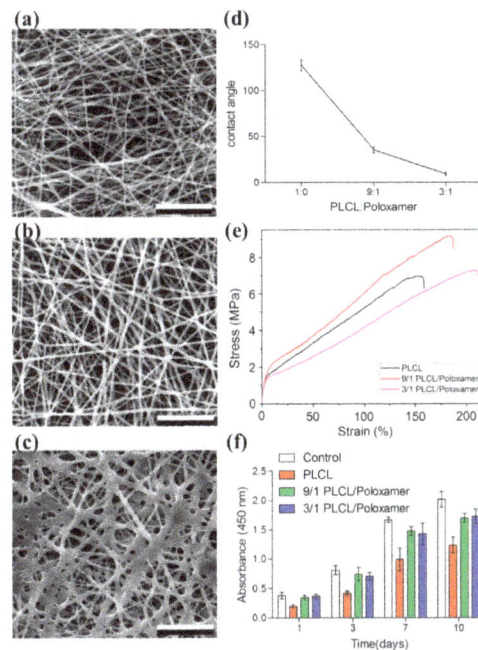

Figure 6. (a) SEM image of fiber structure from PLCL; (b) SEM image of fiber structure from PLCL/poloxamer (9/1 w/w); (c) SEM image of fiber structure from PLCL/poloxamer (3/1 w/w); (d) Water contact angle of PLCL/poloxamer fibers; (e) Stress strain curves of PLCL/poloxamer fibers; (f) Adipose-derived stem cell proliferation on PLCL/poloxamer fibers [150].

7.2. Fiber-Nanoparticle Composites

Similarly, nanoparticles containing electrospun fibers have shown promising potential to make a significant impact in drug release and wound healing research. Studies demonstrated that gold, copper, titanium, and zinc have therapeutic effects during wound healing [153]. In addition, others have incorporated silver nanoparticles into electrospun fibers as an antibacterial agent [154,155]. For example, silver nanoparticles when incorporated with the bipolymer guar gum alkylamine exhibited faster wound healing rates and improved cosmetic attributes [156], whereas gold nanoparticles showed reduction in inflammatory response during the wound healing process [157]. In particular, Leu et al. showed that gold nanoparticles increased cell proliferation resulting in the reduction of wound healing time in mice [158]. Zinc nanoparticles exhibited antibacterial and positive effects on wound healing processes. Raguvaran et al. loaded zinc oxide nanoparticles into sodium alginate/gum acacia hydrogels and showed that zinc oxide at high levels can become toxic but at low levels had antibacterial and healing effects on wounds [159]. Martinez et al. demonstrated the beneficial effects of nitric oxide nanoparticles in wound healing, including antibacterial efficacy that promoted the regeneration of dermal architecture through protection of collagen from bacteria [160]. These examples suggest that nanoparticle provide therapeutic effect in wound healing.

Metallic nanoparticles have been combined with electrospun fibers for the purpose of wound healing. Rather et al. fabricated cerium oxide nanoparticles loaded PCL/Gelatin fibers to investigate the effects in wound healing [161]. The study focused on reducing levels of reactive oxygen species that may hinder proper wound healing when at an elevated level. In a similarly study, adhesive nanocomposite through immobilizing ultrasmall ceria nanocrystals onto the surface of uniform mesoporous silica nanoparticles showed the proper controlling of the reactive oxygen species and the ability to stimulate proliferation and cell migration using an in vivo mouse model [162]. The nanoparticle composite not only decreased healing time, but also reduced scar formation as well [163]. Polymeric nanoparticles have also been chosen for drug delivery for wound healing (Table 6). For example, chitosan based polymeric nanoparticles were fabricated in combination with PLLA-CL electrospun fibers to provide a dual delivery system for Nel-like molecule-1 growth factor [164]. Results indicated that the dual release system prolonged the release of growth factor when compared to plain PLLA-CL fibers. In vitro cell proliferation studies showed that human bone mesenchymal stem cells proliferated better on the dual delivery system than the fibers alone. In another study, chitosan, PVA, and zinc oxide composite nanoparticles displayed a much shorter healing time using an in vivo mouse model, whereas the composite nanoparticles showed almost no bacterial growth in antibacterial activity assay assessed by culturing pus from the wounds after three days of treatment [165]. Lipid nanocarriers have also been used to deliver drugs to wound sites. Sanad et al. used lipid nanocarriers in conjunction with a blended hylaruonic acid/chitosan fiber scaffold for the delivery of the natural diterpene lactone Andrographolide [166]. A prolonged release of Andrographolide, which has anti-inflammatory and antioxidant properties, was observed due to the effects of lipid nanocarrier coupled with the depolymerization of chitosan resulting in the reduction of wound healing time significantly.

Table 6. Summary of nanoparticles used in wound healing on their effects and functions.

Nanoparticles	Effects	Functions	Ref.
Silver/guar gum alkylamine	Antibacterial	Exhibited faster would healing rates and improved cosmetic attributes.	[156]
Gold	Anti-Inflammatory	Wounds exhibited reduction in inflammatory response. Increase in cell proliferation resulting in reduction of wound healing time in mice.	[157]

Table 6. *Cont.*

Nanoparticles	Effects	Functions	Ref.
Zinc Oxide loaded alginate/gun acacia	Antibacterial	Showed that Zinc nanoparticles have antibacterial effects at low levels but can become toxic at high levels.	[159]
Nitric Oxide	Antibacterial	Promoted regeneration of dermal architecture through protection of collagen from bacteria.	[160]
Cerum Oxide loaded PCL/Gelatin fibers	Reduction of reactive oxygen levels, decreased healing time	Lowered the level of reactive oxygen levels that hinder proper wound healing.	[161]
Adhesive nanocomposite made of ultrasmall ceria nanocrystals adhered to the surface of mesoporous silica nanoparticles	Reduction of reactive oxygen levels, decreased healing time	Reduced healing time and scar formation. Stimulated proliferation and cell migration in vivo.	[163]
Chitosan nanoparticles with PLLA-CL fibers	Nel-like mlecule-1 growth factor delivery	Dual release system prolonged release of growth factor when compared to plain PLLA-CL fibers. Dual release system Increased cell proliferation in human bone mesenchymal stem cells.	[164]
Chitosan/PVA/Zinc Oxide	Decreased wound healing time/Antibacterial	Displayed shorter healing time and almost no bacterial growth in cultured pus from wounds.	[165]
Lipid nanocarrier/Hyaluronic Acid/Chitosan	Drug delivery	Prolonged release of Andrographolide combined with depolymerization of chitosan resulted in the reduction of wound healing time.	[166]

8. Conclusions and Future Directions

Non-healing wounds remain a challenge for the development of dressing materials. Advances in nanotechnology enable the production of electrospun fibers, which have the potential to become the ideal candidate for encapsulation and delivery of small molecule drugs and/or large macromolecules to the wound site. In particular, electrospinning accepts most of the polymer and drugs where the interactions between them play in important role in drug release rates. For example, delivery of the coagulation factors and anti-inflammation drugs may be required for early stages of wound healing. The choice of using water-soluble (dissolution mechanism) polymers as well as those with minimal drug–polymer interactions (diffusion mechanism) facilitates the fast release of the drugs. In contrast, the proliferation and remodeling processes in late wound healing stages require the sustained delivery of the growth factors and genes. The use of blend fibers to enhance drug–polymer interactions, coaxial fibers to encapsulate drug in the core, and fiber composites will benefit the prolonged delivery of the biological agents. Therefore, the choice of using particular polymers and architectures in electrospun fibers will depend on the types of drugs and the stage of wound so that the healing process can be improved.

In this review, we explore the incorporation of various wound healing drugs, including small molecules, macromolecules, and gene vectors, in electrospun fibers and their release behaviors for wound healing. In addition, our review suggests that electrospun fibers are capable of integrating with typical small molecules, growth factors, and gene vectors to provide a sustained release behavior, depending on the polymer used. Furthermore, the incorporation of micelles/nanoparticles in fibers allows the formation of a composite material for multifunctional delivery purpose. While electrospinning possesses many advantages in drug delivery and tissue engineering, which are beneficial for wound healing, concerns over the use of harsh chemicals (cytotoxicity) may limit its use in pharmaceutical applications for dressing materials. In such case, exhaustion of the organic solvents under vacuum is required to eliminate residual chemicals that remain in the fibers after electrospinning. This is a costly and time-consuming step. Furthermore, low production rate (e.g., approximately 1~1.5 g/h via uniaxial electrospinning) can be another issue that limits the use of electrospun fibers in clinical aspect. This limitation has been improved by free-surface electrospinning process [167,168], whereas the production rates may be 5–10 fold higher than typical uniaxial electrospinning. In general, electrospun fibers demonstrate the ability of sustained release of small molecule drugs, macromolecules, and genes. This drug delivery platform is especially ideal for the use of topical dressing materials in wound healing applications.

Conflicts of Interest: The authors declare no conflict of interest.

References

1. Sidgwick, G.P.; McGeorge, D.; Bayat, A. A comprehensive evidence-based review on the role of topicals and dressings in the management of skin scarring. *Arch. Dermatol. Res.* **2015**, *307*, 461–477. [CrossRef] [PubMed]
2. Chou, S.-F.; Carson, D.; Woodrow, K.A. Current strategies for sustaining drug release from electrospun nanofibers. *J. Control. Release* **2015**, *220*, 584–591. [CrossRef] [PubMed]
3. Strodtbeck, F. Physiology of wound healing. *Newborn Infant Nurs. Rev.* **2001**, *1*, 43–52. [CrossRef]
4. Merlin-Manton, E. Wound care: Selecting the right dressings. *Pract. Nurse* **2017**, *47*, 28–32.
5. Tejiram, S.; Kavalukas, S.L.; Shupp, J.W.; Barbul, A. 1-Wound healing. In *Wound Healing Biomaterials*; Ågren, M.S., Ed.; Woodhead Publishing: Sawston, Cambridge, UK, 2016; pp. 3–39. ISBN 978-1-78242-455-0.
6. Velnar, T.; Bailey, T.; Smrkolj, V. The wound healing process: An overview of the cellular and molecular mechanisms. *J. Int. Med. Res.* **2009**, *37*, 1528–1542. [CrossRef] [PubMed]
7. Abrigo, M.; McArthur, S.L.; Kingshott, P. Electrospun nanofibers as dressings for chronic wound care: Advances, challenges, and future prospects. *Macromol. Biosci.* **2014**, *14*, 772–792. [CrossRef] [PubMed]
8. Whitney, J.D. Overview: Acute and Chronic Wounds. *Nurs. Clin. N. Am.* **2005**, *40*, 191–205. [CrossRef] [PubMed]
9. Boateng, J.S.; Matthews, K.H.; Stevens, H.N.E.; Eccleston, G.M. Wound healing dressings and drug delivery systems: A review. *J. Pharm. Sci.* **2008**, *97*, 2892–2923. [CrossRef] [PubMed]
10. Arasteh, S.; Kazemnejad, S.; Khanjani, S.; Heidari-Vala, H.; Akhondi, M.M.; Mobini, S. Fabrication and characterization of nano-fibrous bilayer composite for skin regeneration application. *Methods* **2016**, *99*, 3–12. [CrossRef] [PubMed]
11. Lee, Y.-H.; Chang, J.-J.; Yang, M.-C.; Chien, C.-T.; Lai, W.-F. Acceleration of wound healing in diabetic rats by layered hydrogel dressing. *Carbohydr. Polym.* **2012**, *88*, 809–819. [CrossRef]
12. Kasuya, A.; Tokura, Y. Attempts to accelerate wound healing. *J. Dermatol. Sci.* **2014**, *76*, 169–172. [CrossRef] [PubMed]
13. Neuenschwander, P.F.; Jesty, J. Blood coagulation. In *Encyclopedia of Life Sciences*; John Wiley & Sons, Ltd.: Chichester, UK, 2011; ISBN 978-0-470-01617-6.
14. Castellanos, G.; Bernabé-García, Á.; Moraleda, J.M.; Nicolás, F.J. Amniotic membrane application for the healing of chronic wounds and ulcers. *Placenta* **2017**, *59*, 146–153. [CrossRef] [PubMed]
15. Hosgood, G. Stages of Wound healing and their clinical relevance. *Vet. Clin. N. Am. Small Anim. Pract.* **2006**, *36*, 667–685. [CrossRef] [PubMed]
16. Reinke, J.M.; Sorg, H. Wound repair and regeneration. *Eur. Surg. Res.* **2012**, *49*, 35–43. [CrossRef] [PubMed]
17. Landén, N.X.; Li, D.; Ståhle, M. Transition from inflammation to proliferation: A critical step during wound healing. *Cell. Mol. Life Sci.* **2016**, *73*, 3861–3885. [CrossRef] [PubMed]
18. Clinical Guidelines (Nursing): Wound Care. Available online: https://www.rch.org.au/rchcpg/hospital_clinical_guideline_index/Wound_care/ (accessed on 19 October 2017).
19. Hanna, J.R.; Giacopelli, J.A. A review of wound healing and wound dressing products. *J. Foot Ankle Surg.* **1997**, *36*, 2–14. [CrossRef]

20. Harding, K.G.; Morris, H.L.; Patel, G.K. Healing chronic wounds. *BMJ Br. Med. J. Lond.* **2002**, *324*, 160–163. [CrossRef]

21. Dovi, J.V.; Szpaderska, A.M.; DiPietro, L.A. Neutrophil function in the healing wound: Adding insult to injury? *Thromb. Haemost.* **2004**, *92*, 275–280. [CrossRef] [PubMed]

22. Mathews, V.; Hanson, P.T.; Ford, E.; Fujita, J.; Polonsky, K.S.; Graubert, T.A. Recruitment of bone marrow-derived endothelial cells to sites of pancreatic β-cell injury. *Diabetes* **2004**, *53*, 91–98. [CrossRef] [PubMed]

23. Velazquez, O.C. Angiogenesis and vasculogenesis: Inducing the growth of new blood vessels and wound healing by stimulation of bone marrow–derived progenitor cell mobilization and homing. *J. Vasc. Surg.* **2007**, *45*, A39–A47. [CrossRef] [PubMed]

24. Aramwit, P. 1-Introduction to biomaterials for wound healing. In *Wound Healing Biomaterials*; Ågren, M.S., Ed.; Woodhead Publishing: Sawston, Cambridge, UK, 2016; pp. 3–38. ISBN 978-1-78242-456-7.

25. Sood, A.; Granick, M.S.; Tomaselli, N.L. Wound dressings and comparative effectiveness data. *Adv. Wound Care* **2014**, *3*, 511–529. [CrossRef] [PubMed]

26. Liu, M.; Duan, X.-P.; Li, Y.-M.; Yang, D.-P.; Long, Y.-Z. Electrospun nanofibers for wound healing. *Mater. Sci. Eng. C* **2017**, *76*, 1413–1423. [CrossRef] [PubMed]

27. Wang, J.; Windbergs, M. Functional electrospun fibers for the treatment of human skin wounds. *Eur. J. Pharm. Biopharm.* **2017**, *119*, 283–299. [CrossRef] [PubMed]

28. Ahmadi Majd, S.; Rabbani Khorasgani, M.; Moshtaghian, S.J.; Talebi, A.; Khezri, M. Application of Chitosan/PVA Nano fiber as a potential wound dressing for streptozotocin-induced diabetic rats. *Int. J. Biol. Macromol.* **2016**, *92*, 1162–1168. [CrossRef] [PubMed]

29. Chou, S.-F.; Woodrow, K.A. Relationships between mechanical properties and drug release from electrospun fibers of PCL and PLGA blends. *J. Mech. Behav. Biomed. Mater.* **2017**, *65*, 724–733. [CrossRef] [PubMed]

30. Chen, S.; Boda, S.K.; Batra, S.K.; Li, X.; Xie, J. Emerging roles of electrospun nanofibers in cancer research. *Adv. Healthc. Mater.* **2017**. [CrossRef] [PubMed]

31. Mogoşanu, G.D.; Grumezescu, A.M. Natural and synthetic polymers for wounds and burns dressing. *Int. J. Pharm.* **2014**, *463*, 127–136. [CrossRef] [PubMed]

32. Huang, S.; Fu, X. Naturally derived materials-based cell and drug delivery systems in skin regeneration. *J. Control. Release* **2010**, *142*, 149–159. [CrossRef] [PubMed]

33. Dai, T.; Tanaka, M.; Huang, Y.-Y.; Hamblin, M.R. Chitosan preparations for wounds and burns: Antimicrobial and wound-healing effects. *Expert Rev. Anti-Infect. Ther.* **2011**, *9*, 857–879. [CrossRef] [PubMed]

34. Bano, I.; Arshad, M.; Yasin, T.; Ghauri, M.A.; Younus, M. Chitosan: A potential biopolymer for wound management. *Int. J. Biol. Macromol.* **2017**, *102*, 380–383. [CrossRef] [PubMed]

35. Pakravan, M.; Heuzey, M.-C.; Ajji, A. A fundamental study of chitosan/PEO electrospinning. *Polymer* **2011**, *52*, 4813–4824. [CrossRef]

36. Lu, J.-W.; Zhu, Y.-L.; Guo, Z.-X.; Hu, P.; Yu, J. Electrospinning of sodium alginate with poly(ethylene oxide). *Polymer* **2006**, *47*, 8026–8031. [CrossRef]

37. Topuz, F.; Uyar, T. Electrospinning of gelatin with tunable fiber morphology from round to flat/ribbon. *Mater. Sci. Eng. C* **2017**, *80*, 371–378. [CrossRef] [PubMed]

38. Ostrovidov, S.; Shi, X.; Zhang, L.; Liang, X.; Kim, S.B.; Fujie, T.; Ramalingam, M.; Chen, M.; Nakajima, K.; Al-Hazmi, F.; et al. Myotube formation on gelatin nanofibers—Multi-walled carbon nanotubes hybrid scaffolds. *Biomaterials* **2014**, *35*, 6268–6277. [CrossRef] [PubMed]

39. Zhang, K.; Li, Z.; Kang, W.; Deng, N.; Yan, J.; Ju, J.; Liu, Y.; Cheng, B. Preparation and characterization of tree-like cellulose nanofiber membranes via the electrospinning method. *Carbohydr. Polym.* **2018**, *183*, 62–69. [CrossRef] [PubMed]

40. Bak, S.Y.; Yoon, G.J.; Lee, S.W.; Kim, H.W. Effect of humidity and benign solvent composition on electrospinning of collagen nanofibrous sheets. *Mater. Lett.* **2016**, *181*, 136–139. [CrossRef]

41. Sadeghi-Avalshahr, A.; Nokhasteh, S.; Molavi, A.M.; Khorsand-Ghayeni, M.; Mahdavi-Shahri, M. Synthesis and characterization of collagen/PLGA biodegradable skin scaffold fibers. *Regen. Biomater.* **2017**, *4*, 309–314. [CrossRef] [PubMed]

42. Kutlusoy, T.; Oktay, B.; Apohan, N.K.; Süleymanoğlu, M.; Kuruca, S.E. Chitosan-co-Hyaluronic acid porous cryogels and their application in tissue engineering. *Int. J. Biol. Macromol.* **2017**, *103*, 366–378. [CrossRef] [PubMed]

43. Liu, Y.; Ma, G.; Fang, D.; Xu, J.; Zhang, H.; Nie, J. Effects of solution properties and electric field on the electrospinning of hyaluronic acid. *Carbohydr. Polym.* **2011**, *83*, 1011–1015. [CrossRef]

44. Brenner, E.K.; Schiffman, J.D.; Thompson, E.A.; Toth, L.J.; Schauer, C.L. Electrospinning of hyaluronic acid nanofibers from aqueous ammonium solutions. *Carbohydr. Polym.* **2012**, *87*, 926–929. [CrossRef]

45. Esparza, Y.; Ullah, A.; Boluk, Y.; Wu, J. Preparation and characterization of thermally crosslinked poly(vinyl alcohol)/feather keratin nanofiber scaffolds. *Mater. Des.* **2017**, *133*, 1–9. [CrossRef]

46. Yukseloglu, S.M.; Sokmen, N.; Canoglu, S. Biomaterial applications of silk fibroin electrospun nanofibres. *Microelectron. Eng.* **2015**, *146*, 43–47. [CrossRef]

47. Dias Antunes, M.; da Silva Dannenberg, G.; Fiorentini, Â.M.; Pinto, V.Z.; Lim, L.-T.; da Rosa Zavareze, E.; Dias, A.R.G. Antimicrobial electrospun ultrafine fibers from zein containing eucalyptus essential oil/cyclodextrin inclusion complex. *Int. J. Biol. Macromol.* **2017**, *104*, 874–882. [CrossRef] [PubMed]

48. Deng, L.; Kang, X.; Liu, Y.; Feng, F.; Zhang, H. Characterization of gelatin/zein films fabricated by electrospinning vs solvent casting. *Food Hydrocoll.* **2018**, *74*, 324–332. [CrossRef]

49. Wongkanya, R.; Chuysinuan, P.; Pengsuk, C.; Techasakul, S.; Lirdprapamongkol, K.; Svasti, J.; Nooeaid, P. Electrospinning of alginate/soy protein isolated nanofibers and their release characteristics for biomedical applications. *J. Sci. Adv. Mater. Devices* **2017**, *2*, 309–316. [CrossRef]

50. Okutan, N.; Terzi, P.; Altay, F. Affecting parameters on electrospinning process and characterization of electrospun gelatin nanofibers. *Food Hydrocoll.* **2014**, *39*, 19–26. [CrossRef]

51. Entekhabi, E.; Haghbin Nazarpak, M.; Moztarzadeh, F.; Sadeghi, A. Design and manufacture of neural tissue engineering scaffolds using hyaluronic acid and polycaprolactone nanofibers with controlled porosity. *Mater. Sci. Eng. C* **2016**, *69*, 380–387. [CrossRef] [PubMed]

52. Ma, H.; Shen, J.; Cao, J.; Wang, D.; Yue, B.; Mao, Z.; Wu, W.; Zhang, H. Fabrication of wool keratin/polyethylene oxide nano-membrane from wool fabric waste. *J. Clean. Prod.* **2017**, *161*, 357–361. [CrossRef]

53. Kishimoto, Y.; Morikawa, H.; Yamanaka, S.; Tamada, Y. Electrospinning of silk fibroin from all aqueous solution at low concentration. *Mater. Sci. Eng. C* **2017**, *73*, 498–506. [CrossRef] [PubMed]

54. Tampau, A.; González-Martinez, C.; Chiralt, A. Carvacrol encapsulation in starch or PCL based matrices by electrospinning. *J. Food Eng.* **2017**, *214*, 245–256. [CrossRef]

55. Ranjbar-Mohammadi, M.; Zamani, M.; Prabhakaran, M.P.; Bahrami, S.H.; Ramakrishna, S. Electrospinning of PLGA/gum tragacanth nanofibers containing tetracycline hydrochloride for periodontal regeneration. *Mater. Sci. Eng. C* **2016**, *58*, 521–531. [CrossRef] [PubMed]

56. Ju, J.; Shi, Z.; Fan, L.; Liang, Y.; Kang, W.; Cheng, B. Preparation of elastomeric tree-like nanofiber membranes using thermoplastic polyurethane by one-step electrospinning. *Mater. Lett.* **2017**, *205*, 190–193. [CrossRef]

57. Dorneanu, P.P.; Cojocaru, C.; Olaru, N.; Samoila, P.; Airinei, A.; Sacarescu, L. Electrospun PVDF fibers and a novel PVDF/CoFe$_2$O$_4$ fibrous composite as nanostructured sorbent materials for oil spill cleanup. *Appl. Surf. Sci.* **2017**, *424*, 389–396. [CrossRef]

58. Yan, S.; Li, X.; Dai, J.; Wang, Y.; Wang, B.; Lu, Y.; Shi, J.; Huang, P.; Gong, J.; Yao, Y. Electrospinning of PVA/sericin nanofiber and the effect on epithelial-mesenchymal transition of A549 cells. *Mater. Sci. Eng. C* **2017**, *79*, 436–444. [CrossRef] [PubMed]

59. Son, W.K.; Youk, J.H.; Lee, T.S.; Park, W.H. The effects of solution properties and polyelectrolyte on electrospinning of ultrafine poly(ethylene oxide) fibers. *Polymer* **2004**, *45*, 2959–2966. [CrossRef]

60. Reksamunandar, R.P.; Edikresnha, D.; Munir, M.M.; Damayanti, S. Khairurrijal Encapsulation of β-carotene in poly(vinylpyrrolidone) (PVP) by Electrospinning Technique. *Procedia Eng.* **2017**, *170*, 19–23. [CrossRef]

61. Chou, S.-F.; Gunaseelan, S.; Kiellani, M.H.H.; Thottempudi, V.V.K.; Neuenschwander, P.; Nie, H. A review of injectable and implantable biomaterials for treatment and repair of soft tissues in wound healing. *J. Nanotechnol.* **2017**, *2017*, 1–15. [CrossRef]

62. Ponjavic, M.; Nikolic, M.S.; Nikodinovic-Runic, J.; Jeremic, S.; Stevanovic, S.; Djonlagic, J. Degradation behaviour of PCL/PEO/PCL and PCL/PEO block copolymers under controlled hydrolytic, enzymatic and composting conditions. *Polym. Test.* **2017**, *57*, 67–77. [CrossRef]

63. Luong-Van, E.; Grøndahl, L.; Chua, K.N.; Leong, K.W.; Nurcombe, V.; Cool, S.M. Controlled release of heparin from poly(ε-caprolactone) electrospun fibers. *Biomaterials* **2006**, *27*, 2042–2050. [CrossRef] [PubMed]

64. Lu, L.; Garcia, C.A.; Mikos, A.G. In vitro degradation of thin poly(DL-lactic-*co*-glycolic acid) films. *J. Biomed. Mater. Res.* **1999**, *46*, 236–244. [CrossRef]

65. Park, T.G. Degradation of poly(lactic-co-glycolic acid) microspheres: Effect of copolymer composition. *Biomaterials* **1995**, *16*, 1123–1130. [CrossRef]

66. Peña, J.; Corrales, T.; Izquierdo-Barba, I.; Doadrio, A.L.; Vallet-Regí, M. Long term degradation of poly(ε-caprolactone) films in biologically related fluids. *Polym. Degrad. Stab.* **2006**, *91*, 1424–1432. [CrossRef]

67. You, Y.; Min, B.-M.; Lee, S.J.; Lee, T.S.; Park, W.H. In vitro degradation behavior of electrospun polyglycolide, polylactide, and poly(lactide-co-glycolide). *J. Appl. Polym. Sci.* **2005**, *95*, 193–200. [CrossRef]

68. Yeganegi, M.; Kandel, R.A.; Santerre, J.P. Characterization of a biodegradable electrospun polyurethane nanofiber scaffold: Mechanical properties and cytotoxicity. *Acta Biomater.* **2010**, *6*, 3847–3855. [CrossRef] [PubMed]

69. Sheikh, F.A.; Zargar, M.A.; Tamboli, A.H.; Kim, H. A super hydrophilic modification of poly(vinylidene fluoride) (PVDF) nanofibers: By in situ hydrothermal approach. *Appl. Surf. Sci.* **2016**, *385*, 417–425. [CrossRef]

70. Li, X.; Kanjwal, M.A.; Lin, L.; Chronakis, I.S. Electrospun polyvinyl-alcohol nanofibers as oral fast-dissolving delivery system of caffeine and riboflavin. *Colloids Surf. B Biointerfaces* **2013**, *103*, 182–188. [CrossRef] [PubMed]

71. Kim, T.G.; Lee, D.S.; Park, T.G. Controlled protein release from electrospun biodegradable fiber mesh composed of poly(ε-caprolactone) and poly(ethylene oxide). *Int. J. Pharm.* **2007**, *338*, 276–283. [CrossRef] [PubMed]

72. Bui, H.T.; Chung, O.H.; Cruz, J.D.; Park, J.S. Fabrication and characterization of electrospun curcumin-loaded polycaprolactone-polyethylene glycol nanofibers for enhanced wound healing. *Macromol. Res.* **2014**, *22*, 1288–1296. [CrossRef]

73. Ahire, J.J.; Robertson, D.D.; van Reenen, A.J.; Dicks, L.M.T. Polyethylene oxide (PEO)-hyaluronic acid (HA) nanofibers with kanamycin inhibits the growth of Listeria monocytogenes. *Biomed. Pharmacother.* **2017**, *86*, 143–148. [CrossRef] [PubMed]

74. Balogh, A.; Farkas, B.; Verreck, G.; Mensch, J.; Borbás, E.; Nagy, B.; Marosi, G.; Nagy, Z.K. AC and DC electrospinning of hydroxypropylmethylcellulose with polyethylene oxides as secondary polymer for improved drug dissolution. *Int. J. Pharm.* **2016**, *505*, 159–166. [CrossRef] [PubMed]

75. Haider, A.; Haider, S.; Kang, I.-K. A comprehensive review summarizing the effect of electrospinning parameters and potential applications of nanofibers in biomedical and biotechnology. *Arab. J. Chem.* **2015**. [CrossRef]

76. Megelski, S.; Stephens, J.S.; Chase, D.B.; Rabolt, J.F. Micro- and nanostructured surface morphology on electrospun polymer fibers. *Macromolecules* **2002**, *35*, 8456–8466. [CrossRef]

77. Rodoplu, D.; Mutlu, M. Effects of electrospinning setup and process parameters on nanofiber morphology intended for the modification of quartz crystal microbalance surfaces. *J. Eng. Fibers Fabr.* **2012**, *7*, 118–123.

78. Matabola, K.P.; Moutloali, R.M. The influence of electrospinning parameters on the morphology and diameter of poly(vinyledene fluoride) nanofibers- effect of sodium chloride. *J. Mater. Sci.* **2013**, *48*, 5475–5482. [CrossRef]

79. Ball, C.; Chou, S.-F.; Jiang, Y.; Woodrow, K.A. Coaxially electrospun fiber-based microbicides facilitate broadly tunable release of maraviroc. *Mater. Sci. Eng. C* **2016**, *63*, 117–124. [CrossRef] [PubMed]

80. Zamani, M.; Prabhakaran, M.P.; Ramakrishna, S. Advances in drug delivery via electrospun and electrosprayed nanomaterials. *Int. J. Nanomed.* **2013**, 2997–3017.

81. Contardi, M.; Heredia-Guerrero, J.A.; Perotto, G.; Valentini, P.; Pompa, P.P.; Spanò, R.; Goldoni, L.; Bertorelli, R.; Athanassiou, A.; Bayer, I.S. Transparent ciprofloxacin-povidone antibiotic films and nanofiber mats as potential skin and wound care dressings. *Eur. J. Pharm. Sci.* **2017**, *104*, 133–144. [CrossRef] [PubMed]

82. Li, H.; Williams, G.R.; Wu, J.; Lv, Y.; Sun, X.; Wu, H.; Zhu, L.-M. Thermosensitive nanofibers loaded with ciprofloxacin as antibacterial wound dressing materials. *Int. J. Pharm.* **2017**, *517*, 135–147. [CrossRef] [PubMed]

83. Kabay, G.; Meydan, A.E.; Kaleli Can, G.; Demirci, C.; Mutlu, M. Controlled release of a hydrophilic drug from electrospun amyloid-like protein blend nanofibers. *Mater. Sci. Eng. C* **2017**, *81*, 271–279. [CrossRef] [PubMed]

84. Sohrabi, A.; Shaibani, P.M.; Etayash, H.; Kaur, K.; Thundat, T. Sustained drug release and antibacterial activity of ampicillin incorporated poly(methyl methacrylate)–nylon6 core/shell nanofibers. *Polymer* **2013**, *54*, 2699–2705. [CrossRef]

85. Sultanova, Z.; Kaleli, G.; Kabay, G.; Mutlu, M. Controlled release of a hydrophilic drug from coaxially electrospun polycaprolactone nanofibers. *Int. J. Pharm.* **2016**, *505*, 133–138. [CrossRef] [PubMed]

86. Xue, J.; He, M.; Niu, Y.; Liu, H.; Crawford, A.; Coates, P.; Chen, D.; Shi, R.; Zhang, L. Preparation and in vivo efficient anti-infection property of GTR/GBR implant made by metronidazole loaded electrospun polycaprolactone nanofiber membrane. *Int. J. Pharm.* **2014**, *475*, 566–577. [CrossRef] [PubMed]

87. He, M.; Jiang, H.; Wang, R.; Xie, Y.; Zhao, C. Fabrication of metronidazole loaded poly (ε-caprolactone)/zein core/shell nanofiber membranes via coaxial electrospinning for guided tissue regeneration. *J. Colloid Interface Sci.* **2017**, *490*, 270–278. [CrossRef] [PubMed]

88. Zupančič, Š.; Potrč, T.; Baumgartner, S.; Kocbek, P.; Kristl, J. Formulation and evaluation of chitosan/polyethylene oxide nanofibers loaded with metronidazole for local infections. *Eur. J. Pharm. Sci.* **2016**, *95*, 152–160. [CrossRef] [PubMed]

89. Sadri, M.; Sorkhi, S.A. Preparation and characterization of CS/PEO/cefazolin nanofibers with in vitro and in vivo testing. *Nanomedicine Res. J.* **2017**, *2*, 100–110.

90. Rath, G.; Hussain, T.; Chauhan, G.; Garg, T.; Goyal, A.K. Development and characterization of cefazolin loaded zinc oxide nanoparticles composite gelatin nanofiber mats for postoperative surgical wounds. *Mater. Sci. Eng. C* **2016**, *58*, 242–253. [CrossRef] [PubMed]

91. Kataria, K.; Gupta, A.; Rath, G.; Mathur, R.B.; Dhakate, S.R. In vivo wound healing performance of drug loaded electrospun composite nanofibers transdermal patch. *Int. J. Pharm.* **2014**, *469*, 102–110. [CrossRef] [PubMed]

92. Zhang, H.; Lou, S.; Williams, G.R.; Branford-White, C.; Nie, H.; Quan, J.; Zhu, L.-M. A systematic study of captopril-loaded polyester fiber mats prepared by electrospinning. *Int. J. Pharm.* **2012**, *439*, 100–108. [CrossRef] [PubMed]

93. Zheng, X.-F.; Lu, X.-Y. Measurement and correlation of solubilities of asiaticoside in water, methanol, ethanol, *n* -propanol, *n* -butanol, and a methanol + water mixture from (278.15 to 343.15) K. *J. Chem. Eng. Data* **2011**, *56*, 674–677. [CrossRef]

94. Zhu, L.; Liu, X.; Du, L.; Jin, Y. Preparation of asiaticoside-loaded coaxially electrospinning nanofibers and their effect on deep partial-thickness burn injury. *Biomed. Pharmacother.* **2016**, *83*, 33–40. [CrossRef] [PubMed]

95. Mutlu, G.; Calamak, S.; Ulubayram, K.; Guven, E. Curcumin-loaded electrospun PHBV nanofibers as potential wound-dressing material. *J. Drug Deliv. Sci. Technol.* **2018**, *43*, 185–193. [CrossRef]

96. Ranjbar-Mohammadi, M.; Rabbani, S.; Bahrami, S.H.; Joghataei, M.T.; Moayer, F. Antibacterial performance and in vivo diabetic wound healing of curcumin loaded gum tragacanth/poly(ε-caprolactone) electrospun nanofibers. *Mater. Sci. Eng. C* **2016**, *69*, 1183–1191. [CrossRef] [PubMed]

97. Basar, A.O.; Castro, S.; Torres-Giner, S.; Lagaron, J.M.; Turkoglu Sasmazel, H. Novel poly(ε-caprolactone)/gelatin wound dressings prepared by emulsion electrospinning with controlled release capacity of Ketoprofen anti-inflammatory drug. *Mater. Sci. Eng. C* **2017**, *81*, 459–468. [CrossRef] [PubMed]

98. Kenawy, E.-R.; Abdel-Hay, F.I.; El-Newehy, M.H.; Wnek, G.E. Controlled release of ketoprofen from electrospun poly(vinyl alcohol) nanofibers. *Mater. Sci. Eng. A* **2007**, *459*, 390–396. [CrossRef]

99. Liu, L.; Bai, S.; Yang, H.; Li, S.; Quan, J.; Zhu, L.; Nie, H. Controlled release from thermo-sensitive PNVCL-*co*-MAA electrospun nanofibers: The effects of hydrophilicity/hydrophobicity of a drug. *Mater. Sci. Eng. C* **2016**, *67*, 581–589. [CrossRef] [PubMed]

100. Yu, D.-G.; Yu, J.-H.; Chen, L.; Williams, G.R.; Wang, X. Modified coaxial electrospinning for the preparation of high-quality ketoprofen-loaded cellulose acetate nanofibers. *Carbohydr. Polym.* **2012**, *90*, 1016–1023. [CrossRef] [PubMed]

101. Hamori, M.; Yoshimatsu, S.; Hukuchi, Y.; Shimizu, Y.; Fukushima, K.; Sugioka, N.; Nishimura, A.; Shibata, N. Preparation and pharmaceutical evaluation of nano-fiber matrix supported drug delivery system using the solvent-based electrospinning method. *Int. J. Pharm.* **2014**, *464*, 243–251. [CrossRef] [PubMed]

102. Lin, X.; Tang, D.; Du, H. Self-assembly and controlled release behaviour of the water-insoluble drug nifedipine from electrospun PCL-based polyurethane nanofibres: Self-assembly and release of drug. *J. Pharm. Pharmacol.* **2013**, *65*, 673–681. [CrossRef] [PubMed]

103. Lin, X.; Tang, D.; Gu, S.; Du, H.; Jiang, E. Electrospun poly(*N*-isopropylacrylamide)/poly(caprolactone)-based polyurethane nanofibers as drug carriers and temperature-controlled release. *New J. Chem.* **2013**, *37*, 2433–2439. [CrossRef]

104. Zahedi, P.; Rezaeian, I.; Jafari, S.H. In vitro and in vivo evaluations of phenytoin sodium-loaded electrospun PVA, PCL, and their hybrid nanofibrous mats for use as active wound dressings. *J. Mater. Sci.* **2013**, *48*, 3147–3159. [CrossRef]

105. Kurczewska, J.; Pecyna, P.; Ratajczak, M.; Gajęcka, M.; Schroeder, G. Halloysite nanotubes as carriers of vancomycin in alginate-based wound dressing. *Saudi Pharm. J.* **2017**, *25*, 911–920. [CrossRef] [PubMed]

106. El-Khordagui, L.; El-Sayed, N.; Galal, S.; El-Gowelli, H.; Omar, H.; Mohamed, M. Photosensitizer-eluting nanofibers for enhanced photodynamic therapy of wounds: A preclinical study in immunocompromized rats. *Int. J. Pharm.* **2017**, *520*, 139–148. [CrossRef] [PubMed]

107. Anstead, G.M.; Hart, L.M.; Sunahara, J.F.; Liter, M.E. Phenytoin in wound healing. *Ann. Pharmacother.* **1996**, *30*, 768–775. [CrossRef] [PubMed]

108. Hokkam, E.; El-Labban, G.; Shams, M.; Rifaat, S.; El-mezaien, M. The use of topical phenytoin for healing of chronic venous ulcerations. *Int. J. Surg.* **2011**, *9*, 335–338. [CrossRef] [PubMed]

109. Lupo, E.; Locher, R.; Weisser, B.; Vetter, W. In vitro antioxidant activity of calcium antagonists against LDL oxidation compared with α-tocopherol. *Biochem. Biophys. Res. Commun.* **1994**, *203*, 1803–1808. [CrossRef] [PubMed]

110. Barrientos, S.; Stojadinovic, O.; Golinko, M.S.; Brem, H.; Tomic-Canic, M. PERSPECTIVE ARTICLE: Growth factors and cytokines in wound healing. *Wound Repair Regen.* **2008**, *16*, 585–601. [CrossRef] [PubMed]

111. Lai, H.-J.; Kuan, C.-H.; Wu, H.-C.; Tsai, J.-C.; Chen, T.-M.; Hsieh, D.-J.; Wang, T.-W. Tailored design of electrospun composite nanofibers with staged release of multiple angiogenic growth factors for chronic wound healing. *Acta Biomater.* **2014**, *10*, 4156–4166. [CrossRef] [PubMed]

112. Choi, J.S.; Choi, S.H.; Yoo, H.S. Coaxial electrospun nanofibers for treatment of diabetic ulcers with binary release of multiple growth factors. *J. Mater. Chem.* **2011**, *21*, 5258–5267. [CrossRef]

113. Phipps, M.; Ma, Y.; Bellis, S. Delivery of platelet-derived growth factor as a chemotactic factor for mesenchymal stem cells by bone-mimetic electrospun scaffolds. *PLoS ONE* **2012**, *7*, e40831. [CrossRef] [PubMed]

114. Kim, I.L.; Pfeifer, C.G.; Fisher, M.B.; Saxena, V.; Meloni, G.R.; Kwon, M.Y.; Kim, M.; Steinberg, D.R.; Mauck, R.L.; Burdick, J.A. Fibrous scaffolds with varied fiber chemistry and growth factor delivery promote repair in a porcine cartilage defect model. *Tissue Eng. Part A* **2015**, *21*, 2680–2690. [CrossRef] [PubMed]

115. Hu, F.; Zhang, X.; Liu, H.; Xu, P.; Doulathunnisa; Teng, G.; Xiao, Z. Neuronally differentiated adipose-derived stem cells and aligned PHBV nanofiber nerve scaffolds promote sciatic nerve regeneration. *Biochem. Biophys. Res. Commun.* **2017**, *489*, 171–178. [CrossRef] [PubMed]

116. Davis, M.E.; Hsieh, P.C.H.; Takahashi, T.; Song, Q.; Zhang, S.; Kamm, R.D.; Grodzinsky, A.J.; Anversa, P.; Lee, R.T. Local myocardial insulin-like growth factor 1 (IGF-1) delivery with biotinylated peptide nanofibers improves cell therapy for myocardial infarction. *Proc. Natl. Acad. Sci. USA* **2006**, *103*, 8155–8160. [CrossRef] [PubMed]

117. Noh, K.H.; Park, Y.M.; Kim, H.S.; Kang, T.H.; Song, K.-H.; Lee, Y.-H.; Byeon, Y.; Jeon, H.N.; Jung, I.D.; Shin, B.C. GM-CSF-loaded chitosan hydrogel as an immunoadjuvant enhances antigen-specific immune responses with reduced toxicity. *BMC Immunol.* **2014**, *15*, 1–7. [CrossRef] [PubMed]

118. Olvera, D.; Sathy, B.N.; Carroll, S.F.; Kelly, D.J. Modulating microfibrillar alignment and growth factor stimulation to regulate mesenchymal stem cell differentiation. *Acta Biomater.* **2017**, *64*, 148–160. [CrossRef] [PubMed]

119. Choi, J.S.; Leong, K.W.; Yoo, H.S. In vivo wound healing of diabetic ulcers using electrospun nanofibers immobilized with human epidermal growth factor (EGF). *Biomaterials* **2008**, *29*, 587–596. [CrossRef] [PubMed]

120. Norouzi, M.; Shabani, I.; Ahvaz, H.H.; Soleimani, M. PLGA/gelatin hybrid nanofibrous scaffolds encapsulating EGF for skin regeneration. *J. Biomed. Mater. Res. A* **2015**, *103*, 2225–2235. [CrossRef] [PubMed]

121. Schneider, A.; Wang, X.Y.; Kaplan, D.L.; Garlick, J.A.; Egles, C. Biofunctionalized electrospun silk mats as a topical bioactive dressing for accelerated wound healing. *Acta Biomater.* **2009**, *5*, 2570–2578. [CrossRef] [PubMed]

122. Gümüşderelioğlu, M.; Dalkıranoğlu, S.; Aydın, R.S.T.; Çakmak, S. A novel dermal substitute based on biofunctionalized electrospun PCL nanofibrous matrix. *J. Biomed. Mater. Res. A* **2011**, *98*, 461–472. [CrossRef] [PubMed]

123. Jin, G.; Prabhakaran, M.P.; Ramakrishna, S. Photosensitive and biomimetic core–shell nanofibrous scaffolds as wound dressing. *Photochem. Photobiol.* **2014**, *90*, 673–681. [CrossRef] [PubMed]

124. Jin, G.; Prabhakaran, M.P.; Kai, D.; Ramakrishna, S. Controlled release of multiple epidermal induction factors through core–shell nanofibers for skin regeneration. *Eur. J. Pharm. Biopharm.* **2013**, *85*, 689–698. [CrossRef] [PubMed]

125. Gil, E.S.; Panilaitis, B.; Bellas, E.; Kaplan, D.L. Functionalized silk biomaterials for wound healing. *Adv. Healthc. Mater.* **2013**, *2*, 206–217. [CrossRef] [PubMed]

126. Mirdailami, O.; Soleimani, M.; Dinarvand, R.; Khoshayand, M.R.; Norouzi, M.; Hajarizadeh, A.; Dodel, M.; Atyabi, F. Controlled release of rhEGF and rhbFGF from electrospun scaffolds for skin regeneration. *J. Biomed. Mater. Res. A* **2015**, *103*, 3374–3385. [CrossRef] [PubMed]

127. Kobsa, S.; Kristofik, N.J.; Sawyer, A.J.; Bothwell, A.L.M.; Kyriakides, T.R.; Saltzman, W.M. An electrospun scaffold integrating nucleic acid delivery for treatment of full thickness wounds. *Biomaterials* **2013**, *34*, 3891–3901. [CrossRef] [PubMed]

128. Zhao, Q.; Lu, W.W.; Wang, M. Modulating the release of vascular endothelial growth factor by negative-voltage emulsion electrospinning for improved vascular regeneration. *Mater. Lett.* **2017**, *193*, 1–4. [CrossRef]

129. Song, D.W.; Kim, S.H.; Kim, H.H.; Lee, K.H.; Ki, C.S.; Park, Y.H. Multi-biofunction of antimicrobial peptide-immobilized silk fibroin nanofiber membrane: Implications for wound healing. *Acta Biomater.* **2016**, *39*, 146–155. [CrossRef] [PubMed]

130. Sebe, I.; Ostorhazi, E.; Fekete, A.; Kovacs, K.N.; Zelko, R.; Kovalszky, I.; Li, W.; Wade, J.D.; Szabo, D.; Otvos, L. Polyvinyl alcohol nanofiber formulation of the designer antimicrobial peptide APO sterilizes Acinetobacter baumannii-infected skin wounds in mice. *Amino Acids Vienna* **2016**, *48*, 203–211. [CrossRef] [PubMed]

131. Lee, Y.J.; Lee, J.-H.; Cho, H.-J.; Kim, H.K.; Yoon, T.R.; Shin, H. Electrospun fibers immobilized with bone forming peptide-1 derived from BMP7 for guided bone regeneration. *Biomaterials* **2013**, *34*, 5059–5069. [CrossRef] [PubMed]

132. Shao, Z.; Zhang, X.; Pi, Y.; Wang, X.; Jia, Z.; Zhu, J.; Dai, L.; Chen, W.; Yin, L.; Chen, H.; et al. Polycaprolactone electrospun mesh conjugated with an MSC affinity peptide for MSC homing in vivo. *Biomaterials* **2012**, *33*, 3375–3387. [CrossRef] [PubMed]

133. Foldvari, M.; Chen, D.W.; Nafissi, N.; Calderon, D.; Narsineni, L.; Rafiee, A. Non-viral gene therapy: Gains and challenges of non-invasive administration methods. *J. Control. Release* **2016**, *240*, 165–190. [CrossRef] [PubMed]

134. Penn, M.; Michler, R.E.; Espinal, E.; McGrath, M.F.; Firstenberg, M.S.; McCarthy, P.M.; Patel, A.N. Stromal cell derived factor-1 over-expression immediately following surgical closure minimizes scar formation. *J. Am. Coll. Surg.* **2014**, *219*, e66. [CrossRef]

135. Ko, J.; Jun, H.; Chung, H.; Yoon, C.; Kim, T.; Kwon, M.; Lee, S.; Jung, S.; Kim, M.; Park, J.H. Comparison of EGF with VEGF non-viral gene therapy for cutaneous wound healing of streptozotocin diabetic mice. *Diabetes Metab. J.* **2011**, *35*, 226–235. [CrossRef] [PubMed]

136. Dou, C.; Lay, F.; Ansari, A.M.; Rees, D.J.; Ahmed, A.K.; Kovbasnjuk, O.; Matsangos, A.E.; Du, J.; Hosseini, S.M.; Steenbergen, C.; et al. Strengthening the skin with topical delivery of keratinocyte growth factor-1 using a novel DNA plasmid. *Mol. Ther.* **2014**, *22*, 752–761. [CrossRef] [PubMed]

137. Lee, S.; Jin, G.; Jang, J.-H. Electrospun nanofibers as versatile interfaces for efficient gene delivery. *J. Biol. Eng.* **2014**, *8*, 30. [CrossRef] [PubMed]

138. Kim, H.S.; Yoo, H.S. MMPs-responsive release of DNA from electrospun nanofibrous matrix for local gene therapy: In vitro and in vivo evaluation. *J. Control. Release* **2010**, *145*, 264–271. [CrossRef] [PubMed]

139. Kim, H.S.; Yoo, H.S. Matrix metalloproteinase-inspired suicidal treatments of diabetic ulcers with siRNA-decorated nanofibrous meshes. *Gene Ther.* **2013**, *20*, 378–385. [CrossRef] [PubMed]

140. Kim, H.S.; Yoo, H.S. In vitro and in vivo epidermal growth factor gene therapy for diabetic ulcers with electrospun fibrous meshes. *Acta Biomater.* **2013**, *9*, 7371–7380. [CrossRef] [PubMed]

141. Luu, Y.K.; Kim, K.; Hsiao, B.S.; Chu, B.; Hadjiargyrou, M. Development of a nanostructured DNA delivery scaffold via electrospinning of PLGA and PLA–PEG block copolymers. *J. Control. Release* **2003**, *89*, 341–353. [CrossRef]

142. Saraf, A.; Baggett, L.S.; Raphael, R.M.; Kasper, F.K.; Mikos, A.G. Regulated non-viral gene delivery from coaxial electrospun fiber mesh scaffolds. *J. Control. Release* **2010**, *143*, 95–103. [CrossRef] [PubMed]

143. Yang, Y.; Xia, T.; Chen, F.; Wei, W.; Liu, C.; He, S.; Li, X. Electrospun fibers with plasmid bFGF polyplex loadings promote skin wound healing in diabetic rats. *Mol. Pharm.* **2012**, *9*, 48–58. [CrossRef] [PubMed]

144. He, S.; Xia, T.; Wang, H.; Wei, L.; Luo, X.; Li, X. Multiple release of polyplexes of plasmids VEGF and bFGF from electrospun fibrous scaffolds towards regeneration of mature blood vessels. *Acta Biomater.* **2012**, *8*, 2659–2669. [CrossRef] [PubMed]

145. Kataoka, K.; Harada, A.; Nagasaki, Y. Block copolymer micelles for drug delivery: Design, characterization and biological significance. *Adv. Drug Deliv. Rev.* **2001**, *47*, 113–131. [CrossRef]

146. Kazunori, K.; Glenn S., K.; Masayuki, Y.; Teruo, O.; Yasuhisa, S. Block copolymer micelles as vehicles for drug delivery. *J. Control. Release* **1993**, *24*, 119–132. [CrossRef]

147. Redhead, H.M.; Davis, S.S.; Illum, L. Drug delivery in poly(lactide-co-glycolide) nanoparticles surface modified with poloxamer 407 and poloxamine 908: In vitro characterisation and in vivo evaluation. *J. Control. Release* **2001**, *70*, 353–363. [CrossRef]

148. Thotakura, N.; Dadarwal, M.; Kumar, R.; Singh, B.; Sharma, G.; Kumar, P.; Katare, O.P.; Raza, K. Chitosan-palmitic acid based polymeric micelles as promising carrier for circumventing pharmacokinetic and drug delivery concerns of tamoxifen. *Int. J. Biol. Macromol.* **2017**, *102*, 1220–1225. [CrossRef] [PubMed]

149. Wang, Y.; Ke, X.; Voo, Z.X.; Yap, S.S.L.; Yang, C.; Gao, S.; Liu, S.; Venkataraman, S.; Obuobi, S.A.O.; Khara, J.S.; et al. Biodegradable functional polycarbonate micelles for controlled release of amphotericin B. *Acta Biomater.* **2016**, *46*, 211–220. [CrossRef] [PubMed]

150. Pan, J.; Liu, N.; Sun, H.; Xu, F. Preparation and characterization of electrospun PLCL/poloxamer nanofibers and dextran/gelatin hydrogels for skin tissue engineering. *PLoS ONE* **2014**, *9*, e112885. [CrossRef] [PubMed]

151. Gong, C.; Wu, Q.; Wang, Y.; Zhang, D.; Luo, F.; Zhao, X.; Wei, Y.; Qian, Z. A biodegradable hydrogel system containing curcumin encapsulated in micelles for cutaneous wound healing. *Biomaterials* **2013**, *34*, 6377–6387. [CrossRef] [PubMed]

152. Kim, J.H.; Ramasamy, T.; Tran, T.H.; Choi, J.Y.; Cho, H.J.; Yong, C.S.; Kim, J.O. Polyelectrolyte complex micelles by self-assembly of polypeptide-based triblock copolymer for doxorubicin delivery. *Asian J. Pharm. Sci.* **2014**, *9*, 191–198. [CrossRef]

153. Oyarzun-Ampuero, F.; Vidal, A.; Concha, M.; Morales, J.; Orellana, S.; Moreno-Villoslada, I. Nanoparticles for the Treatment of Wounds. *Curr. Pharm. Des.* **2015**, *21*, 4329–4341. [CrossRef] [PubMed]

154. Kalwar, K.; Sun, W.-X.; Li, D.-L.; Zhang, X.-J.; Shan, D. Coaxial electrospinning of polycaprolactone@chitosan: Characterization and silver nanoparticles incorporation for antibacterial activity. *React. Funct. Polym.* **2016**, *107*, 87–92. [CrossRef]

155. Chen, C.-H.; Chen, S.-H.; Shalumon, K.T.; Chen, J.-P. Dual functional core–sheath electrospun hyaluronic acid/polycaprolactone nanofibrous membranes embedded with silver nanoparticles for prevention of peritendinous adhesion. *Acta Biomater.* **2015**, *26*, 225–235. [CrossRef] [PubMed]

156. Ghosh Auddy, R.; Abdullah, M.F.; Das, S.; Roy, P.; Datta, S.; Mukherjee, A. New guar biopolymer silver nanocomposites for wound healing applications. *BioMed Res. Int.* **2013**, *2013*, 1–8. [CrossRef] [PubMed]

157. Hazer, D.B.; Hazer, B.; Dinçer, N. Soft tissue response to the presence of polypropylene-*g*-poly(ethylene glycol) comb-type graft copolymers containing gold nanoparticles. *J. Biomed. Biotechnol.* **2011**, *2011*, 1–7. [CrossRef] [PubMed]

158. Leu, J.-G.; Chen, S.-A.; Chen, H.-M.; Wu, W.-M.; Hung, C.-F.; Yao, Y.-D.; Tu, C.-S.; Liang, Y.-J. The effects of gold nanoparticles in wound healing with antioxidant epigallocatechin gallate and α-lipoic acid. *Nanomed. Nanotechnol. Biol. Med.* **2012**, *8*, 767–775. [CrossRef] [PubMed]

159. Raguvaran, R.; Manuja, B.K.; Chopra, M.; Thakur, R.; Anand, T.; Kalia, A.; Manuja, A. Sodium alginate and gum acacia hydrogels of ZnO nanoparticles show wound healing effect on fibroblast cells. *Int. J. Biol. Macromol.* **2017**, *96*, 185–191. [CrossRef] [PubMed]

160. Martinez, L.R.; Han, G.; Chacko, M.; Mihu, M.R.; Jacobson, M.; Gialanella, P.; Friedman, A.J.; Nosanchuk, J.D.; Friedman, J.M. Antimicrobial and healing efficacy of sustained release nitric oxide nanoparticles against *Staphylococcus aureus* skin infection. *J. Investig. Dermatol.* **2009**, *129*, 2463–2469. [CrossRef] [PubMed]

161. Rather, H.A.; Thakore, R.; Singh, R.; Jhala, D.; Singh, S.; Vasita, R. Antioxidative study of cerium oxide nanoparticle functionalised PCL-Gelatin electrospun fibers for wound healing application. *Bioact. Mater.* **2017**. [CrossRef]

162. Quignard, S.; Coradin, T.; Powell, J.J.; Jugdaohsingh, R. Silica nanoparticles as sources of silicic acid favoring wound healing in vitro. *Colloids Surf. B Biointerfaces* **2017**, *155*, 530–537. [CrossRef] [PubMed]

163. Wu, H.; Li, F.; Wang, S.; Lu, J.; Li, J.; Du, Y.; Sun, X.; Chen, X.; Gao, J.; Ling, D. Ceria nanocrystals decorated mesoporous silica nanoparticle based ROS-Scavenging tissue adhesive for highly efficient regenerative wound healing. *Biomaterials* **2018**, *151*, 66–77. [CrossRef] [PubMed]

164. Wang, C.; Hou, W.; Guo, X.; Li, J.; Hu, T.; Qiu, M.; Liu, S.; Mo, X.; Liu, X. Two-phase electrospinning to incorporate growth factors loaded chitosan nanoparticles into electrospun fibrous scaffolds for bioactivity retention and cartilage regeneration. *Mater. Sci. Eng. C* **2017**, *79*, 507–515. [CrossRef] [PubMed]

165. Gutha, Y.; Pathak, J.L.; Zhang, W.; Zhang, Y.; Jiao, X. Antibacterial and wound healing properties of chitosan/poly(vinyl alcohol)/zinc oxide beads (CS/PVA/ZnO). *Int. J. Biol. Macromol.* **2017**, *103*, 234–241. [CrossRef] [PubMed]

166. Sanad, R.A.-B.; Abdel-Bar, H.M. Chitosan–hyaluronic acid composite sponge scaffold enriched with Andrographolide-loaded lipid nanoparticles for enhanced wound healing. *Carbohydr. Polym.* **2017**, *173*, 441–450. [CrossRef] [PubMed]

167. Blakney, A.; Jiang, Y.; Woodrow, K.; Krogstad, E. Delivery of multipurpose prevention drug combinations from electrospun nanofibers using composite microarchitectures. *Int. J. Nanomed.* **2014**, 2967–2978. [CrossRef] [PubMed]

168. Krogstad, E.A.; Woodrow, K.A. Manufacturing scale-up of electrospun poly(vinyl alcohol) fibers containing tenofovir for vaginal drug delivery. *Int. J. Pharm.* **2014**, *475*, 282–291. [CrossRef] [PubMed]

4

Metabolic Reprogramming and the Recovery of Physiological Functionality in 3D Cultures in Micro-Bioreactors

Krzysztof Wrzesinski [1,2] ⓘ and **Stephen J. Fey** [1,2,*] ⓘ

1   Tissue Culture Engineering Laboratory, Department of Biochemistry and Molecular Biology, University of Southern Denmark, 5230 Odense, Denmark; kwr@celvivo.com
2   CelVivo IVS, 5491 Blommenslyst, Denmark
*   Correspondence: sjf@celvivo.com

Abstract: The recovery of physiological functionality, which is commonly seen in tissue mimetic three-dimensional (3D) cellular aggregates (organoids, spheroids, acini, etc.), has been observed in cells of many origins (primary tissues, embryonic stem cells (ESCs), induced pluripotent stem cells (iPSCs), and immortal cell lines). This plurality and plasticity suggest that probably several basic principles promote this recovery process. The aim of this study was to identify these basic principles and describe how they are regulated so that they can be taken in consideration when micro-bioreactors are designed. Here, we provide evidence that one of these basic principles is hypoxia, which is a natural consequence of multicellular structures grown in microgravity cultures. Hypoxia drives a partial metabolic reprogramming to aerobic glycolysis and an increased anabolic synthesis. A second principle is the activation of cytoplasmic glutaminolysis for lipogenesis. Glutaminolysis is activated in the presence of hypo- or normo-glycaemic conditions and in turn is geared to the hexosamine pathway. The reducing power needed is produced in the pentose phosphate pathway, a prime function of glucose metabolism. Cytoskeletal reconstruction, histone modification, and the recovery of the physiological phenotype can all be traced to adaptive changes in the underlying cellular metabolism. These changes are coordinated by mTOR/Akt, p53 and non-canonical Wnt signaling pathways, while myc and NF-kB appear to be relatively inactive. Partial metabolic reprogramming to aerobic glycolysis, originally described by Warburg, is independent of the cell's rate of proliferation, but is interwoven with the cells abilities to execute advanced functionality needed for replicating the tissues physiological performance.

Keywords: bioreactors; 3D cell culture; spheroids; organoids; hypoxia; aerobic glycolysis; glutaminolysis; metabolic reprogramming; physiological performance; Warburg

1. Introduction

Three-dimensional (3D) cell culture offers a glimpse into tissue function that is only recently being appreciated. This is true whether primary, immortal or stem cells are used and it appears to apply to all types of tissue.

What drives this process? Often the growth conditions used for these two-dimensional (2D) or 3D cell culture studies are identical: the same growth media, temperature, and atmosphere. Often the same cells are used and yet the performance of the cells is radically different. Here we examine the role of metabolic reprogramming and factors that induce the recovery or development of mimetic tissues.

Primary cells retain their physiological behaviour longer when grown in 3D culture conditions (e.g., astrocytes [1], prostate [2], and microvascular networks [3]). Numerous different colorectal cancer-derived tumour spheroids retain characteristics of original tumours [4,5] and lung cancer spheroids their chemosensitivity [6] but cardiac pluripotent cells do adapt to growth in 3D culture [7]. Immortal cell lines recover ('re-differentiate') structural and functional features of their parental tissue and regain in vivo drug sensitivity (e.g., breast, MCF-7 [8,9]; pancreatic β-cell β-TC6 [10] and RIN 5F [11]; glial-like GL15, neuronal-like SH-SY5Y [12]; ovarian, OV-MZ-6 and SKOV-3 [13]; trophoblast BeWo, Jeg-3 and JAr [14]; and liver, HepG2 [15], HuH-6 [16], HepG2/C3A [17–19]. Stem cells, whether embryonic or induced pluripotent, differentiate into 3D organoids when provided with appropriate molecular guidelines (e.g., ECM and growth factors) [20,21]. These recapitulate differentiation into a wide variety of mimetic tissues. (e.g., optic cup [22], pancreas [23] & gastric [24]). Growth in 3D microenvironments boosts the induction of pluripotency [25]. Transplantation of 3D structures into animals induces them to differentiate further, in some cases, to tissues that are almost indistinguishable from the native organ [26].

While the term 'spheroid' is usually used to indicate a mimetic tissue that is constructed from immortal cells and the term 'organoid' used to indicate a mimetic tissue derived from primary cells (including stem cells of any derivation) the basic principles driving recovery will be the same and so for the purposes of this article, the word 'spheroid' is used for both.

Clinostat rotating vessels (Figure 1, also known as rotating wall vessels (RWV), rotating cell culture systems (RCCS) or high aspect rotating wall vessels (HARV)), are commonly used to generate a 'microgravity' environment that is conducive to the production of highly reproducible long lived 3D cultures which allow for the investigation and manipulation of mimetic tissues. Strictly speaking, they produce omnidirectional gravity—i.e., the tissue is influenced by gravity from all sides, effectively neutralizing directional gravity effects. In practice, the gravity is slightly larger than 1G: a clinostat running at 20 rpm will generate a G-force of 1.0089 at 2 cm from the axis of rotation.

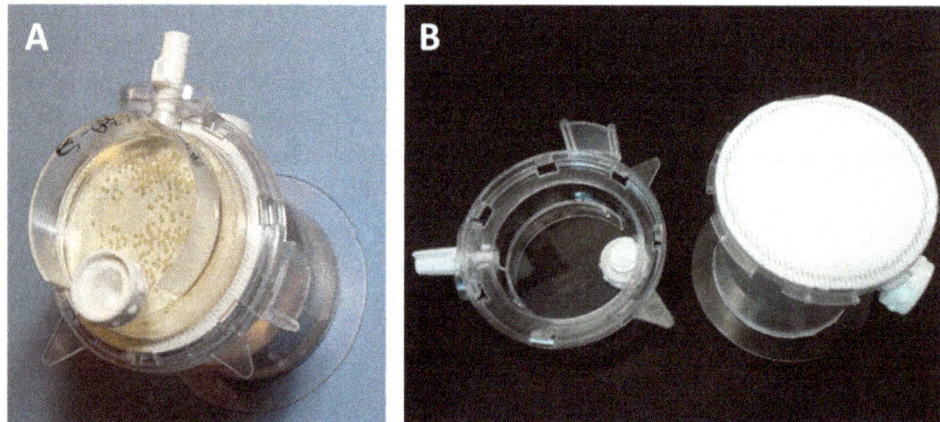

Figure 1. (**A**) an assembled bioreactor containing >300 21 day old spheroids. (**B**) The open bioreactor with (left) a 10 mL petri-dish like culture chamber and (right) the gas exchange membrane. Behind the membrane is a water reservoir and humidification labyrinth. White stoppers allow access for media change or filling the reservoir. This type of bioreactor has a gas membrane exchange area of 13.2 cm^2 and a fixed volume (nominally 10 mL) and is available from CelVivo (Denmark).

The clinostat bioreactor has a number of advantages. Most cells in tissues, with the obvious exception of endothelial cells, experience little or no shear forces. Critical/lethal shear stresses for different mammalian cell types are in the range of 0.3–1.7 Pascal (1 Pa = 10 dyne/cm^2) (Croughan and Wang, 1991) [27].

Mimetic tissue culture in a clinostat bioreactor provides very low shear forces (at 20 rpm, ca. 0.01 Pa [28] on the suspended spheroids, similar to rocking platform 'wave' bioreactors (set at an oscillation of 7° at 20 rpm) [29,30] and 0.02–0.064 Pa for micro-fluidic devices (flow rate 650 μL/min) [31,32].

Higher shear forces (and cellular effects) are seen for stirred suspension bioreactors (100–200 rpm, 0.3–0.66 Pa) [33,34] and for orbital shakers (20–60 rpm, 0.6–1.6 Pa) [35,36].

In rocking platform bioreactors and micro-fluidic devices, the mimetic tissue is usually in contact with plastic surfaces and the shear forces vary either with time, location, or both. This will induce differential growth rates, sizes and biochemical properties of the spheroids [37]. These differences will affect, for example, drug response [38]. The clinostat bioreactor exposes all spheroids to an equal and very low shear force and has been show to result in uniform spheroids that even after 21 days in culture have a standard deviation in their size of only ±21%. Clinostat spheroids are therefore the best suited for studies of metabolism or pharmacology, especially for kinetic measurements.

This recovery of the physiological phenotype in 3D culture suggests that either there is a common driving mechanism or that it occurs spontaneously. Strangely enough, clues as to why this happens can be found in ideas that have been around for almost 100 years.

1.1. The Relationship between Oxidative Phosphorylation and Aerobic Glycolysis

Rapid cancer cell proliferation favours reprogramming from oxidative phosphorylation to aerobic glycolysis. This concept was described initially by Warburg in the 1920's [39,40]. He observed that proliferating ascites tumour cells convert most of their glucose to lactate and referred to this process as aerobic glycolysis because it occurred regardless of whether oxygen was present or absent [41].

In dynamic nuclear polarization (DNP) spectroscopic techniques, hyperpolarized [^{13}C]-labelled pyruvate or other glycolytic intermediates have shown that tumours in situ produce lactate at levels which correlate with their degree of tumour progression or response to treatment [42,43]. Similar studies using hyperpolarized [^{13}C]-labelled glucose revealed increased lactate production in mouse lymphoma and lung tumours, but not in healthy tissues [44]. Other whole-body approaches, such as PET and MRI brain-imaging techniques, have measured that 10–15% of the glucose that is used by the healthy brain is metabolised by aerobic glycolysis [45]. Similar results have been obtained in vitro using perfused heart and liver tissues [46] and in cell culture, comparing non-proliferating myocytes with proliferating Rh30 cell line. The proportion of aerobic glycolysis to oxidative phosphorylation varies between different tumour types: rapidly growing tumours tending to utilise a higher degree of aerobic glycolysis, in keeping with their reduced reliance on oxygen and their need to synthesise larger amounts of precursors faster [40,47].

Warburg hypothesized that metabolic reprogramming was specific to cancer cells, and that it arose from mitochondrial defects [39]. While his observations have been corroborated many times, the hypothesis that cancer growth was driven by these defects has been disproven [40,48,49]. Weinberg et al. demonstrated that mitochondrially generated reactive oxygen species are essential for Kras-induced proliferation and tumourigenesis of HCT116 colon cancer cells [50].

The current viewpoint is that, in aerobic glycolysis, glucose is metabolised via the glycolytic pathway to produce lactic acid, nucleotides, amino acids, and other metabolites. Simultaneously, glutamine is converted via glutaminolysis to citrate for cholesterol and lipid production. In contrast, non-proliferating, differentiated cells in healthy tissues efficiently produce ATP through oxidative phosphorylation. In oxidative phosphorylation, glucose is metabolised via the glycolytic pathway and then the pyruvate produced is oxidised to CO_2 and water in the tricarboxylic (TCA) cycle. The electrochemical gradient generated is used to produce ATP [41].

1.2. Are Growth Rates Inversely Related to Functionality?

Although generally accepted, there is a flaw in the argument that rapidly growing cells preferentially utilise aerobic metabolism. Most tumours and healthy tissues proliferate relatively slowly, doubling their numbers every 20–60 days [51]. In contrast, cultured cells double every 1–8 days and therefore would be expected to exhibit strong aerobic metabolism. This is not the case. In a painstaking review, using oxygen consumption and lactate production to define oxidative and glycolytic ATP production, Zu and Guppy showed that there is no evidence indicating that cancer cells or cells in culture are inherently glycolytic [52]. Despite considerable variation, both normal and cancer cells produce about 19% of their ATP using aerobic glycolysis and the rest by oxidative phosphorylation. They concluded that cancers tend to be glycolytic because they are hypoxic [52]. Metabolic reprogramming to aerobic glycolysis has been called a hallmark of cancer [41,53,54], but in practice, it is a consequence of hypoxia.

There is another paradox. By definition, healthy tissues exhibit full physiological functionality. Tumours tend to lose these functionalities in a reciprocal proportion to their proliferation rate [51,55,56]. Immortal cell lines are often considered a 'terminal condition' for tumour cells where they proliferate rapidly but have lost much of their in vivo functionality. However, when immortal cells are given sufficient time to adapt to 3D conditions, their proliferation rate slows to that seen in tumours and healthy tissues in vivo and they regain physiological functionality [18,19].

So, the question becomes how do growth rate, metabolic reprogramming, and physiological functionality relate to each other. Is proliferation linked to metabolic reprogramming or is it independent? Is there a 'spectrum' between, on the one extreme, healthy tissue and at the other, immortal cell lines grown in 2D? Or are the 'axes' of normal to transformed independent of the axis of hyperoxic to hypoxic? Where should 3D cultures 'be placed' in this spectrum? Does metabolic reprogramming to aerobic glycolysis invariably lead to the loss of physiological function, as seen in the transformation of normal tissue to cancer, or are these phenomena independent? Can cells reprogram between oxidative phosphorylation and aerobic glycolysis based on their growth requirements or does it have consequences? We have addressed these questions by reviewing what is known about metabolic reprograming.

2. Materials and Methods

This manuscript is based on the deeper evaluation of raw data published previously as supplementary data [57]. For convenience, we describe here in brief the methods used. A full description can be found in our previous manuscript.

2.1. Cell Culture

HepG2/C3A cells were grown in DMEM containing 1 g glucose/L. In 2D conditions, they were left until they were nearly confluent (day 5) before they were collected for mass spectrometry. HepG2/C3A cell spheroids were prepared using AggreWell™ 400 plates (Stemcell Technologies, Vancouver, Canada) and left to mature in a rotating 'microgravity' micro-bioreactors on a BioArray Matrix drive (CelVivo IVS, Blommenslyst, Denmark) for at least 21 days to reach a dynamic equilibrium [57]. The clinostat 3D culture reduces shear forces on the cells to a minimum while increasing nutrient and gas exchange (see Supplementary video).

2.2. Determination of Protein Content of Spheroids

Spheroids were washed with PBS, collected, photographed to calculate their individual protein content using a look-up table derived, as described previously [17].

2.3. Determination of Glucose and Glycogen Content of Spheroids

Glucose in the media was measured using a Onetouch Vita glucose meter' and test strips, (LifeScan, Inc., Cat Nos. 6407078 and 6407079 respectively, Milpitas, CA, USA). Total glycogen was measured in individual spheroids using a flurometric assay kit (Sigma cat. No. MAK016, Merck KGaA, Darmstadt, Germany). Glycogen is hydrolysed to glucose and measured in a fluorometer. Amounts of glucose in the spheroids prior to glycogen hydrolysis were negligible.

2.4. Mass Spectroscopy

Protein samples were collected from classical 2D cell culture five days after trypsinisation and from 3D spheroid culture 21 days after spheroid culture initiation. The proteins were quantitated, alkylated, digested with trypsin and washed, stable-isotope dimethyl labelled, and electrosprayed into the LTQ Orbitrap Velos (Thermo Scientific, Waltham, MA, USA). Data on 1346 proteins was deemed statistically reliable and was analysed with reference to multiple programs and information sources including MedLine, SwissProt, Kegg, Ingenuity™ and Go protein annotations [57].

3. Results and Discussion

We previously catalogued a plethora of differences between growing HepG2/C3A cells in 2D and 3D conditions. These include changes in the cell architecture (actin, microtubules, intermediate filaments) and metabolism (glycolysis, fatty acid metabolism, cholesterol and urea synthesis, DNA repair, RNA processing, protein folding and degradation, cell cycle arrest, transport around the cell) [57]. In this manuscript we take the analysis of the raw data a step further and use it to describe a coherent model of the driving force behind these differences between 2D and 3D culture. Where the cell type used to construct the spheroids is not defined in this article, the results refer to the raw data generated with HepG2/C3A.

3.1. Adaptation to Growth in 3D Culture

Cells need time to adapt to growing in 3D cultures and implement the changes catalogued above: to slow growth rates, reorganise their cytoskeleton, establish tight junctions, polarity, relocate membrane transporters and secrete tissue specific extracellular matrix components. Establishment coincides with the time needed for spheroids to reach a radius where their core is severely hypoxic. In a passive diffusion culture system (ultra-low attachment dishes or hanging drop) this requires about 8 days. Using NIH 3T3 fibroblasts or HepG2/C3A hepatocytes, the spheroid's radius is about 160 μm [58,59]. In irrigated spheroids, (i.e., where media flows past them in microgravity cultures), this occurs after 18 days (radius is 450 μm) [18].

Evidence for hypoxia-induced metabolic reprogramming to aerobic glycosylation has been provided by reanalysing proteomic studies of cells grown in 2D and clinostat 3D conditions. Protein abundance has been measured in mature, 21 day old spheroids and compared to that seen in 80% confluent, five day old 2D cultures of the same cells (human hepatocellular carcinoma HepG2/C3A) using quantitative proteomics [57] (the raw data is presented in supplementary information) and [60]. For convenience, the reciprocal has been taken of all values below 1 (i.e., where the protein level is lower in 3D than in 2D), and this is indicated with a negative sign. In this way, equally significant changes, for example, a doubling or a halving of the amount of protein would be indicated by '2.00' or '−2.00' respectively.

The central metabolic pathways are illustrated in Figure 2. Enzyme expression levels illustrate a clear increase in glycolytic, glutaminolytic, hexosamine, and pentose phosphate pathways, as well as increased nucleotide, amino acid, and lipid synthesis. In contrast, enzymes of the TCA cycle are essentially unchanged. This is consistent with metabolic reprogramming [58,61]. If enzyme levels can be used to roughly indicate enzymatic activity, then one third of the glucose is metabolised by oxidative phosphorylation, corresponding to that seen in a rapidly growing tumour.

Metabolic reprogramming renders cancer cells susceptible to growth suppression because of their increased dependence on glucose for these anabolic pathways. Tumour cells predominantly express the embryonic M2 splicing isoform of pyruvate kinase (PKM2) [62]. Short hairpin RNA knockdown of PKM2 leads to its replacement by the adult PKM1 form, reverses metabolic reprogramming, increasing oxygen consumption and reducing lactate production and tumourigenicity in nude mouse xenografts [63]. PKM1 and PKM2 switching is regulated by three heterogeneous nuclear ribonucleoproteins hnRNPA1, hnRNPA2 and the polypyrimidine tract binding protein PTBP1 (or hnRNPI). These proteins bind flanking regions around exon 9, and in doing so, promote PKM2 expression [64]. In 3D spheroids, their expression is reduced (hnRNPA1 −2.20; hnRMPA2 −2.29; PTBP1 −1.23) favouring PKM1 and oxidative phosphorylation. However, no PKM1-specific peptides were detected in the original mass spectrometry data making it impossible to differentiate them [57].

Metabolic reprogramming results in the cell becoming increasingly dependent on glutaminolysis for fatty acid synthesis [41,58,61]. The key enzyme, ATP citrate lyase (ACL) shows increased expression in tumours (lung, prostate, bladder, breast, liver, stomach and colon) [65] and in spheroids (Figure 2). Cytoplasmic isocitrate dehydrogenase 1 produces isocitrate from α-ketoglutarate. IDH-1 mutations have been associated with gliomas via nuclear factor-κB activation in an hypoxia-inducible-factor (HIF1-α) dependent manner [66]. Inhibition of the IDH-1 or ACL inhibits A549 metabolism in vitro and, when injected into nude mice, reduces tumour growth [49,58].

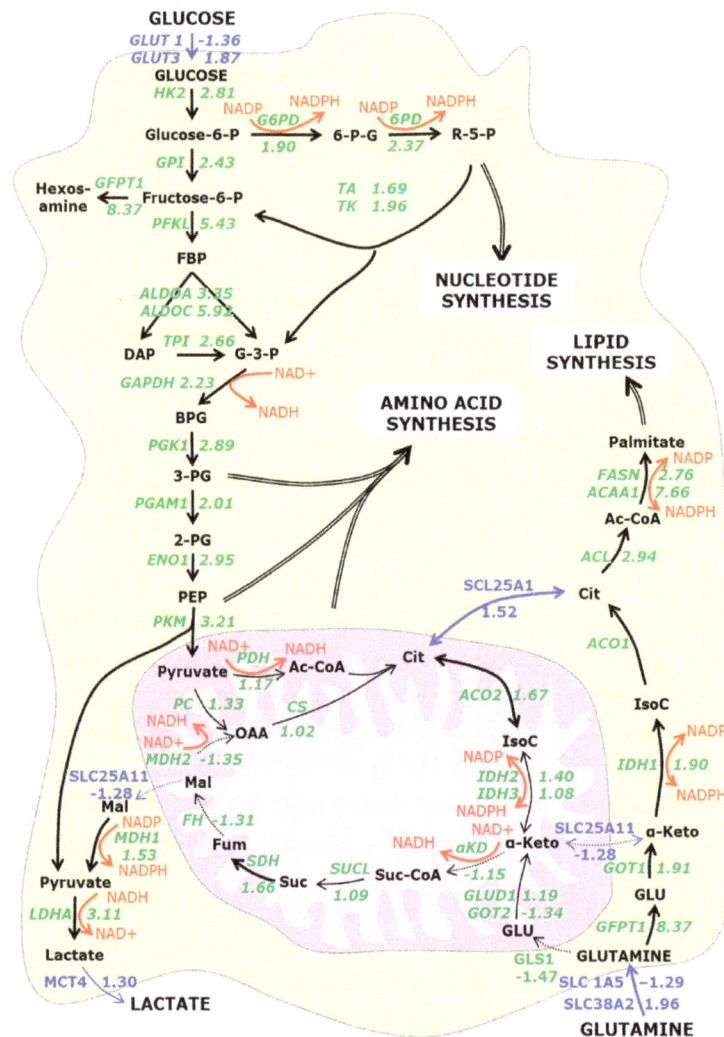

Figure 2. Ratios of the protein abundance of central metabolic enzymes and membrane transporter in 3D compared to two-dimensional (2D) cultures. Metabolites are marked in black, enzymes in green, transporters in blue and selected cofactors in red. Negative ratios indicate that the protein is present in higher amounts in 2D cultures. Arrows connecting metabolites are marked in bold if the enzyme expression is increased by a factor of 1.5 or greater. Arrows connecting metabolites are dotted if the enzyme expression is decreased. (Raw data taken from [57]).

Interestingly, activated effector T cells (TE) utilise anabolic aerobic glycolysis, while memory T cells (TM) use catabolic pathways. Microscopy shows that TE cells have punctate mitochondria, while

TM cells maintain fused networks. The protein Opa1 is required for maintaining fused networks [67]. Opa1 is reduced in 3D spheroids (OPA1 −2.71), also indicating that they utilise aerobic glycolysis [18].

Evidence for metabolic reprogramming of cells in culture has also been found using hyperpolarized [^{13}C] spectroscopy. Two cell lines (Huh-7 hepatocellular carcinoma cells and SF188-derived glioblastoma cells) have been cultivated in 2D and pulse-labelled with hyperpolarized [1-^{13}C] pyruvate to determine the activities of pyruvate dehydrogenase (PDH, as a surrogate indicator of oxphos) and pyruvate carboxylase (PKM, for lactate). While both enzymes were active, supplementation with glucose favoured lactate production. Inhibition of glycolysis using an Akt inhibitor reversed this effect [68]. This illustrates that cells can adapt their metabolic activity to their environment. Rat hepatoma cells (JM1) have been probed using [^{13}C]-labelled glucose while cultivated in either 2D or 3D conditions (encapsulated in alginate beads). These studies showed that in both conditions, 85% of [^{13}C]-glucose was converted either to lactate or alanine by aerobic glycolysis [69].

Jiang et al., compared the metabolic activity of human H460, A549, MCF7, and HT-29 cells grown in 2D and 3D cultures [1-^{13}C] glutamine or [5-^{13}C] glutamine tracers clearly illustrated that citrate and lipids predominantly were synthesised via reductive glutaminolysis. In particular, while neither isocitrate dehydrogenase-1 nor -2 (cytosolic or mitochondrial, respectively) were necessary for monolayer growth, spheroids were dependent on the cytosolic IDH1 for glutaminolysis. Many cell lines (lung, mammary, colon, embryonic fibroblasts, squamous cell carcinoma, melanoma, glioblastoma, and leukaemia) use glutamine as their primary source of acetyl-CoA for lipogenesis [50,70]. Glycolytic ATP is only necessary in hypoxic conditions [37] and glutamine consumption is increased by reducing the available oxygen to 1% [61]. In these conditions, the major role of glucose metabolism is to drive the pentose phosphate pathway to generate NADPH. Glutaminolysis drives acyl-CoA production and lipidogenesis [50].

3.2. Is Metabolic Reprogramming Driven by Oxygen or Glucose Insufficiency?

Given that spheroids constructed from many types of cell exhibit metabolic reprogramming, the question arises as to what causes the switch?

3.2.1. Diffusion Gradients and the Importance of Irrigation

Mammalian cells need oxygen and nutrients. In tissues, they are normally located within 100 to 200 μm of capillaries [71]. This corresponds to roughly 10–40 cell layers thick. An experimentally derived diffusion limit (i.e., where the PO_2 level falls to 0) of 232 ± 22 μm agrees well with this [72]. Many cells in tissues experience low oxygen tension (e.g., 1% O_2) [73] and it is alternative stressors, such as serum deprivation or acidosis, which induce cell death [74]. Only severe hypoxia (<0.01%) O_2 is capable of inducing apoptosis [75].

The most significant differences between 2D and 3D culture are diffusion gradients. Several types exist including gasses, nutrients, metabolites, signalling molecules, secondary messengers and growth factors. Here we will only consider oxygen, CO_2 and glucose, since they have been suggested to drive metabolic reprogramming and need to be taken into consideration when designing micro-bioreactors (Figure 3).

The existence and depth of the hypoxic zone depends on several factors, including radius, cell type, and media flow rate. Measured diffusion gradients for oxygen follow smooth sigmoidal curves (with no difference in shape outside or inside EMT6 spheroids), suggesting that the presence of cells has little influence on its diffusability. Atmospheric oxygen (21%) provides a partial pressure (PO_2) of about 145 mm Hg (ca. 190 μM) in media. In static cultures where there is no flow of plasma or media, the PO_2 falls rapidly towards the spheroid's centre (Figure 3A). Small radius spheroids (25–50 μm) have about 3.3% PO_2 in their core, close to physiological levels in the brain. 100 μm radius spheroids have about 1.6% PO_2 and larger spheroids less [76]. In irrigated spheroids, where media flows past the spheroid, the PO_2 falls to about 13% measured in the media at the surface of the spheroid [77]. Thus, even when irrigated, there is a 'diffusion-depleted zone' in the media surrounding each spheroid (grey zones in Figure 3). The PO_2 reaches a minimal plateau of 3.3% at about 150 µm into a 352 µm radius irrigated EMT6 spheroid, and 1.6% at about 225 µm into a 480 µm radius spheroid. In the latter case, stopping the media flow causes the core PO_2 to quickly fall to 0. Doubling the flow rate had a marginal effect and was confined to the spheroid surface [78]. Core PO_2 reached 0% in irrigated spheroids with radii greater than 600 µm.

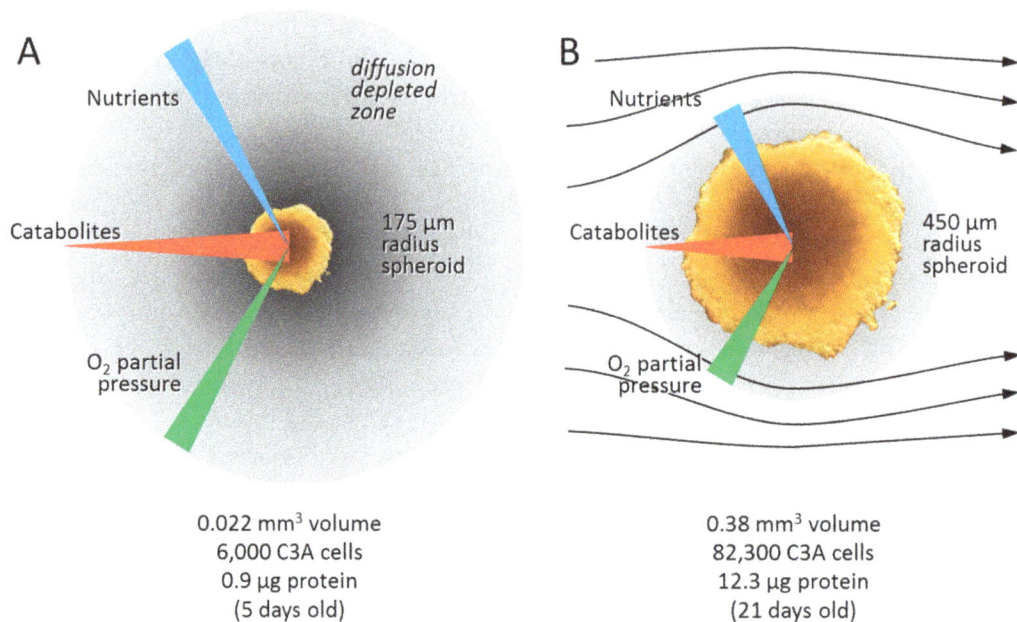

Figure 3. The diffusion depleted zone and its consequences for the oxygen, nutrient and catabolite gradients in (**A**) passive and (**B**) irrigated 3D culture. The size shown indicates the approximate maximum size above which anoxia develops in the spheroid core. Volumes, number of cells and amounts of protein are indicated for each. The apparent increase in cell volume is attributed in part to increased ECM and the development of sinusoidal and bile cannalicular spaces between the cells.

Haemoglobin is normally found in hepatocytes and several cell lines. Hepatocarcinoma spheroids increase their haemoglobin content by a factor of thirty to actively alleviate low PO_2 oxygen concentrations [57].

Experiments using pH as a spatial readout have demonstrated that the diffusivity of CO_2 through spheroids (colorectal HCT116 and HT29, breast MDA-MB-468, pancreatic MiaPaca2, cervical squamous cell carcinomas HeLa and SiHa and ovarian clear cell adenocarcinoma OVTOKO) is exactly the same as its diffusivity through water (2.5×103 µm^2/s) [79]. Usually, the spheroids' cores are slightly more acidic than the surrounding media, possibly due to increased CO_2 or lactate amounts [80,81].

No data is available describing a glucose gradient in or around spheroids. When considering that the glucose molecule is larger than oxygen or CO_2, the gradient would intrinsically be expected to be steeper, but glucose transporters may alleviate this.

3.2.2. Hypoxia Affects Glycolysis and Oxidative Phosphorylation

Hypoxia has numerous effects on mammalian cells. One is the activation of the constitutively expressed hypoxia-inducible factor (HIF-1α) (Figure 4).

Under normoxia, defined as the PO_2 levels normally seen in healthy tissues (usually 1–5% [73]), HIF-1α subunits are hydroxylated by prolyl hydroxylases (PHD1-3 including the TCA cycle enzyme α-KD). The modified HIF-1α is recognized and targeted for proteasomal degradation by the VHL-E3-ubiquitin ligase complex.

When oxygen concentrations decrease, the oxygen-dependent PHDs are inactivated, allowing for the HIF-1α protein to accumulate. This promotes HIF-1α translocation to the nucleus where it interacts with HIF-1β/ARNT and p300. This complex binds hypoxia-response elements (HREs) in promoter regions of numerous target genes, including glucose transporters and glycolytic enzymes [82].

The translationally controlled tumour protein (TCTP 4.49) binds competitively to VHL, reducing PHD binding and accelerating HIF-1α accumulation, nuclear translocation, and transcription reprogramming [83].

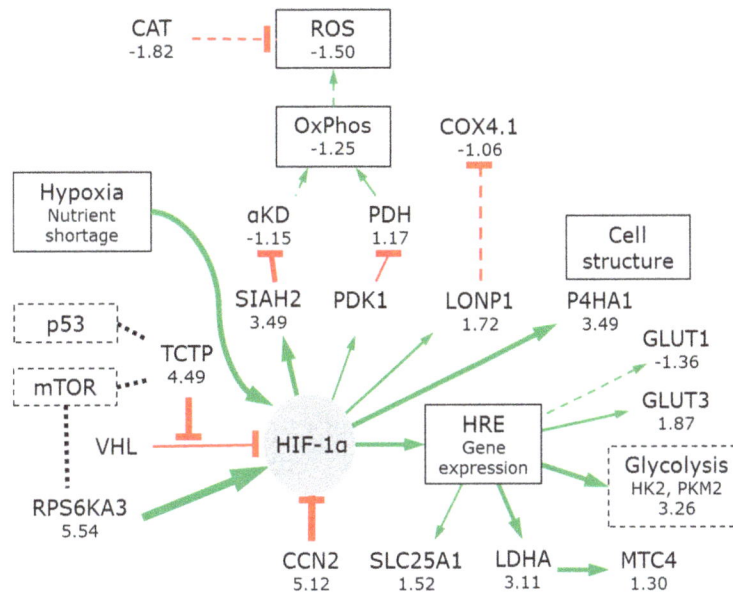

Figure 4. Effects of hypoxia on HIF-1α. The thickness of the line indicates the ratio of protein amount in thre—dimensional (3D) cultures compared to 2D cultures. Dotted lines indicate reduced expression. Green lines ending in arrowheads indicate activators, while red lines ending in a bar indicate inhibitory activity. Dotted-boxes indicate links to other pathway figures.

HIF-1α induces pyruvate dehydrogenase kinase 1 (PDK1) expression. PDK1 inhibits the mitochondrial pyruvate dehydrogenase (PDH) [84]. This reduces pyruvate flux into the TCA cycle and lowers the mitochondrial oxygen requirements. This switch increases lactate production and secretion, as observed by Warburg. Differentially transformed rat embryo fibroblasts showed increasing levels of lactate content and unchanged or decreasing lactate secretion in irrigated spheroids with increasing radii of up to about 450 μM. Above this radius, lactate content and secretion stabilised, illustrating that hypoxia-induced glycolysis need not lead to lactate secretion [58,81] suggesting that most of the glycolytic metabolites are utilised in anabolic processes.

Liver cells can convert lactate back to pyruvate. Despite this, the lactate transporter (MCT4 or SLC16A3) is increased (1.30) suggesting that the cells might 'pump' the lactate towards the spheroid surface.

Spheroids show the increase in glucose transporters, glycolytic enzymes and lactate dehydrogenase by on average about a factor of 3.26. This metabolic reprogramming is partial: spheroids do not show a significant decrease in the pyruvate dehydrogenase or of any enzymes of the TCA cycle (Figure 2).

HIF-1α also induces E3-ubiquitin ligase SIAH2 synthesis. This mediates the proteasomal degradation of the OGDH subunit of α-KD and forms part of the feedback control of HIF-1α. A modest reduction of the α-KD 3 enzyme complex is observed in spheroids (DLD -1.26, DLST -1.11, OGDH -1.09). This will slow the TCA cycle and allow for more citrate to be transported into the cytoplasm by an upregulated citrate transporter protein (SLC25A1, 1.52), supporting the metabolomics [81] and isotope analyses [58,61].

Interestingly HIF-1α also promotes extracellular matrix remodelling via collagen hydroxylases (P4HA1 3.49), a facility useful for cancer cell morphology, adhesion, and motility [85].

Part of the indirect negative feedback regulatory circuit for HIF-1α is the connective tissue growth factor (CCN family member 2) or insulin-like growth factor-binding protein 8, (IBP-8). It is strongly upregulated in spheroids (5.12) illustrating strong positive and negative regulatory mechanisms are active.

HIF-1α can also induce the mitochondrial protease LONP1, which degrades the less efficient cytochrome C oxidase 4 subunit 1 (COX4-1) from the complex IV of the electron transport chain and allows it to be replaced by the more efficient COX4-2 [82]. Although LONP1 was increased (1.72), there was no change in the level of COX4-1 (−1.06). LONP1 is an essential central regulator of mitochondrial activity and is overexpression in oncogenesis [86]. Despite that spheroids contain higher levels of ATP, subunit IV and the ATP synthase (subunit V) are reduced by −1.12 and −1.37 respectively [18,19]. Reduced mitochondrial respiration will result in fewer reactive oxygen species correlating with reduced levels of catalase (CAT −1.82) [87] resulting in diminishes hydrogen peroxide damage and 50% less oxidised proteins.

3.3. Glucose Starvation Has Little Effect on Metabolic Reprogramming

The feature that Warburg noticed—that cancer cells rapidly use glucose and convert it to lactate would suggest that glucose availability might also play a central role.

Liver cell spheroids are known to rapidly import glucose and convert it to glycogen. When cultured in bioreactors with physiological amounts of glucose (5.5 mM), the media glucose is typically exhausted 8 h after media exchange (Figure 5). Thereafter, the spheroids experience 'glucose starvation' and catabolise the glycogen they have synthesised.

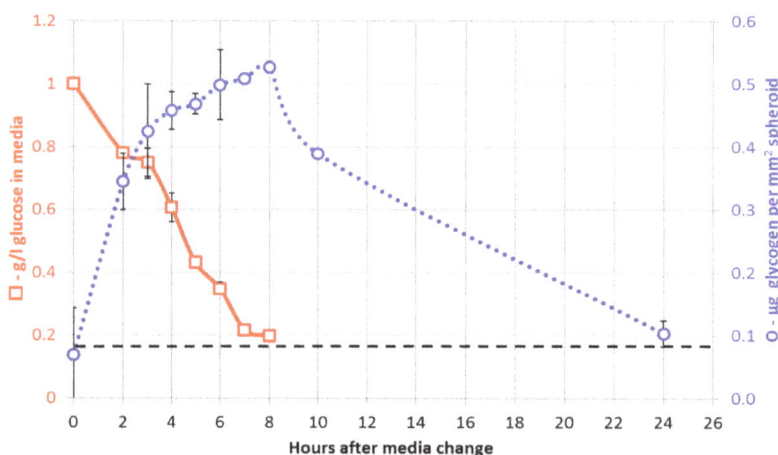

Figure 5. Relationship of glucose consumption from the media with the level of glycogen present in the spheroids. The amount of glucose in the spheroids prior to glycogen hydrolysis was negligible.

Glucose starvation could unleash a number of changes in the cell, as initiated by the Glucose Regulated Proteins (GRP). These are typically found in the ER, often overexpressed in cancers and associated with aggressive growth and invasion [88]. Their first effect would be to increase ER stress and initiate the unfolded (or misfolded) protein response (UPR) [89]. In the UPR, GRP78 dissociates from three protein-folding quality sensors (IRE1, PERK, and ATF6) embedded in the ER membrane. These sensors activate the UPR signal transduction program, a negative feedback loop that alters gene expression to slow protein synthesis and the cell cycle (eventually leading to arrest in G1 [90,91]).

Surprisingly, the amount of GRP78 is unchanged between 2D and 3D (1.05) and there are only weak changes in GRP58 (1.34) and GRP60 (1.37), suggesting that there is no UPR or ER stress.

The mitochondrial GRP75 can inactivate p53 and induce apoptosis [88], but it is only slightly elevated (1.36), suggesting that the mitochondria also suffer very little stress.

GRP94 and GRP170 show the strongest responses (2.09 and 2.15, respectively). GRP 94 plays critical roles in folding and exporting proteins in the secretory pathway (e.g., insulin-like growth factors IGF-1 and 2), which could activate the PI3K-Akt pathway. GRP170 is a glycosylated protein also known as the hypoxia up-regulated protein 1, HYOU1. It plays a role in suppressing apoptosis and is up-regulated in invasive tumours [88]. Considering the relatively little stress caused by prolonged glucose starvation, the lack of glucose appears to play a minor role in the metabolic reprogramming. What effects there are, appear to stabilise cellular metabolism and are anti-apoptotic.

3.4. Metabolic Reprogramming 'Links' Glutamine Metabolism to the Hexosamine Pathway

Metabolic reprogramming results in an increased reliance on glutamine. Intracellular levels are regulated by plasma membrane transporters SLC1A5 and SLC38A2 [92]. ER stress would induce their degradation and ultimately to autophagy and cell death [92,93]. In spheroids, SLC38A2 is increased (1.96) while SLC1A5 is decreased (-1.29), suggesting that they play subtly different roles. In agreement with this, net glutamine uptake in HeLa cells was not dependent on SLC1A5 but required SLC38A1 or 2 [94].

3.4.1. Conversion of Glutamine to Glutamate

The cells' glutamate demand is probably supplied by the highly upregulated GFPT1 which is the first, and rate-limiting step, of the hexosamine pathway (8.37). This enzyme catalyses the conversion of fructose 6-phosphate and glutamine to glucosamine 6-phosphate and glutamate. Activation of glutaminolysis was necessary for adaptive cell survival in the mouse model of pancreatic ductal adenocarcinoma [95]. Hypoxia is considered to drive this adaptive process, which, amongst other things, leads to increased amounts of O-linked N-acetylglucosaminylated proteins. In agreement with this, several polysaccharide, proteoglycan and glycosylation synthetic pathway enzymes are strongly upregulated in 3D spheroids (UDP-glucose pyrophosphorylase UGP2 6.59; UDP-glucose 6-dehydrogenase UGD 7.46; UDP-glucose 4-epimerase GALE 12.71; and sialic acid synthase NANS 5.37).

3.4.2. α-Ketoglutarate

Glutamate can be converted to α-ketoglutarate by the mitochondrial GLS1 (-1.47) and GLUD (1.19). It can also be converted by cytoplasmic or mitochondrial alanine or aspartate aminotransferases [50] and it is the cytoplasmic enzyme that is upregulated (GOT1 1.91, GOT2 -1.34).

There are three possible routes by which α-ketoglutarate can be converted to citrate (Figure 2). Firstly, it can be converted around the TCA cycle. Secondly it could be converted via isocitrate to citrate (by IDH2 1.40 and ACO2 1.67), reversing the normal TCA cycle flux by reductive glutamate metabolism [61]. IDH3 is not increased (1.08) because it can only catalyse the 'forward' reaction. Finally cytoplasmic α-ketoglutarate, produced via the upregulated GFPT1 and GOT1, can be converted, by IDH1 (1.90) and ACO1. While all three processes probably occur, both isotope tracing and enzyme abundance suggests that the latter route is the most active [57,61].

3.4.3. NADH

Conversion of α-ketoglutarate to isocitrate requires the cofactor NADPH. The reduction in the mitochondrial MDH2 (-1.35) and the essentially unchanged abundance of its NAD(P) transhydrogenase (NNT 1.13) suggest that the mitochondrial source is of low significance. MDH1 is increased (1.53) but lacks a malate source (the SLC25A11 transporter is reduced -1.28) and the conversion of cytoplasmic pyruvate to lactate would actually consume the NADH that is produced. The richest source of NADPH is the pentose phosphate pathway where G6PD and 6PD are both upregulated (1.90 and 2.37 fold respectively), in agreement with isotope tracing data [57,61].

3.4.4. Citrate

Citrate is used for fatty acid synthesis. ATP-citrate synthase (ACL 2.94) uses citrate to generate cytosolic acetyl-CoA. Acetyl-CoA is used for: histone acetylation by acetyl-CoA acyltransferase (ACAA1 7.66); palmitate synthesis by fatty acid synthase (FASN 2.76); cholesterol, steroid hormones, haem and a plethora of other biomolecules. Glutamine is as important as glucose in metabolic reprogramming and blocking glutamate-dependent cellular pathways (at either IDH1 or ACL) limits tumorigenic growth [49,58].

3.5. Metabolic Reprogramming Is Associated with Chromatin Remodelling

Conversion between transcriptionally active euchromatin and inactive heterochromatin is brought about by processes, including acetylation, methylation, and clipping of histones. Hypoxia can change these epigenetic markings. HIF-1α stabilisation leads to increases in histone lysine demethylases (KDM3A, KDM4B, KDM4C, and KDM6B) [96]. While C3A cells that are grown in 2D culture essentially show little epigenetic marking, spheroids recover extensive histone methylation, acetylation, and clipping on both H2B and H3 [97]. Hypoxia also upregulates the arginine N-methyltransferase PRMT1 (2.66), increasing methylation of arginine 3 of H4 [98]. PRMT1 can asymmetrically methylate the ReIA subunit [99] inhibiting its binding to DNA and repressing NF-κB target genes. The 'Chromatin target of PRMT1' protein, (CHTOP) [100] which promotes cell cycle progression, is strongly reduced in spheroids (-8.09), resulting in few cells in the G2/M phase. Histone deacetylases do not appear to be affected (HDAC1 1.00).

3.6. The Switch to Anabolic Metabolism

Spheroids composed of either OVTOKO or SIHA cell lines have been shown to contain higher levels of serine, glutamine and other amino acids as well as citrate [81]. The amounts of all anabolic rate-limiting enzymes are increased while catabolic enzymes are unchanged (Table 1) in concordance with metabolic reprogramming to aerobic glycolysis.

Table 1. Rate limiting enzymes for central metabolic pathways and the ration of their expression in 3D spheroids compared to 2D exponential growth. n.d. not detected.

Pathway	Gene	Fold Change
Glucose phosphorylation	HK2	2.81
Glycogenolysis	PYGB	4.06
Glycolysis	PFKL	5.43
Glycolysis	PKM	3.21
Pentose Phosphate	G6PD	1.90
Hexose	GFPT1	8.37
TCA Cycle	IDH2 & 3	1.40 & 1.08
Pyrimidine synthesis	CAD	3.49
Purine synthesis	PRPS1	3.74
Fatty acid synthesis	FASN	2.76
Fatty acid synthesis	ACAA1	7.66
Fatty acid oxidation	CRAT	1.17
Alanine synthesis	ALT	n.d.
Asparagine synthesis	ASNS	5.58
Aspartate synthesis	GOT1	1.91
Cysteine synthesis	MAT1	5.62
Glutamine-glutamate conversion	GLUD	1.19
	GFPT1	8.37
Glycine synthesis	SHMT	1.47
Methionine synthesis	MTR	n.d.
Proline synthesis	PYCR1 & 2	1.06 & 1.03
Serine synthesis	PHGDH	7.67
Tyrosine synthesis	PAH	3.80
Urea synthesis	CPS	3.49
Folate synthesis	MDHFD1	2.49

The three rate-limiting glycolytic pathway steps (HK2, PFKL, and PKM) are three of the four most increased enzymes of the pathway (the 4th being aldolase). Interestingly, PFKL is repressed by high ATP/AMP ratios [101]. Since spheroids have high ATP amounts [19], high PFK levels suggest that AMP levels are also high.

The glutamine-dependent cytosolic carbamoyl-phosphate synthetase 2, is upregulated in spheroids (CAD 3.49). CAD is the rate-limiting enzyme carrying out the first three steps in pyrimidine synthesis. CAD is essential for uridine diphosphate (UDP) synthesis, which in turn, is essential for glycogenesis. This correlates with the appearance of glycogen granules in hepatocyte spheroids and with protein glycosylation and the hexosamine pathway [102].

3.7. Signal Pathways Involved in Orchestrating Metabolic Reprogramming

All of the adaptations seen in glycolysis and glutaminolysis, pentose phosphate pathway, TCA cycle, and fatty acid synthesis indicate that spheroids, grown in a wide variety of 3D culture systems, are utilising a significant degree of metabolic reprogramming to aerobic glycolysis.

The typical features of 3D culture—diffusion gradients resulting in hypoxia (and to a less extent glucose starvation) clearly drive metabolic reprogramming. Warburg saw this phenomenon as a hallmark of cancer. In order to investigate how metabolic reprogramming is orchestrated, we reviewed the status of pathways that are often associated with tumour development: PI3K/Akt/mTOR, Myc, p53, nuclear factor kappa-B (NF-κB), and Wnt [54].

3.7.1. PIK3/AKT/mTOR

The PI3K/AKT/mTOR pathway (Figure 6) plays a key integrating role, sensing concentrations of nutrients (including glucose, oxygen, amino acids and ATP levels) and regulating the anabolic processes of the cell for growth and maintenance [103].

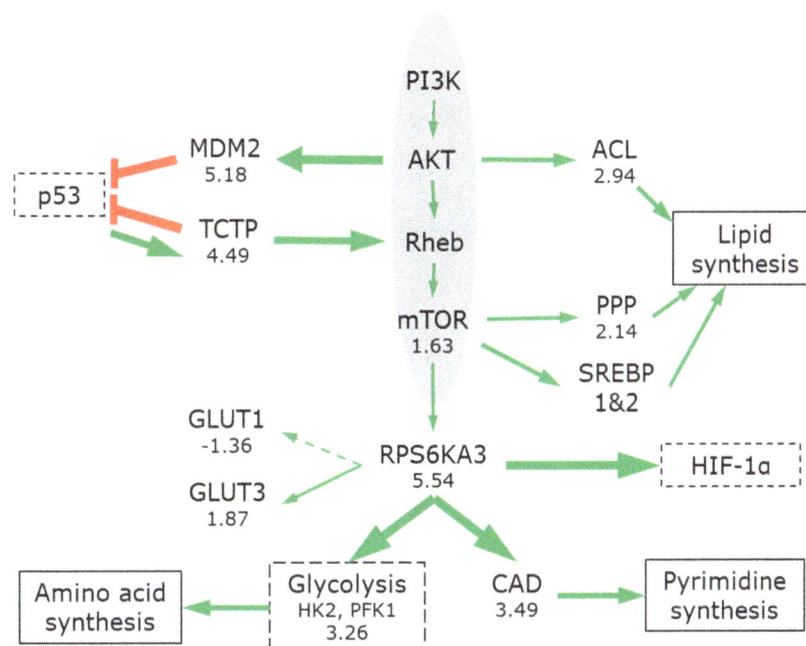

Figure 6. mTOR signalling in 3D spheroids. See legend to Figure 4 for nomenclature.

While only two key proteins from this pathway were detected (mTOR, 1.63; and ribosomal protein S6 kinase RPS6KA3, 5.54), strong downstream effects are clearly visible showing that pathway is activated in 3D (Table 1). mTOR signalling increases translation of hypoxia-inducible factor 1α (HIF-1α), glucose transporters and glycolytic enzymes, and promotes metabolic reprogramming [101,104] (Figure 6).

mTOR promotes pentose phosphate pathway (PPP) enzyme expression (on average by 2.14) and channels metabolic flux into its oxidative, NADPH-producing branch [91]. mTOR strongly stimulates pyrimidine synthesis via the RPS6KA-mediated phosphorylation of CAD (3.49), thereby increasing the pool of nucleotides available [105]. AKT can phosphorylate ACL, enhancing its lipogenic activities and mTOR signalling promotes NADPH-requiring lipid synthesis by activating sterol regulatory element-binding proteins (SREBP1 and 2) [106].

3.7.2. Myc

Myc has the potential to play a key role in metabolic reprogramming. Myc is central to growth regulation and is one of the most frequently deregulated oncogene transcription factors seen in a wide variety of cancers [107,108]. Myc directly transactivates gene expression of GLUT1, phosphofructokinase (PFK), enolase (ENO) and LDHA and indirectly increases phosphoglucose isomerase (GPI), glyceraldehyde-3-phosphate dehydrogenase (GAPDH) and phosphoglycerate kinase (PGK1) [109] (Figure 7). This is consistent (with the exception of GLUT1) with their increased levels in spheroids. However, as described above, PIK3/AKT/mTOR can also induce these proteins (via HIF-1α) and so this effect need not be attributed to Myc. HIF-1α can inactivate Myc [110], and in doing so, induce cell cycle arrest [111].

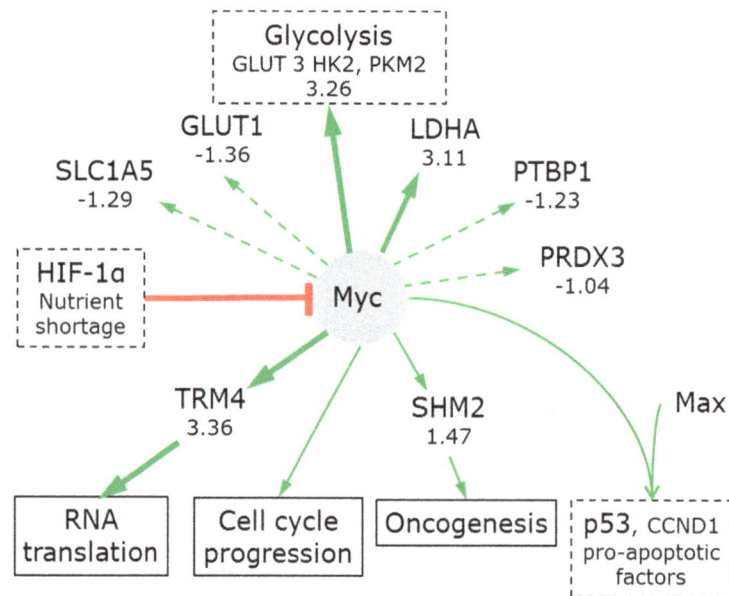

Figure 7. Myc signalling in 3D spheroids. See legend to Figure 4 for nomenclature.

Low expression levels of several proteins normally induced by Myc suggest that Myc is not particularly active in 3D spheroids. Examples include: PTBP1 [112] (−1.23); GLUT1 (−1.36); SLC1A5 [107] (−1.29); and, PRDX3's [113] (−1.04). Myc regulates serine hydroxymethyl transferases and pathway hyperactivation is a driver of oncogenesis [107]. However, the moderate increase of SHMT2 (1.47) cannot qualify as hyperactivation. One exception may be tRNA (cytosine34-C5)-methyltransferase (TRM4 which methylates the first position of the cytosine anticodon). Myc enhances TRM4 expression (3.36). The formation of a covalent complex between dual-cysteine RNA:m5C methyltransferases and methylated RNA has been proposed to provide a unique mechanism by which metabolic factors can influence RNA translation, in particular the processing and utilisation of m5C-containing RNAs [114].

Nutrient shortage and/or hypoxia can inhibit Myc translation; reduce its stability and its ability to dimerise with another transcription factor MAX. Inhibition of Myc/MAX dimerization prevents specific gene expression, most significantly of p53, cyclin D1 and pro-apoptotic factors [115]. Therefore, while Myc regulates many proteins in cancer [107], it appears that the slow proliferation of cells in spheroids is a result of low myc activity.

3.7.3. p53

The tumour suppressor p53 can transactivate a broad array of target genes that are involved in redox maintenance, DNA repair, cell cycle checkpoints, and can thus affect cellular senescence, proliferation, and apoptosis. Mutations of p53 are found in over 50% of human tumours and disturb the IGF1-AKT branch of the mTOR pathway [116].

p53 activity is tightly linked to the oncogene protein DJ-1 (1.83). In a self-regulating loop, DJ-1 is necessary for hypoxic stress-induced p53 activation, while p53 prevents the accumulation of the DJ-1 protein (Figure 8). DJ-1 can bind the ubiquitin-independent 20S proteasomal core and its quantitative increase mirrors the increase in the core (1.92) and in NADPH:quinone oxidoreductase 1 (NQO1, by 2.42) (which protects p53 from proteasomal degradation).

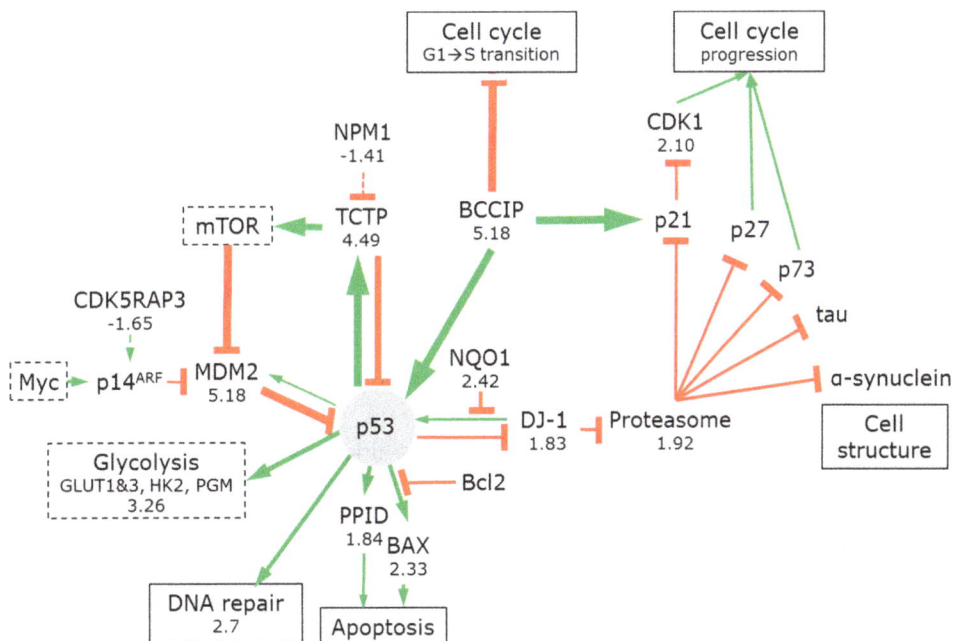

Figure 8. p53 signalling in 3D spheroids. See legend to Figure 4 for nomenclature.

Many key regulatory proteins, including tumour suppressors p53 and p73, tau, α-synuclein and the cell cycle regulators p21 and p27 are degraded by the proteasome core. DJ-1 binding inhibits the activity of the core and by slowing their degradation, leads to an increase in their abundance and activity [117]. DJ-1 can also activate the AKT/mTOR pathway [116].

In addition, the transcriptional suppressor CDK5RAP3 is reduced (−1.65). This will allow for the synthesis of p14ARF and its binding to MDM2. This releases p53 from inhibition. This results in the stabilization, accumulation, and activation of p53 [118].

Activation of p53 is consistent with the increase in DNA repair enzyme expression [57] (on average by 2.7). Interestingly, both the positive (BCCIP) and negative (TCTP) p53 regulators are strongly increased (5.18 and 4.49, respectively), illustrating that p53 is subjected to a tight feedback regulation. BCCIPβ plays a role in cell growth regulation [119]. Overexpression of the BCCIPβ splices variant delays the G1-to-S cell cycle transition and elevates p21 expression. Elevated p21 expression would inhibit cyclin dependent kinase 1 (CDK1 2.10) induction of cell cycle progression.

The evolutionarily conserved TCTP is emerging as a pleiotropic key to phenotypic reprogramming through its ability to regulate the mTOR pathway [120], as well as being an upstream activator of OCT4 and NANOG transcription factors (which play essential roles in nuclear reprogramming). p53 induces TCTP, reducing oxidative stress and minimizing apoptosis [121]. Forming another negative feedback loop, TCTP can inhibit both transcription and function of p53 [122]. The activation of TCTP (4.49) suggests reduced proliferative drive [123]. This is confirmed by the reduction in nucleoplasmin (NPM1 −1.41), which would otherwise complex with TCTP during mitosis to promote cell proliferation.

p53 directly regulates cellular redox homeostasis by modifying expression of pro- and anti-oxidant enzymes peroxiredoxins and thioredoxins. Peroxiredoxins (that act as both sensors and barriers to MAPK activation) are upregulated (PRDX1-6 of 1.74, 2.27, −1.04, 1.86, 1.42, and 2.75, respectively), as is thioredoxin (TXN 1.84). Their upregulation contributes to Myc regulation [124]. The modest changes in mitochondrial peroxiredoxins (PRDX 3 and 5: −1.04 and 1.42) suggest that mitochondrial ROS are insignificant 'stress factors' in keeping with the relatively reduced mitochondrial activity.

p53 is activated, but is exposed to tight feedback control. Together with the low activity of Myc, p53 and associated pathways arrest the cells predominantly in G1 or Go [19].

3.7.4. Wnt GSK-3β/β-Catenin

In the canonical Wnt pathway, the Wnt ligand can bind to a Frizzled family receptor, causing a deactivation of the β-catenin destruction complex. This leads to the dephosphorylation of β-catenin, its accumulation and migration to the nucleus where it acts as a coactivator of TCF/LEF transcription factors. Activation of the Wnt/β-catenin pathway activates cell proliferation and the homeostatic renewal of the liver from pericentral hepatocytes [125]. In 3D spheroids, this pathway is inactive: the amount of β-catenin is reduced (CTNNB1 −1.57), and the protein phosphorylase 2A, although present (PPP2RA1 (the constant regulatory subunit core of the PP2A) 1.02), is strongly inhibited by I1PP2A and I2PP2A (5.77 and 2.18). mTOR also negatively regulates PP2A, allowing for the integration of these two pathways. Reverse regulation occurs in amino-acid depleted conditions: PP2A can inhibit mTOR via dephosphorylation of p170 [126].

The histidine triad nucleotide-binding protein 1 is significantly increased (HINT1, 5.99). It keeps the Wnt GSK-3β/β-catenin pathway inactive, limits cell growth [127], and promotes apoptosis via p53 and Bax [128]. The mitochondrial HINT2 may also promote apoptosis (1.32).

In contrast, the non-canonical Wnt pathway appears to be active in 3D spheroids. Binding of Wnt to Frizzled recruits Dsh, which then binds directly to RAC1 (1.40) and indirectly to profilin (2.75) amongst others. Both of these and numerous other upregulated actin-structure modifying proteins lead to the dramatic restructuring seen in spheroids [57].

3.7.5. NF-κB

NF-κB is a rapid-acting primary transcription factor well suited to respond to harmful stimuli like cell stress, cytokines and free radicals. Many different types of human tumours have constitutively active NF-κB [129].

In its inactive state, the NF-κB heterodimer (composed of p50 and ReiA) is complexed with its inhibitor IκBα. In spheroids, HINT1 (5.99) promotes IκBα stability maintaining NF-κB inactive [127]. Hypoxia upregulates PRMT1 (2.66) [57], which asymmetrically methylates ReIA inhibiting ReIA's binding to DNA, and further repressing NF-κB [99]. The type III transforming growth factor β receptor (TGFβR3) regulates both the epithelial-mesenchymal transition and cell invasion during development via NF-κB activation [130]. Deactivation of TGFβR3 (−4.58) is consistent with NF-κB inactivity. The extracellular matrix TGFβ-induced protein (TGFBI), involved in tissue remodelling and found in liver metastases stroma, is very highly upregulated (TGFBI 8.83) [131]. TGFBI reduces NF-κB activation [132]. Deactivation of NF-κB in 3D sensitises the cell to apoptosis or necrosis by allowing for TNF-α to active the JNK pathway and lead to cell death.

3.7.6. Cell Death

Many of the pathways described above influence necrosis and apoptosis. The fundamental difference between them is that bioenergetic failure in necrosis leads to free radical damage, swelling, rupture, and cytolysis, while apoptosis is ATP-requiring and leads to shrinkage, caspase activation, DNA fragmentation, and retention of the plasma membrane [133].

Apoptosis is often 'defeated' as a cell is transformed from healthy to tumourigenic. Many specific mechanisms operating in many organelles can lead to apoptosis [133]. The apoptotic potential is a balance between pro- and anti-apoptotic signals, which are integrated in mitochondria. The decreased amounts of NF-κB and other factors noted above, result in the under expression of anti-apoptotic proteins including Bcl-2; Bcl-XL; NR13; Bcl-2 inhibitor of transcription 1 (PTRH2 −1.37); Bcl-2-associated transcription factor 1 (BCLAF1 −2.61); Bcl-2-associated athanogene 2 (BAG2 −1.18); Bcl-XL-binding protein v68 (PGAM5 −1.78) and the 'defender against apoptotic cell death' (DAD1 −1.37). These anti-apoptotoic proteins would otherwise bind and inactivate pro-apoptotic proteins. The only pro-apoptotic protein detected, BAX (Bcl-2-like protein 4), was increased (2.33). The net result in spheroids is to increase their apoptotic sensitivity, but without activating apoptosis.

Necrosis, as judged by the microscopic appearance of core cells, has often been reported for spheroids. Activated p53 would interact directly with PPID and push the cell towards necrosis [134]. This interaction may be enhanced by increased BAX abundance (2.33), especially when anti-apoptotic Bcl2 proteins are depleted. Thus, both apoptotic and necrotic processes are sensitised.

3D spheroid cultures have illustrated that the serine protease tumour suppressor MASPIN facilitates the mitochondrial permeability transition (MPT). However, since ATP levels are high, neither process opens the MPT pore. Its opening would initiate a collapse of the transmembrane proton gradient and lead to apoptotic or necrotic cell death (depending on the initiating factors). The essential component of the MPT pore, the peptidyl-prolyl isomerase D, located in the mitochondrial matrix is increased (PPID 1.84). The non-essential components, VDAC, (Voltage Dependent Anion Channel, which spans the outer membrane) and ANT (Adenine Nucleotide Translocase which spans the inner membrane) are either unchanged or are decreased (VDAC1 1.02; VDAC2 −1.16; VDAC3 −1.23; ANT1, (ATP/ADP antiporter SLC25A4) −1.01; ANT2, −1.98; ANT3 −1.25). ANT1 can interact with BAX. ANT2 is anti-apoptotic and it's reduction matches other anti-apoptotic BCl-2 proteins. On the balance, necrosis may be favoured over apoptosis due to the reduction in the chromatinolytic activity of AIFM1 (apoptosis-inducing factor mitochondrion-associated 1, −1.90) [135].

4. Conclusions

The most widely used approach to reproducibly produce 3D spheroids or organoids that are stable for long periods of time are clinostat 'microgravity' cultures in micro-bioreactors. In these spheroids, the majority of cells experience hypoxia and glucose starvation. These conditions are certainly closer to those present in tissues than those experienced by cells in classical 2D cultures (which typically experience hyperoxia and hyperglycaemia), and are therefore critical to take into account when designing a micro-bioreactor. The recovery of physiological behavior stems from:

1. Oxygen limitations (and to a less extent glucose) induce metabolic reprogramming from oxidative phosphorylation to aerobic glycolysis and result in a strong anabolic phenotype.
2. The metabolic reprogramming includes an activation of glutaminolysis (via extra-mitochondrial pathways) (consistent with physiological increases in lipid and cholesterol synthesis).
3. Glutamine conversion to the lipid 'precursor' glutamate is linked to the hexosamine pathway activation. This correlates to increased glycogen production and protein glycosylation.

4. The additional NADPH needed for citrate and lipid synthesis is mainly generated by pentose phosphate pathway activation. Increases in acetyl-CoA also provide precursors for the observed histone acetylation.

5. Signalling pathway activities (activation of mTOR and p53, repression of NF-κB and canonical Wnt) are consistent with significant retardation of proliferation and the accumulation of cells in G1/G0, (resulting in a rate resembling that seen in both healthy and transformed cells in tissues and tumours).

6. The reduction in proliferation rate allows the cell to achieve higher ATP levels.

7. Activation of the non-canonical Wnt signalling pathway orchestrates the significant ultrastructural changes.

8. The rate of proliferation is not coupled to aerobic glycolysis.

9. Metabolic reprogramming underpins the recovery of traits mimicking in vivo physiology.

3D tissues offer an exciting model to investigate in vivo-like functionality where cells are grown in conditions that are not drastically different to those seen in vivo. Given the right growth conditions, cells 'spontaneously' revert to an in vivo mimetic physiological performance.

Acknowledgments: This work was supported in part by a grant from MC2 Therapeutics, Hørsholm, Denmark and from the University of Southern Denmark. We would also like to acknowledge the support of the COST actions CM1407 (Challenging organic synthesis inspired by nature—from natural products chemistry to drug discovery) and CA16119 (In vitro 3-D total cell guidance and fitness). The funding sponsors had no role in the design of the study; in the collection, analyses, or interpretation of data; in the writing of the manuscript, and in the decision to publish the results. We would like to thank Kira Eyd Joensen for excellent technical assistance, Aleksandra Amaladas for help with preparation of glucose/glycogen utilization experiment and Adelina Rogowska-Wrzesinska for critical review of the manuscript.

Author Contributions: K.W. and S.J.F. conceived and designed the experiments, analyzed the data and wrote the paper.

Conflicts of Interest: K.W. and S.J.F. are owners of CelVivo IVS, a company producing equipment and micro-bioreactors for 3D cell culture.

References

1. Balasubramanian, S.; Packard, J.A.; Leach, J.B.; Powell, E.M. Three-dimensional environment sustains morphological heterogeneity and promotes phenotypic progression during astrocyte development. *Tissue Eng. Part A* **2016**, *22*, 885–898. [CrossRef] [PubMed]

2. Drost, J.; Karthaus, W.R.; Gao, D.; Driehuis, E.; Sawyers, C.L.; Chen, Y.; Clevers, H. Organoid culture systems for prostate epithelial and cancer tissue. *Nat. Protoc.* **2016**, *11*, 347–358. [CrossRef] [PubMed]

3. Bersini, S.; Moretti, M. 3D functional and perfusable microvascular networks for organotypic microfluidic models. *J. Mater. Sci. Mater. Med.* **2015**, *26*, 180. [CrossRef] [PubMed]

4. Ashley, N.; Jones, M.; Ouaret, D.; Wilding, J.; Bodmer, W.F. Rapidly derived colorectal cancer cultures recapitulate parental cancer characteristics and enable personalized therapeutic assays. *J. Pathol.* **2014**, *234*, 34–45. [CrossRef] [PubMed]

5. Lee, S.H.; Hong, J.H.; Park, H.K.; Park, J.S.; Kim, B.K.; Lee, J.Y.; Jeong, J.Y.; Yoon, G.S.; Inoue, M.; Choi, G.S.; et al. Colorectal cancer-derived tumor spheroids retain the characteristics of original tumors. *Cancer Lett.* **2015**, *367*, 34–42. [CrossRef] [PubMed]

6. Ruppen, J.; Wildhaber, F.D.; Strub, C.; Hall, S.R.; Schmid, R.A.; Geiser, T.; Guenat, O.T. Towards personalized medicine: Chemosensitivity assays of patient lung cancer cell spheroids in a perfused microfluidic platform. *Lab Chip* **2015**, *15*, 3076–3085. [CrossRef] [PubMed]

7. Rajcevic, U.; Knol, J.C.; Piersma, S.; Bougnaud, S.; Fack, F.; Sundlisaeter, E.; Sondenaa, K.; Myklebust, R.; Pham, T.V.; Niclou, S.P.; et al. Colorectal cancer derived organotypic spheroids maintain essential tissue characteristics but adapt their metabolism in culture. *Proteome Sci.* **2014**, *12*, 39. [CrossRef] [PubMed]

8. Horning, J.L.; Sahoo, S.K.; Vijayaraghavalu, S.; Dimitrijevic, S.; Vasir, J.K.; Jain, T.K.; Panda, A.K.; Labhasetwar, V. 3-D tumor model for in vitro evaluation of anticancer drugs. *Mol. Pharm.* **2008**, *5*, 849–862. [CrossRef] [PubMed]

9. Vantangoli, M.M.; Wilson, S.; Madnick, S.J.; Huse, S.M.; Boekelheide, K. Morphologic effects of estrogen stimulation on 3D MCF-7 microtissues. *Toxicol. Lett.* **2016**, *248*, 1–8. [CrossRef] [PubMed]

10. Samuelson, L.; Gerber, D.A. Improved function and growth of pancreatic cells in a three-dimensional bioreactor environment. *Tissue Eng. Part C Methods* **2013**, *19*, 39–47. [CrossRef] [PubMed]

11. Joo, D.J.; Kim, J.Y.; Lee, J.I.; Jeong, J.H.; Cho, Y.; Ju, M.K.; Huh, K.H.; Kim, M.S.; Kim, Y.S. Manufacturing of insulin-secreting spheroids with the RIN-5F cell line using a shaking culture method. *Transplant. Proc.* **2010**, *42*, 4225–4227. [CrossRef] [PubMed]

12. Morabito, C.; Steimberg, N.; Mazzoleni, G.; Guarnieri, S.; Fano-Illic, G.; Mariggio, M.A. RCCS bioreactor-based modelled microgravity induces significant changes on in vitro 3D neuroglial cell cultures. *BioMed Res. Int.* **2015**, *2015*, 754283. [CrossRef] [PubMed]

13. Loessner, D.; Stok, K.S.; Lutolf, M.P.; Hutmacher, D.W.; Clements, J.A.; Rizzi, S.C. Bioengineered 3D platform to explore cell-ECM interactions and drug resistance of epithelial ovarian cancer cells. *Biomaterials* **2010**, *31*, 8494–8506. [CrossRef] [PubMed]

14. Grummer, R.; Hohn, H.P.; Mareel, M.M.; Denker, H.W. Adhesion and invasion of three human choriocarcinoma cell lines into human endometrium in a three-dimensional organ culture system. *Placenta* **1994**, *15*, 411–429. [CrossRef]

15. Ramaiahgari, S.C.; den Braver, M.W.; Herpers, B.; Terpstra, V.; Commandeur, J.N.; van de Water, B.; Price, L.S. A 3D in vitro model of differentiated HepG2 cell spheroids with improved liver-like properties for repeated dose high-throughput toxicity studies. *Arch. Toxicol.* **2014**, *88*, 1083–1095. [CrossRef] [PubMed]

16. Kosaka, T.; Tsuboi, S.; Fukaya, K.; Pu, H.; Ohno, T.; Tsuji, T.; Miyazaki, M.; Namba, M. Spheroid cultures of human hepatoblastoma cells (HuH-6 line) and their application for cytotoxicity assay of alcohols. *Acta Med. Okayama* **1996**, *50*, 61–66. [PubMed]

17. Fey, S.J.; Wrzesinski, K. Determination of acute lethal and chronic lethal dose thresholds of valproic acid using 3D spheroids constructed from the immortal human hepatocyte cell line HepG2/C3A. In *Valproic Acid*; Boucher, A., Ed.; Nova Science Publishers, Inc.: New York, NY, USA, 2013; pp. 141–165.

18. Wrzesinski, K.; Fey, S.J. After trypsinisation, 3D spheroids of C3A hepatocytes need 18 days to re-establish similar levels of key physiological functions to those seen in the liver. *Toxicol. Res.* **2013**, *2*, 123–135. [CrossRef]

19. Wrzesinski, K.; Magnone, M.C.; Visby Hansen, L.; Kruse, M.E.; Bergauer, T.; Bobadilla, M.; Gubler, M.; Mizrahi, J.; Zhang, K.; Andreasen, C.M.; et al. HepG2/C3a 3D spheroids exhibit stable physiological functionality for at least 24 days after recovering from trypsinisation. *Toxicol. Res.* **2013**, *2*, 163–172. [CrossRef]

20. Fatehullah, A.; Tan, S.H.; Barker, N. Organoids as an in vitro model of human development and disease. *Nat. Cell Biol.* **2016**, *18*, 246–254. [CrossRef] [PubMed]

21. Clevers, H. Modeling development and disease with organoids. *Cell* **2016**, *165*, 1586–1597. [CrossRef] [PubMed]

22. Nakano, T.; Ando, S.; Takata, N.; Kawada, M.; Muguruma, K.; Sekiguchi, K.; Saito, K.; Yonemura, S.; Eiraku, M.; Sasai, Y. Self-formation of optic cups and storable stratified neural retina from human ESCS. *Cell Stem Cell* **2012**, *10*, 771–785. [CrossRef] [PubMed]

23. Greggio, C.; De Franceschi, F.; Figueiredo-Larsen, M.; Gobaa, S.; Ranga, A.; Semb, H.; Lutolf, M.; Grapin-Botton, A. Artificial three-dimensional niches deconstruct pancreas development in vitro. *Development* **2013**, *140*, 4452–4462. [CrossRef] [PubMed]

24. McCracken, K.W.; Cata, E.M.; Crawford, C.M.; Sinagoga, K.L.; Schumacher, M.; Rockich, B.E.; Tsai, Y.H.; Mayhew, C.N.; Spence, J.R.; Zavros, Y.; et al. Modelling human development and disease in pluripotent stem-cell-derived gastric organoids. *Nature* **2014**, *516*, 400–404. [CrossRef] [PubMed]

25. Caiazzo, M.; Okawa, Y.; Ranga, A.; Piersigilli, A.; Tabata, Y.; Lutolf, M.P. Defined three-dimensional microenvironments boost induction of pluripotency. *Nat. Mater.* **2016**, *15*, 344–352. [CrossRef] [PubMed]

26. Peloso, A.; Dhal, A.; Zambon, J.P.; Li, P.; Orlando, G.; Atala, A.; Soker, S. Current achievements and future perspectives in whole-organ bioengineering. *Stem Cell Res. Ther.* **2015**, *6*, 107. [CrossRef] [PubMed]

27. Croughan, M.S.; Wang, D.I. Hydrodynamic effects on animal cells in microcarrier bioreactors. *Biotechnology* **1991**, *17*, 213–249. [PubMed]

28. Cinbiz, M.N.; Tigli, R.S.; Beskardes, I.G.; Gumusderelioglu, M.; Colak, U. Computational fluid dynamics modeling of momentum transport in rotating wall perfused bioreactor for cartilage tissue engineering. *J. Biotechnol.* **2010**, *150*, 389–395. [CrossRef] [PubMed]

29. Tsai, A.C.; Liu, Y.; Yuan, X.; Chella, R.; Ma, T. Aggregation kinetics of human mesenchymal stem cells under wave motion. *Biotechnol. J.* **2017**, *12*, 1600448. [CrossRef] [PubMed]

30. Kalmbach, A.; Bordas, R.; Oncul, A.A.; Thevenin, D.; Genzel, Y.; Reichl, U. Experimental characterization of flow conditions in 2- and 20-L bioreactors with wave-induced motion. *Biotechnol. Prog.* **2011**, *27*, 402–409. [CrossRef] [PubMed]

31. Jang, K.J.; Mehr, A.P.; Hamilton, G.A.; McPartlin, L.A.; Chung, S.; Suh, K.Y.; Ingber, D.E. Human kidney proximal tubule-on-a-chip for drug transport and nephrotoxicity assessment. *Integr. Biol.* **2013**, *5*, 1119–1129. [CrossRef] [PubMed]

32. Esch, M.B.; Prot, J.M.; Wang, Y.I.; Miller, P.; Llamas-Vidales, J.R.; Naughton, B.A.; Applegate, D.R.; Shuler, M.L. Multi-cellular 3D human primary liver cell culture elevates metabolic activity under fluidic flow. *Lab Chip* **2015**, *15*, 2269–2277. [CrossRef] [PubMed]

33. Sousa, M.F.; Silva, M.M.; Giroux, D.; Hashimura, Y.; Wesselschmidt, R.; Lee, B.; Roldao, A.; Carrondo, M.J.; Alves, P.M.; Serra, M. Production of oncolytic adenovirus and human mesenchymal stem cells in a single-use, vertical-wheel bioreactor system: Impact of bioreactor design on performance of microcarrier-based cell culture processes. *Biotechnol. Prog.* **2015**, *31*, 1600–1612. [CrossRef] [PubMed]

34. Ismadi, M.Z.; Gupta, P.; Fouras, A.; Verma, P.; Jadhav, S.; Bellare, J.; Hourigan, K. Flow characterization of a spinner flask for induced pluripotent stem cell culture application. *PLoS ONE* **2014**, *9*, e106493. [CrossRef] [PubMed]

35. Dardik, A.; Chen, L.; Frattini, J.; Asada, H.; Aziz, F.; Kudo, F.A.; Sumpio, B.E. Differential effects of orbital and laminar shear stress on endothelial cells. *J. Vasc. Surg.* **2005**, *41*, 869–880. [CrossRef] [PubMed]

36. Gareau, T.; Lara, G.G.; Shepherd, R.D.; Krawetz, R.; Rancourt, D.E.; Rinker, K.D.; Kallos, M.S. Shear stress influences the pluripotency of murine embryonic stem cells in stirred suspension bioreactors. *J. Tissue Eng. Regen. Med.* **2014**, *8*, 268–278. [CrossRef] [PubMed]

37. Bai, G.; Bee, J.S.; Biddlecombe, J.G.; Chen, Q.; Leach, W.T. Computational fluid dynamics (CFD) insights into agitation stress methods in biopharmaceutical development. *Int. J. Pharm.* **2012**, *423*, 264–280. [CrossRef] [PubMed]

38. Filipovic, N.; Ghimire, K.; Saveljic, I.; Milosevic, Z.; Ruegg, C. Computational modeling of shear forces and experimental validation of endothelial cell responses in an orbital well shaker system. *Comput. Methods Biomech. Biomed. Eng.* **2016**, *19*, 581–590. [CrossRef] [PubMed]

39. Warburg, O. On the origin of cancer cells. *Science* **1956**, *123*, 309–314. [CrossRef] [PubMed]

40. Moreno-Sanchez, R.; Rodriguez-Enriquez, S.; Marin-Hernandez, A.; Saavedra, E. Energy metabolism in tumor cells. *FEBS J.* **2007**, *274*, 1393–1418. [CrossRef] [PubMed]

41. Vander Heiden, M.G.; Cantley, L.C.; Thompson, C.B. Understanding the warburg effect: The metabolic requirements of cell proliferation. *Science* **2009**, *324*, 1029–1033. [CrossRef] [PubMed]

42. Day, S.E.; Kettunen, M.I.; Gallagher, F.A.; Hu, D.E.; Lerche, M.; Wolber, J.; Golman, K.; Ardenkjaer-Larsen, J.H.; Brindle, K.M. Detecting tumor response to treatment using hyperpolarized ^{13}C magnetic resonance imaging and spectroscopy. *Nat. Med.* **2007**, *13*, 1382–1387. [CrossRef] [PubMed]

43. Albers, M.J.; Bok, R.; Chen, A.P.; Cunningham, C.H.; Zierhut, M.L.; Zhang, V.Y.; Kohler, S.J.; Tropp, J.; Hurd, R.E.; Yen, Y.F.; et al. Hyperpolarized ^{13}C lactate, pyruvate, and alanine: Noninvasive biomarkers for prostate cancer detection and grading. *Cancer Res.* **2008**, *68*, 8607–8615. [CrossRef] [PubMed]

44. Rodrigues, T.B.; Serrao, E.M.; Kennedy, B.W.; Hu, D.E.; Kettunen, M.I.; Brindle, K.M. Magnetic resonance imaging of tumor glycolysis using hyperpolarized ^{13}C-labeled glucose. *Nat. Med.* **2014**, *20*, 93–97. [CrossRef] [PubMed]

45. Shannon, B.J.; Vaishnavi, S.N.; Vlassenko, A.G.; Shimony, J.S.; Rutlin, J.; Raichle, M.E. Brain aerobic glycolysis and motor adaptation learning. *Proc. Natl. Acad. Sci. USA* **2016**, *113*, E3782–E3791. [CrossRef] [PubMed]

46. Lumata, L.; Yang, C.; Ragavan, M.; Carpenter, N.; DeBerardinis, R.J.; Merritt, M.E. Hyperpolarized (13)C magnetic resonance and its use in metabolic assessment of cultured cells and perfused organs. *Methods Enzymol.* **2015**, *561*, 73–106. [PubMed]

47. Fan, T.W.; Kucia, M.; Jankowski, K.; Higashi, R.M.; Ratajczak, J.; Ratajczak, M.Z.; Lane, A.N. Rhabdomyosarcoma cells show an energy producing anabolic metabolic phenotype compared with primary myocytes. *Mol. Cancer* **2008**, *7*, 79. [CrossRef] [PubMed]

48. Fantin, V.R.; St-Pierre, J.; Leder, P. Attenuation of LDH-A expression uncovers a link between glycolysis, mitochondrial physiology, and tumor maintenance. *Cancer Cell* **2006**, *9*, 425–434. [CrossRef] [PubMed]

49. Hatzivassiliou, G.; Zhao, F.; Bauer, D.E.; Andreadis, C.; Shaw, A.N.; Dhanak, D.; Hingorani, S.R.; Tuveson, D.A.; Thompson, C.B. ATP citrate lyase inhibition can suppress tumor cell growth. *Cancer Cell* **2005**, *8*, 311–321. [CrossRef] [PubMed]

50. Weinberg, F.; Hamanaka, R.; Wheaton, W.W.; Weinberg, S.; Joseph, J.; Lopez, M.; Kalyanaraman, B.; Mutlu, G.M.; Budinger, G.R.; Chandel, N.S. Mitochondrial metabolism and ROS generation are essential for Kras-mediated tumorigenicity. *Proc. Natl. Acad. Sci. USA* **2010**, *107*, 8788–8793. [CrossRef] [PubMed]

51. Nakajima, T.; Moriguchi, M.; Mitsumoto, Y.; Katagishi, T.; Kimura, H.; Shintani, H.; Deguchi, T.; Okanoue, T.; Kagawa, K.; Ashihara, T. Simple tumor profile chart based on cell kinetic parameters and histologic grade is useful for estimating the natural growth rate of hepatocellular carcinoma. *Hum. Pathol.* **2002**, *33*, 92–99. [CrossRef] [PubMed]

52. Zu, X.L.; Guppy, M. Cancer metabolism: Facts, fantasy, and fiction. *Biochem. Biophys. Res. Commun.* **2004**, *313*, 459–465. [CrossRef] [PubMed]

53. Ward, P.S.; Thompson, C.B. Metabolic reprogramming: A cancer hallmark even warburg did not anticipate. *Cancer Cell* **2012**, *21*, 297–308. [CrossRef] [PubMed]

54. DeBerardinis, R.J.; Lum, J.J.; Hatzivassiliou, G.; Thompson, C.B. The biology of cancer: Metabolic reprogramming fuels cell growth and proliferation. *Cell Metab.* **2008**, *7*, 11–20. [CrossRef] [PubMed]

55. Klein, C.A. Selection and adaptation during metastatic cancer progression. *Nature* **2013**, *501*, 365–372. [CrossRef] [PubMed]

56. Sabo, A.; Kress, T.R.; Pelizzola, M.; de Pretis, S.; Gorski, M.M.; Tesi, A.; Morelli, M.J.; Bora, P.; Doni, M.; Verrecchia, A.; et al. Selective transcriptional regulation by Myc in cellular growth control and lymphomagenesis. *Nature* **2014**, *511*, 488–492. [CrossRef] [PubMed]

57. Wrzesinski, K.; Rogowska-Wrzesinska, A.; Kanlaya, R.; Borkowski, K.; Schwammle, V.; Dai, J.; Joensen, K.E.; Wojdyla, K.; Carvalho, V.B.; Fey, S.J. The cultural divide: Exponential growth in classical 2d and metabolic equilibrium in 3D environments. *PLoS ONE* **2014**, *9*, e106973. [CrossRef] [PubMed]

58. Jiang, L.; Shestov, A.A.; Swain, P.; Yang, C.; Parker, S.J.; Wang, Q.A.; Terada, L.S.; Adams, N.D.; McCabe, M.T.; Pietrak, B.; et al. Reductive carboxylation supports redox homeostasis during anchorage-independent growth. *Nature* **2016**, *532*, 255–258. [CrossRef] [PubMed]

59. Wrzesinski, K.; Fey, S.J. From 2D to 3D—A new dimension for modelling the effect of natural products on human tissue. *Curr. Pharm. Des.* **2015**, *21*, 5605–5616. [CrossRef] [PubMed]

60. Rogowska-Wrzesinska, A.; Wrzesinski, K.; Fey, S.J. Heteromer score-using internal standards to assess the quality of proteomic data. *Proteomics* **2014**, *14*, 1042–1047. [CrossRef] [PubMed]

61. Metallo, C.M.; Gameiro, P.A.; Bell, E.L.; Mattaini, K.R.; Yang, J.; Hiller, K.; Jewell, C.M.; Johnson, Z.R.; Irvine, D.J.; Guarente, L.; et al. Reductive glutamine metabolism by IDH1 mediates lipogenesis under hypoxia. *Nature* **2012**, *481*, 380–384. [CrossRef] [PubMed]

62. Mazurek, S.; Boschek, C.B.; Hugo, F.; Eigenbrodt, E. Pyruvate kinase type M2 and its role in tumor growth and spreading. *Semin. Cancer Biol.* **2005**, *15*, 300–308. [CrossRef] [PubMed]

63. Christofk, H.R.; Vander Heiden, M.G.; Harris, M.H.; Ramanathan, A.; Gerszten, R.E.; Wei, R.; Fleming, M.D.; Schreiber, S.L.; Cantley, L.C. The M2 splice isoform of pyruvate kinase is important for cancer metabolism and tumour growth. *Nature* **2008**, *452*, 230–233. [CrossRef] [PubMed]

64. Chaneton, B.; Gottlieb, E. Rocking cell metabolism: Revised functions of the key glycolytic regulator PKM2 in cancer. *Trends Biochem. Sci.* **2012**, *37*, 309–316. [CrossRef] [PubMed]

65. Zaidi, N.; Swinnen, J.V.; Smans, K. ATP-citrate lyase: A key player in cancer metabolism. *Cancer Res.* **2012**, *72*, 3709–3714. [CrossRef] [PubMed]

66. Wang, G.; Sai, K.; Gong, F.; Yang, Q.; Chen, F.; Lin, J. Mutation of isocitrate dehydrogenase 1 induces glioma cell proliferation via nuclear factor-kappaB activation in a hypoxia-inducible factor 1-alpha dependent manner. *Mol. Med. Rep.* **2014**, *9*, 1799–1805. [CrossRef] [PubMed]

67. Buck, M.D.; O'Sullivan, D.; Klein Geltink, R.I.; Curtis, J.D.; Chang, C.H.; Sanin, D.E.; Qiu, J.; Kretz, O.; Braas, D.; van der Windt, G.J.; et al. Mitochondrial dynamics controls T cell fate through metabolic programming. *Cell* **2016**, *166*, 63–76. [CrossRef] [PubMed]

68. Yang, C.; Harrison, C.; Jin, E.S.; Chuang, D.T.; Sherry, A.D.; Malloy, C.R.; Merritt, M.E.; DeBerardinis, R.J. Simultaneous steady-state and dynamic ^{13}C NMR can differentiate alternative routes of pyruvate metabolism in living cancer cells. *J. Biol. Chem.* **2014**, *289*, 6212–6224. [CrossRef] [PubMed]

69. Keshari, K.R.; Kurhanewicz, J.; Jeffries, R.E.; Wilson, D.M.; Dewar, B.J.; Van Criekinge, M.; Zierhut, M.; Vigneron, D.B.; Macdonald, J.M. Hyperpolarized (13)C spectroscopy and an NMR-compatible bioreactor system for the investigation of real-time cellular metabolism. *Magn. Reson. Med.* **2010**, *63*, 322–329. [CrossRef] [PubMed]

70. DeBerardinis, R.J.; Mancuso, A.; Daikhin, E.; Nissim, I.; Yudkoff, M.; Wehrli, S.; Thompson, C.B. Beyond aerobic glycolysis: Transformed cells can engage in glutamine metabolism that exceeds the requirement for protein and nucleotide synthesis. *Proc. Natl. Acad. Sci. USA* **2007**, *104*, 19345–19350. [CrossRef] [PubMed]

71. Carmeliet, P.; Jain, R.K. Angiogenesis in cancer and other diseases. *Nature* **2000**, *407*, 249–257. [CrossRef] [PubMed]

72. Grimes, D.R.; Kelly, C.; Bloch, K.; Partridge, M. A method for estimating the oxygen consumption rate in multicellular tumour spheroids. *J. R. Soc. Interface* **2014**, *11*, 20131124. [CrossRef] [PubMed]

73. Carreau, A.; El Hafny-Rahbi, B.; Matejuk, A.; Grillon, C.; Kieda, C. Why is the partial oxygen pressure of human tissues a crucial parameter? Small molecules and hypoxia. *J. Cell. Mol. Med.* **2011**, *15*, 1239–1253. [CrossRef] [PubMed]

74. Lenihan, C.R.; Taylor, C.T. The impact of hypoxia on cell death pathways. *Biochem. Soc. Trans.* **2013**, *41*, 657–663. [CrossRef] [PubMed]

75. Papandreou, I.; Krishna, C.; Kaper, F.; Cai, D.; Giaccia, A.J.; Denko, N.C. Anoxia is necessary for tumor cell toxicity caused by a low-oxygen environment. *Cancer Res.* **2005**, *65*, 3171–3178. [CrossRef] [PubMed]

76. Dmitriev, R.I.; Zhdanov, A.V.; Nolan, Y.M.; Papkovsky, D.B. Imaging of neurosphere oxygenation with phosphorescent probes. *Biomaterials* **2013**, *34*, 9307–9317. [CrossRef] [PubMed]

77. Sutherland, R.M. Cell and environment interactions in tumor microregions: The multicell spheroid model. *Science* **1988**, *240*, 177–184. [CrossRef] [PubMed]

78. Mueller-Klieser, W.F.; Sutherland, R.M. Influence of convection in the growth medium on oxygen tensions in multicellular tumor spheroids. *Cancer Res.* **1982**, *42*, 237–242. [PubMed]

79. Hulikova, A.; Swietach, P. Rapid CO_2 permeation across biological membranes: Implications for CO_2 venting from tissue. *FASEB J.* **2014**, *28*, 2762–2774. [CrossRef] [PubMed]

80. Hulikova, A.; Vaughan-Jones, R.D.; Swietach, P. Dual role of CO_2/HCO_3(-) buffer in the regulation of intracellular pH of three-dimensional tumor growths. *J. Biol. Chem.* **2011**, *286*, 13815–13826. [CrossRef] [PubMed]

81. Sato, M.; Kawana, K.; Adachi, K.; Fujimoto, A.; Yoshida, M.; Nakamura, H.; Nishida, H.; Inoue, T.; Taguchi, A.; Takahashi, J.; et al. Spheroid cancer stem cells display reprogrammed metabolism and obtain energy by actively running the tricarboxylic acid (TCA) cycle. *Oncotarget* **2016**, *7*, 33297–33305. [CrossRef] [PubMed]

82. Fukuda, R.; Zhang, H.; Kim, J.W.; Shimoda, L.; Dang, C.V.; Semenza, G.L. HIF-1 regulates cytochrome oxidase subunits to optimize efficiency of respiration in hypoxic cells. *Cell* **2007**, *129*, 111–122. [CrossRef] [PubMed]

83. Chen, K.; Chen, S.; Huang, C.; Cheng, H.; Zhou, R. TCTP increases stability of hypoxia-inducible factor 1alpha by interaction with and degradation of the tumour suppressor VHL. *Biol. Cell* **2013**, *105*, 208–218. [CrossRef] [PubMed]

84. Lum, J.J.; Bui, T.; Gruber, M.; Gordan, J.D.; DeBerardinis, R.J.; Covello, K.L.; Simon, M.C.; Thompson, C.B. The transcription factor HIF-1alpha plays a critical role in the growth factor-dependent regulation of both aerobic and anaerobic glycolysis. *Genes Dev.* **2007**, *21*, 1037–1049. [CrossRef] [PubMed]

85. Gilkes, D.M.; Bajpai, S.; Chaturvedi, P.; Wirtz, D.; Semenza, G.L. Hypoxia-inducible factor 1 (HIF-1) promotes extracellular matrix remodeling under hypoxic conditions by inducing P4HA1, P4HA2, and PLOD2 expression in fibroblasts. *J. Biol. Chem.* **2013**, *288*, 10819–10829. [CrossRef] [PubMed]

86. Quiros, P.M.; Espanol, Y.; Acin-Perez, R.; Rodriguez, F.; Barcena, C.; Watanabe, K.; Calvo, E.; Loureiro, M.; Fernandez-Garcia, M.S.; Fueyo, A.; et al. ATP-dependent lon protease controls tumor bioenergetics by reprogramming mitochondrial activity. *Cell Rep.* **2014**, *8*, 542–556. [CrossRef] [PubMed]

87. Wojdyla, K.; Wrzesinski, K.; Williamson, J.; Fey, S.J.; Rogowska-Wrzesinska, A. Acetaminophen-induced *S*-nitrosylation and *S*-sulfenylation changes in 3D cultured hepatocarcinoma cell spheroids. *Toxicol. Res.* **2016**, *5*, 905–920. [CrossRef]

88. Lee, A.S. Glucose-regulated proteins in cancer: Molecular mechanisms and therapeutic potential. *Nat. Rev. Cancer* **2014**, *14*, 263–276. [CrossRef] [PubMed]

89. Bravo, R.; Parra, V.; Gatica, D.; Rodriguez, A.E.; Torrealba, N.; Paredes, F.; Wang, Z.V.; Zorzano, A.; Hill, J.A.; Jaimovich, E.; et al. Endoplasmic reticulum and the unfolded protein response: Dynamics and metabolic integration. *Int. Rev. Cell Mol. Biol.* **2013**, *301*, 215–290. [PubMed]

90. Korennykh, A.; Walter, P. Structural basis of the unfolded protein response. *Annu. Rev. Cell Dev. Biol.* **2012**, *28*, 251–277. [CrossRef] [PubMed]

91. Behnke, J.; Feige, M.J.; Hendershot, L.M. BiP and its nucleotide exchange factors Grp170 and Sil1: Mechanisms of action and biological functions. *J. Mol. Biol.* **2015**, *427*, 1589–1608. [CrossRef] [PubMed]

92. Bhutia, Y.D.; Ganapathy, V. Glutamine transporters in mammalian cells and their functions in physiology and cancer. *Biochim. Biophys. Acta* **2016**, *1863*, 2531–2539. [CrossRef] [PubMed]

93. Jeon, Y.J.; Khelifa, S.; Ratnikov, B.; Scott, D.A.; Feng, Y.; Parisi, F.; Ruller, C.; Lau, E.; Kim, H.; Brill, L.M.; et al. Regulation of glutamine carrier proteins by RNF5 determines breast cancer response to ER stress-inducing chemotherapies. *Cancer Cell* **2015**, *27*, 354–369. [CrossRef] [PubMed]

94. Broer, A.; Rahimi, F.; Broer, S. Deletion of amino acid transporter ASCT2 (SLC1A5) reveals an essential role for transporters SNAT1 (SLC38A1) and SNAT2 (SLC38A2) to sustain glutaminolysis in cancer cells. *J. Biol. Chem.* **2016**, *291*, 13194–13205. [CrossRef] [PubMed]

95. Guillaumond, F.; Leca, J.; Olivares, O.; Lavaut, M.N.; Vidal, N.; Berthezene, P.; Dusetti, N.J.; Loncle, C.; Calvo, E.; Turrini, O.; et al. Strengthened glycolysis under hypoxia supports tumor symbiosis and hexosamine biosynthesis in pancreatic adenocarcinoma. *Proc. Natl. Acad. Sci. USA* **2013**, *110*, 3919–3924. [CrossRef] [PubMed]

96. Salminen, A.; Kaarniranta, K.; Kauppinen, A. Hypoxia-inducible histone lysine demethylases: Impact on the aging process and age-related diseases. *Aging Dis.* **2016**, *7*, 180–200. [PubMed]

97. Tvardovskiy, A.; Schwammle, V.; Kempf, S.J.; Rogowska-Wrzesinska, A.; Jensen, O.N. Accumulation of histone variant $H_{3.3}$ with age is associated with profound changes in the histone methylation landscape. *Nucleic Acids Res.* **2017**, *45*, 9272–9289. [CrossRef] [PubMed]

98. Lim, S.K.; Jeong, Y.W.; Kim, D.I.; Park, M.J.; Choi, J.H.; Kim, S.U.; Kang, S.S.; Han, H.J.; Park, S.H. Activation of PRMT1 and PRMT5 mediates hypoxia- and ischemia-induced apoptosis in human lung epithelial cells and the lung of miniature pigs: The role of p38 and JNK mitogen-activated protein kinases. *Biochem. Biophys. Res. Commun.* **2013**, *440*, 707–713. [CrossRef] [PubMed]

99. Reintjes, A.; Fuchs, J.E.; Kremser, L.; Lindner, H.H.; Liedl, K.R.; Huber, L.A.; Valovka, T. Asymmetric arginine dimethylation of RelA provides a repressive mark to modulate TNFalpha/NF-kappaB response. *Proc. Natl. Acad. Sci. USA* **2016**, *113*, 4326–4331. [CrossRef] [PubMed]

100. Takai, H.; Masuda, K.; Sato, T.; Sakaguchi, Y.; Suzuki, T.; Suzuki, T.; Koyama-Nasu, R.; Nasu-Nishimura, Y.; Katou, Y.; Ogawa, H.; et al. 5-hydroxymethylcytosine plays a critical role in glioblastomagenesis by recruiting the CHTOP-methylosome complex. *Cell Rep.* **2014**, *9*, 48–60. [CrossRef] [PubMed]

101. Lunt, S.Y.; Vander Heiden, M.G. Aerobic glycolysis: Meeting the metabolic requirements of cell proliferation. *Annu. Rev. Cell Dev. Biol.* **2011**, *27*, 441–464. [CrossRef] [PubMed]

102. Ng, B.G.; Wolfe, L.A.; Ichikawa, M.; Markello, T.; He, M.; Tifft, C.J.; Gahl, W.A.; Freeze, H.H. Biallelic mutations in cad, impair de novo pyrimidine biosynthesis and decrease glycosylation precursors. *Hum. Mol. Genet.* **2015**, *24*, 3050–3057. [CrossRef] [PubMed]

103. Dibble, C.C.; Cantley, L.C. Regulation of mTORC1 by PI3K signaling. *Trends Cell Biol.* **2015**, *25*, 545–555. [CrossRef] [PubMed]

104. Duvel, K.; Yecies, J.L.; Menon, S.; Raman, P.; Lipovsky, A.I.; Souza, A.L.; Triantafellow, E.; Ma, Q.; Gorski, R.; Cleaver, S.; et al. Activation of a metabolic gene regulatory network downstream of mTOR complex 1. *Mol. Cell* **2010**, *39*, 171–183. [CrossRef] [PubMed]

105. Robitaille, A.M.; Christen, S.; Shimobayashi, M.; Cornu, M.; Fava, L.L.; Moes, S.; Prescianotto-Baschong, C.; Sauer, U.; Jenoe, P.; Hall, M.N. Quantitative phosphoproteomics reveal mTORC1 activates de novo pyrimidine synthesis. *Science* **2013**, *339*, 1320–1323. [CrossRef] [PubMed]

106. Peterson, T.R.; Sengupta, S.S.; Harris, T.E.; Carmack, A.E.; Kang, S.A.; Balderas, E.; Guertin, D.A.; Madden, K.L.; Carpenter, A.E.; Finck, B.N.; et al. mTOR complex 1 regulates lipin 1 localization to control the SREBP pathway. *Cell* **2011**, *146*, 408–420. [CrossRef] [PubMed]

107. Stine, Z.E.; Walton, Z.E.; Altman, B.J.; Hsieh, A.L.; Dang, C.V. Myc, metabolism, and cancer. *Cancer Discov.* **2015**, *5*, 1024–1039. [CrossRef] [PubMed]

108. Locasale, J.W. Serine, glycine and one-carbon units: Cancer metabolism in full circle. *Nat. Rev. Cancer* **2013**, *13*, 572–583. [CrossRef] [PubMed]

109. Osthus, R.C.; Shim, H.; Kim, S.; Li, Q.; Reddy, R.; Mukherjee, M.; Xu, Y.; Wonsey, D.; Lee, L.A.; Dang, C.V. Deregulation of glucose transporter 1 and glycolytic gene expression by c-Myc. *J. Biol. Chem.* **2000**, *275*, 21797–21800. [CrossRef] [PubMed]

110. Kim, J.W.; Gao, P.; Liu, Y.C.; Semenza, G.L.; Dang, C.V. Hypoxia-inducible factor 1 and dysregulated c-Myc cooperatively induce vascular endothelial growth factor and metabolic switches hexokinase 2 and pyruvate dehydrogenase kinase 1. *Mol. Cell. Biol.* **2007**, *27*, 7381–7393. [CrossRef] [PubMed]

111. Koshiji, M.; Kageyama, Y.; Pete, E.A.; Horikawa, I.; Barrett, J.C.; Huang, L.E. HIF-1alpha induces cell cycle arrest by functionally counteracting Myc. *EMBO J.* **2004**, *23*, 1949–1956. [CrossRef] [PubMed]

112. David, C.J.; Chen, M.; Assanah, M.; Canoll, P.; Manley, J.L. HnRNP proteins controlled by c-Myc deregulate pyruvate kinase mRNA splicing in cancer. *Nature* **2010**, *463*, 364–368. [CrossRef] [PubMed]

113. Wonsey, D.R.; Zeller, K.I.; Dang, C.V. The c-Myc target gene PRDX3 is required for mitochondrial homeostasis and neoplastic transformation. *Proc. Natl. Acad. Sci. USA* **2002**, *99*, 6649–6654. [CrossRef] [PubMed]

114. Moon, H.J.; Redman, K.L. Trm4 and Nsun2 RNA:m^{5}c methyltransferases form metabolite-dependent, covalent adducts with previously methylated RNA. *Biochemistry* **2014**, *53*, 7132–7144. [CrossRef] [PubMed]

115. Yang, H.; Li, T.W.; Ko, K.S.; Xia, M.; Lu, S.C. Switch from Mnt-Max to Myc-Max induces p53 and cyclin D1 expression and apoptosis during cholestasis in mouse and human hepatocytes. *Hepatology* **2009**, *49*, 860–870. [CrossRef] [PubMed]

116. Vasseur, S.; Afzal, S.; Tomasini, R.; Guillaumond, F.; Tardivel-Lacombe, J.; Mak, T.W.; Iovanna, J.L. Consequences of DJ-1 upregulation following p53 loss and cell transformation. *Oncogene* **2012**, *31*, 664–670. [CrossRef] [PubMed]

117. Moscovitz, O.; Ben-Nissan, G.; Fainer, I.; Pollack, D.; Mizrachi, L.; Sharon, M. The Parkinson's-associated protein DJ-1 regulates the 20S proteasome. *Nat. Commun.* **2015**, *6*, 6609. [CrossRef] [PubMed]

118. Mak, G.W.; Lai, W.L.; Zhou, Y.; Li, M.; Ng, I.O.; Ching, Y.P. CDK5RAP3 is a novel repressor of p14ARF in hepatocellular carcinoma cells. *PLoS ONE* **2012**, *7*, e42210. [CrossRef] [PubMed]

119. Meng, W.; Ellsworth, B.A.; Nirschl, A.A.; McCann, P.J.; Patel, M.; Girotra, R.N.; Wu, G.; Sher, P.M.; Morrison, E.P.; Biller, S.A.; et al. Discovery of dapagliflozin: A potent, selective renal sodium-dependent glucose cotransporter 2 (SGLT2) inhibitor for the treatment of type 2 diabetes. *J. Med. Chem.* **2008**, *51*, 1145–1149. [CrossRef] [PubMed]

120. Amson, R.; Pece, S.; Marine, J.C.; Di Fiore, P.P.; Telerman, A. TPT1/TCTP-regulated pathways in phenotypic reprogramming. *Trends Cell Biol.* **2013**, *23*, 37–46. [CrossRef] [PubMed]

121. Chen, W.; Wang, H.; Tao, S.; Zheng, Y.; Wu, W.; Lian, F.; Jaramillo, M.; Fang, D.; Zhang, D.D. Tumor protein translationally controlled 1 is a p53 target gene that promotes cell survival. *Cell Cycle* **2013**, *12*, 2321–2328. [CrossRef] [PubMed]

122. Amson, R.; Pece, S.; Lespagnol, A.; Vyas, R.; Mazzarol, G.; Tosoni, D.; Colaluca, I.; Viale, G.; Rodrigues-Ferreira, S.; Wynendaele, J.; et al. Reciprocal repression between p53 and TCTP. *Nat. Med.* **2012**, *18*, 91–99. [CrossRef] [PubMed]

123. Johansson, H.; Vizlin-Hodzic, D.; Simonsson, T.; Simonsson, S. Translationally controlled tumor protein interacts with nucleophosmin during mitosis in ES cells. *Cell Cycle* **2010**, *9*, 2160–2169. [CrossRef] [PubMed]

124. Graves, J.A.; Metukuri, M.; Scott, D.; Rothermund, K.; Prochownik, E.V. Regulation of reactive oxygen species homeostasis by peroxiredoxins and c-myc. *J. Biol. Chem.* **2009**, *284*, 6520–6529. [CrossRef] [PubMed]

125. Wang, B.; Zhao, L.; Fish, M.; Logan, C.Y.; Nusse, R. Self-renewing diploid Axin2(+) cells fuel homeostatic renewal of the liver. *Nature* **2015**, *524*, 180–185. [CrossRef] [PubMed]

126. Wlodarchak, N.; Xing, Y. PP2A as a master regulator of the cell cycle. *Crit. Rev. Biochem. Mol. Biol.* **2016**, *51*, 162–184. [CrossRef] [PubMed]

127. Wang, L.; Li, H.; Zhang, Y.; Santella, R.M.; Weinstein, I.B. HINT1 inhibits beta-catenin/TCF4, USF2 and NFkappab activity in human hepatoma cells. *Int. J. Cancer* **2009**, *124*, 1526–1534. [CrossRef] [PubMed]

128. Weiske, J.; Huber, O. The histidine triad protein hint1 triggers apoptosis independent of its enzymatic activity. *J. Biol. Chem.* **2006**, *281*, 27356–27366. [CrossRef] [PubMed]

129. Wu, D.; Wu, P.; Zhao, L.; Huang, L.; Zhang, Z.; Zhao, S.; Huang, J. NF-kappaB expression and outcomes in solid tumors: A systematic review and meta-analysis. *Medicine* **2015**, *94*, e1687. [CrossRef] [PubMed]

130. Clark, C.R.; Robinson, J.Y.; Sanchez, N.S.; Townsend, T.A.; Arrieta, J.A.; Merryman, W.D.; Trykall, D.Z.; Olivey, H.E.; Hong, C.C.; Barnett, J.V. Common pathways regulate type III TGFbeta receptor-dependent cell invasion in epicardial and endocardial cells. *Cell. Signal.* **2016**, *28*, 688–698. [CrossRef] [PubMed]

131. Turtoi, A.; Blomme, A.; Debois, D.; Somja, J.; Delvaux, D.; Patsos, G.; Di Valentin, E.; Peulen, O.; Mutijima, E.N.; De Pauw, E.; et al. Organized proteomic heterogeneity in colorectal cancer liver metastases and implications for therapies. *Hepatology* **2014**, *59*, 924–934. [CrossRef] [PubMed]

132. Yang, Y.; Sun, H.; Li, X.; Ding, Q.; Wei, P.; Zhou, J. Transforming growth factor beta-induced is essential for endotoxin tolerance induced by a low dose of lipopolysaccharide in human peripheral blood mononuclear cells. *Iran. J. Allergy Asthma Immunol.* **2015**, *14*, 321–330. [PubMed]

133. Galluzzi, L.; Bravo-San Pedro, J.M.; Kepp, O.; Kroemer, G. Regulated cell death and adaptive stress responses. *Cell. Mol. Life Sci. CMLS* **2016**, *73*, 2405–2410. [CrossRef] [PubMed]

134. Vaseva, A.V.; Marchenko, N.D.; Ji, K.; Tsirka, S.E.; Holzmann, S.; Moll, U.M. P53 opens the mitochondrial permeability transition pore to trigger necrosis. *Cell* **2012**, *149*, 1536–1548. [CrossRef] [PubMed]

135. Galluzzi, L.; Bravo-San Pedro, J.M.; Kroemer, G. Organelle-specific initiation of cell death. *Nat. Cell Biol.* **2014**, *16*, 728–736. [CrossRef] [PubMed]

5

Microscale 3D Liver Bioreactor for In Vitro Hepatotoxicity Testing under Perfusion Conditions

Nora Freyer [1,*,†], Selina Greuel [1,†], Fanny Knöspel [1], Florian Gerstmann [1], Lisa Storch [1], Georg Damm [2], Daniel Seehofer [2], Jennifer Foster Harris [3], Rashi Iyer [3], Frank Schubert [4] and Katrin Zeilinger [1]

[1] Berlin-Brandenburg Center for Regenerative Therapies (BCRT), Charité–Universitätsmedizin Berlin, 13353 Berlin, Germany; selina.greuel@charite.de (S.G.); fanny.knoespel@gmx.de (F.K.); f.gerstmann@mailbox.tu-berlin.de (F.G.); lisa.storch@gmx.net (L.S.); katrin.zeilinger@charite.de (K.Z.)
[2] Department of Hepatobiliary Surgery and Visceral Transplantation, University of Leipzig, 04103 Leipzig, Germany; georg.damm@medizin.uni-leipzig.de (G.D.); daniel.seehofer@medizin.uni-leipzig.de (D.S.)
[3] Los Alamos National Laboratory, Los Alamos, NM 87545, USA; jfharris@lanl.gov (J.F.H.); rashi@lanl.gov (R.I.)
[4] StemCell Systems GmbH, Berlin 12101, Germany; frank.schubert@stemcell-systems.com
* Correspondence: nora.freyer@charite.de
† These authors contributed equally to this work.

Abstract: The accurate prediction of hepatotoxicity demands validated human in vitro models that can close the gap between preclinical animal studies and clinical trials. In this study we investigated the response of primary human liver cells to toxic drug exposure in a perfused microscale 3D liver bioreactor. The cellularized bioreactors were treated with 5, 10, or 30 mM acetaminophen (APAP) used as a reference substance. Lactate production significantly decreased upon treatment with 30 mM APAP ($p < 0.05$) and ammonia release significantly increased in bioreactors treated with 10 or 30 mM APAP ($p < 0.0001$), indicating APAP-induced dose-dependent toxicity. The release of prostaglandin E2 showed a significant increase at 30 mM APAP ($p < 0.05$), suggesting an inflammatory reaction towards enhanced cellular stress. The expression of genes involved in drug metabolism, antioxidant reactions, urea synthesis, and apoptosis was differentially influenced by APAP exposure. Histological examinations revealed that primary human liver cells in untreated control bioreactors were reorganized in tissue-like cell aggregates. These aggregates were partly disintegrated upon APAP treatment, lacking expression of hepatocyte-specific proteins and transporters. In conclusion, our results validate the suitability of the microscale 3D liver bioreactor to detect hepatotoxic effects of drugs in vitro under perfusion conditions.

Keywords: microscale 3D liver bioreactor; in vitro perfusion; primary human liver cells; hepatotoxicity; acetaminophen

1. Introduction

The evaluation of hepatotoxicity of pharmaceutical substances is one major aspect of drug development. For in vivo hepatotoxicity testing, animal models, especially rats and mice, are currently the method of choice [1,2]. However, the use of such models is controversial as they often do not accurately represent the human metabolism due to differences in pharmacokinetics, pharmacodynamics, and species-specific genetic variations [3,4]. The accurate prediction of potential hepatotoxicity demands validated human in vitro models that can close the gap between preclinical animal studies and clinical trials in drug toxicity testing.

Currently applied in vitro toxicity testing models, especially in earlier developmental stages, are mostly based on 2D cell cultures, which offer several advantages including low costs, high throughput, and reproducibility. However, conventional 2D models using primary human hepatocytes are impeded by a rapid decrease of hepatic function and by cell dedifferentiation [5]. This phenomenon is partly due to loss of the original 3D architecture of the organ, which is characterized by organ-specific cell–cell and cell–extracellular matrix contacts. A promising approach to create a physiologically relevant surrounding in vitro can be seen in the development of 3D culture models [6]. Evidence shows that 3D models better reflect the microcellular environment than 2D cultures and thereby enable a more realistic prediction of in vivo drug effects [7–9]. Moreover, an even closer approximation of the in vivo situation is provided by perfused culture platforms that mimic the in vivo hemodynamics and enhance nutrient supply of the cells [10,11]. In addition, perfused culture models enable a constant exposure to test compounds with simultaneous removal of metabolites, in contrast to static 2D cultures with discontinuous medium exchange. In this context, microfluidic culture systems gain increasing importance, since they allow minimization of the amounts of cells and reagents needed while providing characteristics of 3D cultures with physiological cell arrangement [12,13].

We have previously shown that a scalable 3D multicompartment bioreactor technology with counter-current medium exchange and decentralized oxygenation supports the reorganization and longevity of primary human liver cells in vitro [14]. It was also demonstrated that the technology is suitable for analysis of hepatic drug metabolism and hepatic toxicity using serum-free conditions [15–17]. Based on the existing technology, a microscale 3D liver bioreactor with a cell compartment volume of 100 μL was constructed for applications in preclinical drug development and toxicity testing.

The goal of the present study was to investigate the potential of the device to detect toxic drug effects, using acetaminophen (APAP) as a reference substance. For this purpose, primary human liver cells were cultured in microscale 3D bioreactors over six days and treated with APAP at final concentrations of 0, 5, 10, or 30 mM. For monitoring the cell viability and functionality, the release of intracellular enzymes, parameters of glucose and nitrogen metabolism, as well as the liberation of inflammatory factors were measured daily. Upon termination of the bioreactor cultures, histological and immunofluorescence as well as mRNA analysis were performed to detect possible effects of APAP on the tissue integrity and gene expression of hepatic markers.

2. Materials and Methods

2.1. Bioreactor System

The microscale 3D bioreactor used in this study is based on a four-compartment hollow-fiber bioreactor technology described previously [15,17] and was further down-scaled resulting in a culture volume of about 100 μL for perfusion cell culture at microscale level. Figure 1 shows the bioreactor structure and the configuration of hollow-fiber capillaries in the device. The bioreactor housing (Figure 1A) is made of polyurethane and sized in credit card format. It disposes of tube connections for medium perfusion via two independent capillary systems (Medium I and Medium II), perfusion with an air/CO_2 mixture (Gas), and cell inoculation. A central cavity scaled at 1 cm in diameter harbors the capillary bed, which is made of four hollow-fiber layers. As shown in Figure 1B, each layer is composed of alternately arranged medium and gas capillaries, which serve for cell nutrition and oxygenation while providing an adhesion scaffold for the cells seeded in the extra-capillary space (cell compartment). The arrangement of capillary layers in a 45° angle to each other allows for counter-current medium perfusion of the capillary bed. Mass exchange between the capillary lumen and the cell compartment occurs via the pores of the used hydrophilic high-flux filtration membranes (3M, Neuss, Germany), while air/CO_2 exchange is mediated via hydrophobic membranes (Mitsubishi, Tokyo, Japan).

A   Bioreactor structure

B   Capillary configuration

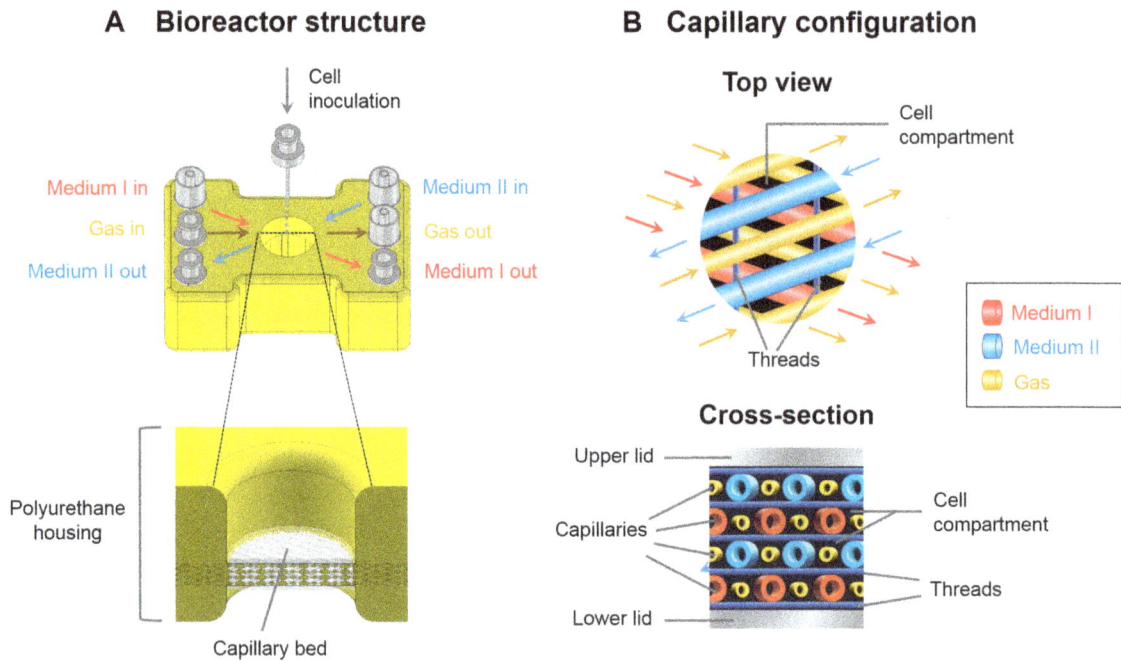

Figure 1. Schematic illustration of the structure of the microscale 3D liver bioreactor. (**A**) Outside view of the bioreactor with tube connections for medium in- and outflow via two independent capillary systems (Medium I and Medium II), perfusion with an air/CO_2 mixture (Gas), and a tube serving for cell inoculation into the extracapillary space (cell compartment); the figure below shows a section of the bioreactor housing with a central cavity containing the capillary bed; (**B**) Capillary structure of the bioreactor shown as top view (upper figure) and cross-section (lower figure). The capillary bed consists of four layers of hollow-fiber capillaries. Cells are seeded in the extra-capillary space (cell compartment). The layers form a 45° angle to each other to enable counter-current medium perfusion of the two capillary systems (red: Medium I, blue: Medium II). Synthetic threads (dark blue) integrated between each layer serve as spacers.

Synthetic threads made of polyethylenterephtalate are placed as spacers between the capillary layers. Thus, direct contacts between the capillaries, which could result in shunt formation and impaired mass exchange, are prevented.

Bioreactors are run in a perfusion circuit consisting of tubing for medium recirculation, medium feed, and medium outflow, as well as gas perfusion lines, as shown schematically in Figure 2. Medium flow rates are regulated by individual pumps for medium recirculation and medium feed, while electronically controlled gas valves (Vögtlin Instruments, Aesch, Switzerland) serve for regulation of air and CO_2 flow rates. The temperature in the bioreactor chamber is maintained at a constant level by means of software-controlled heating cartridges (HS Heizelemente GmbH, Fridingen, Germany).

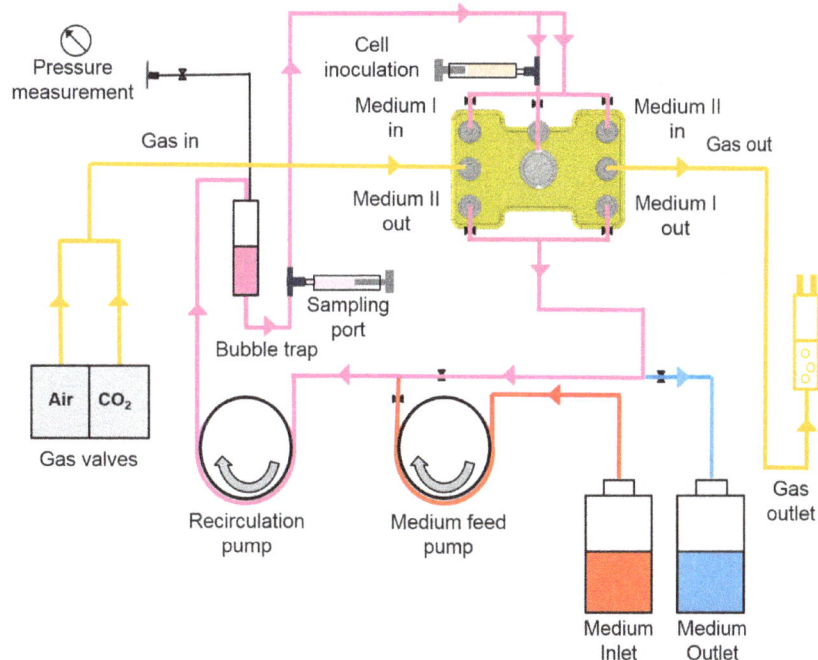

Figure 2. Schematic illustration of the bioreactor perfusion circuit. The bioreactor is integrated into a tubing circuit with continuously recirculating medium (purple). Fresh medium (red) is infused into the circuit via the medium feed pump while used medium (blue) is rinsed out from the circuit upon hydrostatic pressure increase. The tubing circuit contains a bubble trap connected with a line for pressure measurement and is equipped with connections for sample taking. Furthermore, the bioreactor is perfused with a defined air/CO_2 mixture (yellow), which is generated by means of electronically controlled gas valves.

2.2. Primary Human Liver Cell Isolation

Primary human liver cells were gained from tissue remaining from clinical partial liver resection. All patients gave their informed written consent before they participated in the study. The study was conducted in accordance with the Declaration of Helsinki, and the protocol was approved by the Ethics Committee of the Charité–Universitätsmedizin Berlin (EA2/026/09, dated 11 April 2013). Cell isolation was performed by collagenase digestion according to previously described protocols [18,19]. The obtained cell suspension was not further purified to avoid a loss of non-parenchymal cell fractions. The cell viability after isolation was $84.6 \pm 7.0\%$ as determined by their capacity to exclude trypan blue.

2.3. Bioreactor Operation

Primary human liver cells were seeded into each bioreactor at 10^7 cells per bioreactor. Liver cell bioreactors were perfused with Heparmed culture medium (Vito 143, Biochrom, Berlin, Germany), a modification of Williams' Medium E specifically developed for serum-free perfusion of high-density 3D liver cell cultures. The medium was supplemented prior to use with 20 IU/L insulin, 5 mg/L transferrin, 3 μg/L glucagon, 100,000 U/L penicillin, and 100 mg/L streptomycin (all purchased from Biochrom). Culture medium recirculated at a rate of 1 mL/min, while fresh medium was continuously fed into the perfusion circuit at a rate of 0.6 mL/h for the first 24 h followed by 0.2 mL/h until the end of culture. Used medium was removed at the same rate and flowed into the outlet vessel. The pH value in the recirculating medium was kept between 7.35 and 7.45 by adjusting the CO_2 concentration in the supplied gas mixture. Perfusion parameters and operation conditions for running the microscale bioreactors are listed in Table 1.

Table 1. Bioreactor perfusion parameters and operation conditions.

Parameter	Set Values during Operation
Recirculation rate	1 mL/min
Feed rate	0.6 mL/h (0–24 h) 0.2 mL/h (from 24 h on)
Gas flow rate	4 mL/min
Concentration of CO_2 in supplied gas mixture	3–6% [1]
Temperature in bioreactor chamber	38 °C
pH value	7.35–7.45

[1] The CO_2 concentration was adjusted on demand to maintain a constant pH value in the system.

2.4. Clinical Chemistry Parameters

Parameters for assessment of the cell viability and functionality in the bioreactors were measured daily in samples from the culture perfusate. Activities of intracellular enzymes, including lactate dehydrogenase (LDH), aspartate aminotransferase (AST), alanine transaminase (ALT), and glutamate dehydrogenase (GLDH), as well as urea and ammonia concentrations were measured at Labor Berlin GmbH, Berlin, Germany, using automated clinical chemistry analyzers (Cobas® 8000, Roche Diagnostics GmbH, Mannheim, Germany). Glucose and lactate levels were analyzed by means of a blood gas analyzer (ABL 700, Radiometer, Copenhagen, Denmark). Concentrations of prostaglandin E2 (PGE2; Life Technologies, Carlsbad, CA, USA) and interleukin-6 (IL-6; Peprotech, Rocky Hill, NJ, USA) were determined using ELISA Kits according to the instructions of the manufacturers. The required sample volumes and variation coefficients are provided in Table 2.

Table 2. Sample volumes and variation coefficients of analyzed clinical chemistry parameters.

Parameter	Required Volume	Variation Coefficient
Glucose	100 µL in total	4%
Lactate		3%
LDH	250 µL in total	1%
AST		4%
ALT		2.9%
GLDH		0.8%
Urea		1%
Ammonia	250 µL	2.2%
PGE2	200 µL	n.a.
IL-6	200 µL	n.a.

2.5. Acetaminophen (APAP) Application

On the third day of culture, APAP (Sigma-Aldrich, St.-Louis, MO, USA) was added at a final concentration of 5, 10, or 30 mM. The drug was dissolved in methanol, followed by methanol evaporation and dissolution of the substance in culture medium. APAP incubation was initiated by adding 1 mL of a 7× concentrated solution of the drug into the perfusion circuit via the bubble trap (bolus application), to reach the desired final solution of the drug in the recirculation circuit. Subsequently, fresh medium containing APAP at the respective concentration of 5, 10, or 30 mM was continuously infused into the perfusion circuit. The control bioreactor was treated equally, but without adding the drug.

2.6. Histological and Immunofluorescence Analysis

Upon termination of bioreactor cultures on day 6 of culture, bioreactors were opened and sections of the capillary bed containing the cell material were taken. Histological slides were prepared from fixed samples and subjected to hematoxylin-eosin (HE) and immunofluorescence staining as described previously [15]. Double-staining of antigens was performed using monoclonal mouse anti-cytochrome P450 (CYP) 1A2 antibodies (AB) provided by Santa Cruz (Santa Cruz, CA, USA), combined with monoclonal rabbit anti-cytokeratin 18 (CK18) AB (Abcam, Cambridge, UK); monoclonal mouse anti-CK18 AB (Santa Cruz), combined with polyclonal rabbit anti-vimentin AB (Santa Cruz); or monoclonal mouse anti-multidrug resistance protein 2 (MRP2) AB (Abcam) in combination with polyclonal rabbit anti-CYP3A4 AB (Abcam). As secondary AB, fluorochrome-coupled goat anti-mouse IgG 488 AB (Life Technologies) and goat anti-rabbit IgG 594 AB (Life Technologies) were used. Counterstaining of nuclei was performed using bisbenzimide H 33342 trihydrochloride (Hoechst 33342, Sigma-Aldrich).

2.7. qRT-PCR

Total RNA was obtained from human liver cells gained from the bioreactors after termination of the cultures on day 6. The RNA was extracted using the TRIzol® Reagent (Life Technologies) according to the manufacturer's instructions. Afterwards, genomic DNA was digested using the RNase-free DNase-Set (Qiagen, Hilden, Germany). Subsequent cDNA synthesis and quantitative real-time PCR (qRT-PCR) were performed as described elsewhere [20] using human-specific primers and probes (TaqMan Gene Expression Assay system, Life Technologies, Table 3). The expression of specific genes was normalized to that of the housekeeping gene glyceraldehyde-3-phosphate dehydrogenase (*GAPDH*) and fold changes of expression levels were calculated with the $\Delta\Delta C_t$ method [21].

Table 3. Applied Biosystems TaqMan Gene Expression Assays®.

Gene Symbol	Gene Name	Assay ID
AIFM1	Apoptosis-inducing factor, mitochondria-associated, 1	Hs00377585_m1
CASP3	Caspase 3, apoptosis-related cysteine peptidase	Hs00234387_m1
CPS1	Carbamoyl-phosphate synthase 1, mitochondrial	Hs00157048_m1
CYP1A2	Cytochrome P450, family 1, subfamily A, polypeptide 2	Hs00167927_m1
CYP2E1	Cytochrome P450, family 2, subfamily E, polypeptide 1	Hs00559368_m1
GAPDH	Glyceraldehyde-3-phosphate dehydrogenase	Hs03929097_g1
GSTO2	Glutathione S-transferase omega 2	Hs01598184_m1

2.8. Statistics

Four independent experiments were performed with cells from different donors ($N = 4$, unless stated otherwise). Statistical analyses were performed using GraphPad Prism 5.0 for Windows (GraphPad Software, San Diego, CA, USA). Results are provided as mean ± standard error of the mean (SEM). The influence of the drug dose (day 3–day 6) on clinical chemistry parameters in comparison to the control was analyzed by calculating the area under curve (AUC) of values during the drug application interval. The AUCs between day 3 and day 6 of the groups treated with different APAP concentrations were compared with those of untreated control cultures by means of one-way ANOVA with Dunnett's multiple comparison test. The same test was used for statistical evaluation of gene expression data. The group treated with 30 mM APAP was not included in the statistical analysis of gene expression data, since RNA in sufficient quality and quantity was only gained from one culture in this group.

3. Results

3.1. Clinical Chemistry Parameters

Clinical chemistry parameters revealed a dose-dependent effect of APAP on metabolic functions of primary human liver cells maintained in perfused microscale bioreactors (Figure 3).

The time-course of glucose production (Figure 3A) showed stable values with some fluctuations in control bioreactors or those treated with 5 mM APAP, while a clear decrease upon drug application from day 3 onwards was observed in bioreactors exposed to 10 or 30 mM APAP. Lactate production rates (Figure 3B) showed a steadily increasing course in the control group, while bioreactors exposed to 5 or 10 mM APAP remained on a constant level, and the group exposed to 30 mM APAP was characterized by a sharp decline. The comparison of AUCs of lactate values following drug application revealed a significant difference between the 30 mM APAP-treated group and the control group ($p < 0.01$).

The time-course of ammonia release was determined as an indicator for the cells' capacity of nitrogen elimination (Figure 3C). After an initial peak on the first culture day, control bioreactors and those treated with 5 mM APAP showed stable values on a basal level. In contrast, a distinct increase was observed in bioreactors upon exposure to 10 or 30 mM APAP, with significantly ($p < 0.0001$) increased AUCs as compared with the control group. Urea production rates showed a mild, but not significant decrease at 30 mM APAP, while lower drug concentrations did not affect urea levels as compared to untreated control bioreactors (Figure 3D).

Release rates of the intracellular enzymes LDH and AST, indicating disturbed cell integrity and membrane leakage, showed a similar time-course in all experimental groups, characterized by a peak immediately after cell inoculation, which was followed by basal levels from day 3 onwards (Figure 3E,F). The enzymes ALT and GLDH showed a similar time course (data available at http://doi.org/10.5281/zenodo.1169306 (clinical chemistry parameters)). The administration of APAP had no effect on enzyme release rates.

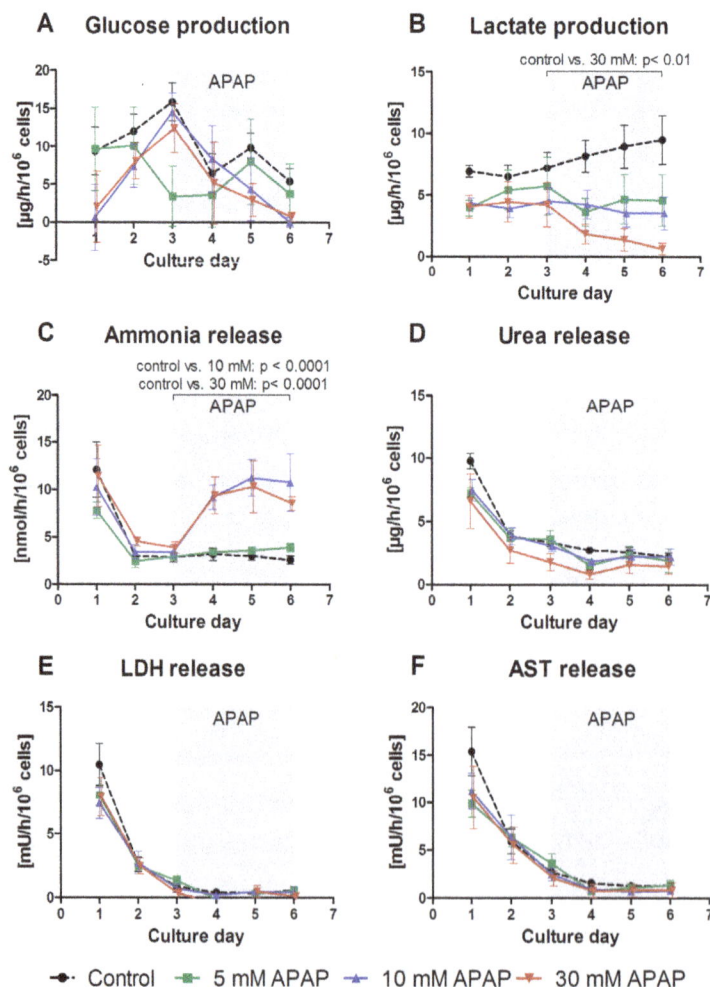

Figure 3. Time-courses of clinical parameters in bioreactors treated with 5 mM, 10 mM or 30 mM acetaminophen (APAP) in comparison to untreated bioreactors used as control group. The figure shows

the course of glucose (**A**) and lactate (**B**) production, ammonia (**C**) and urea (**D**) release, as well as liberation of lactate dehydrogenase (LDH, (**E**)) and aspartate aminotransferase (AST, (**F**)). APAP was continuously introduced from day 3 throughout day 6 of culture. Values were normalized to 10^6 inoculated cells. Data are shown as means $\pm$ SEM (n = 4; control n = 6). The influence of the drug dose (day 3–day 6) on the metabolic activity of the cells in comparison to the control was analyzed by means of one-way ANOVA with Dunnett's multiple comparison test, using the AUCs from day 3 until day 6. Significant changes are indicated in the graphs. Underlying data are available at http://doi.org/10.5281/zenodo.1169306(Clinical_chemistry_parameters).

The release of inflammatory factors was selectively affected by different APAP concentrations (Figure 4).

Release rates of the prostaglandin PGE2 (Figure 4A) showed a peak immediately after cell isolation followed by a rapid decline to basal levels. Upon APAP exposure, PGE2 release rates were characterized by a significant ($p < 0.05$) increase in bioreactors treated with 30 mM APAP as compared with the control group, indicating induction of PGE2 secretion at high APAP doses. In contrast, the cytokine IL-6 showed a general decrease in values in all groups during the culture course, and AUCs upon APAP incubation were similar in APAP-treated and untreated cultures (Figure 4B).

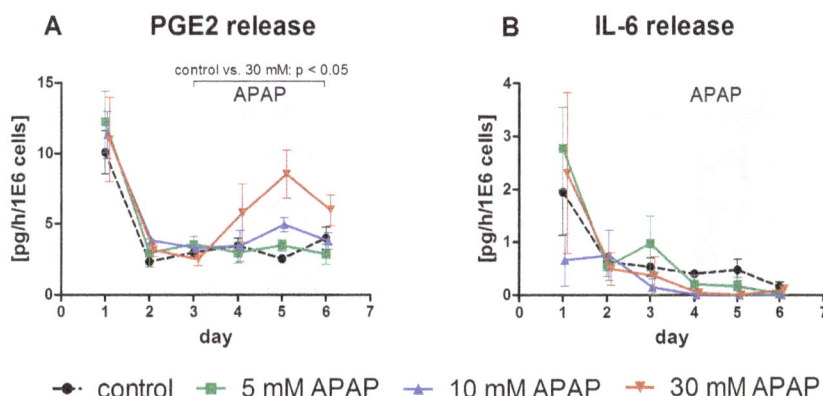

Figure 4. Time-courses of inflammatory factors in bioreactors treated with 5 mM, 10 mM, or 30 mM acetaminophen (APAP) in comparison to untreated bioreactors used as control group. The figure shows release rates of prostaglandin E2 (PGE2, (**A**)) and interleukin 6 (IL-6, (**B**)). APAP was continuously introduced from day 3 throughout day 6 of culture. Values were normalized to 10^6 inoculated cells. Data are shown as means $\pm$ SEM (n = 3; control n = 5). The influence of the drug dose (day 3–day 6) on the metabolic activity of the cells in comparison to the control was analyzed by means of one-way ANOVA with Dunnett's multiple comparison test, using the AUCs from day 3 until day 6. Significant changes are indicated in the graphs. Underlying data are available at http://doi.org/10.5281/zenodo.1169306 (Clinical_chemistry_parameters).

3.2. Gene Expression Analysis

The expression of genes involved in drug metabolism, antioxidant reactions, urea synthesis, and apoptosis was influenced individually by APAP exposure (Figure 5).

The genes encoding for *CYP1A2* (Figure 5A) and *CYP2E1* (Figure 5B) were strongly reduced in cultures exposed to 5 mM APAP as compared to the untreated control bioreactors, followed by a stepwise increase at higher APAP concentrations. While *CYP1A2* expression was lower in all APAP-treated groups than in the control, the expression of *CYP2E1* increased by 20-fold in the group treated with 30 mM APAP. The genes encoding for carbamoyl phosphate synthetase I (*CPS1*, Figure 5C) and glutathione S-transferase omega 2 (*GSTO2*, Figure 5D) showed a similar expression pattern, characterized by a 25-fold (*CPS1*, $p < 0.05$) resp. 4.5-fold (*GSTO2*) reduction in bioreactors treated with 5 mM APAP, and a successive increase at higher drug concentrations, without reaching the values of the control group. A different effect of APAP on gene expression was observed for the genes associated with apoptosis, namely caspase 3, apoptosis-related cysteine peptidase (*CASP3*, Figure 5E), and apoptosis-inducing factor, mitochondria-associated, 1 (*AIFM1*, Figure 5F). *CASP3* showed a slight increase in expression at 5 mM APAP, which was followed by a successive decline in the groups exposed to 10 or 30 mM APAP, whereas *AIFM1* expression was reduced in all APAP-treated groups, with significantly lower values in bioreactors treated with 5 or 10 mM APAP ($p < 0.0001$).

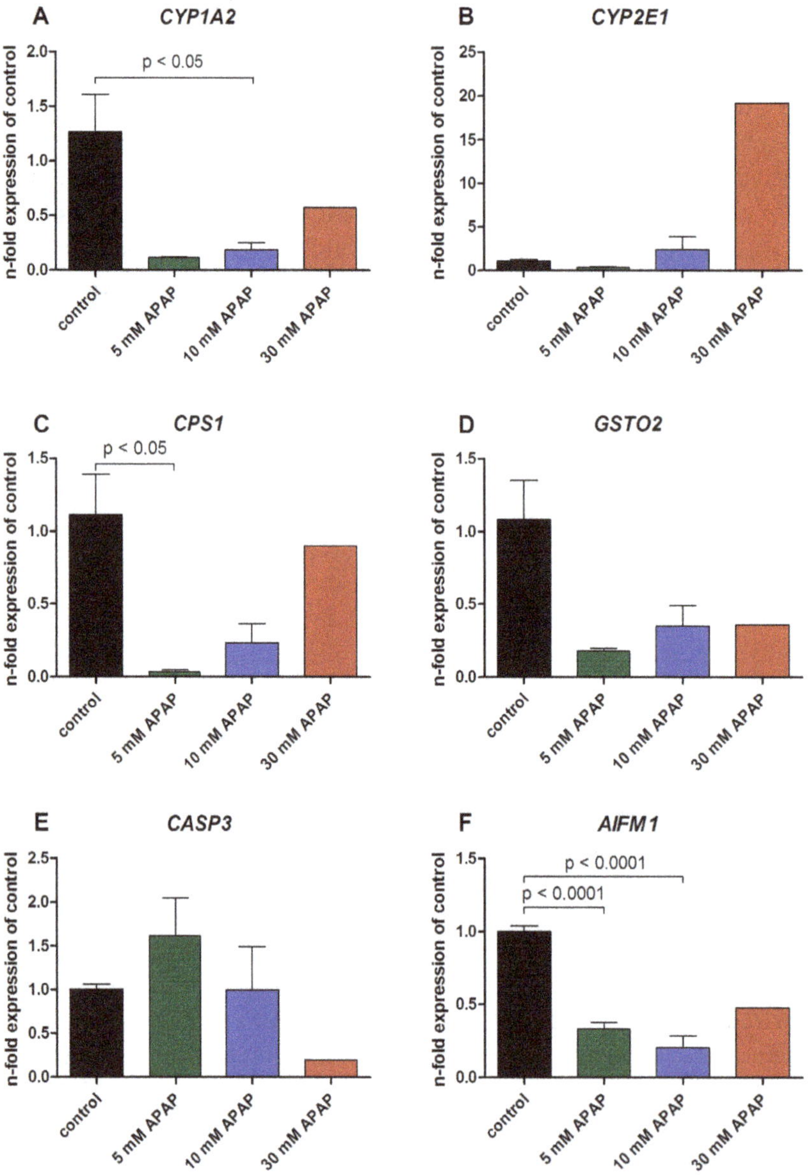

Figure 5. Gene expression analysis of primary human liver cells after culture in bioreactors treated with 5 mM, 10 mM or 30 mM acetaminophen (APAP) in comparison to untreated bioreactors used as control group. The figure shows the gene expression of cytochrome P450 family 1, subfamily A, polypeptide 2 (*CYP1A2*, (**A**)), cytochrome P450 family 2, subfamily E, polypeptide 1 (*CYP2E1*, (**B**)), carbamoyl phosphate synthetase I (*CPS1*, (**C**)), glutathione S-transferase omega 2 (*GSTO2*, (**D**)), caspase 3, apoptosis-related cysteine peptidase (*CASP3*, (**E**)) and apoptosis-inducing factor, mitochondria-associated, 1 (*AIFM1*, (**F**)). Expression data were normalized to the house-keeping gene glyceraldehyde-3-phosphate dehydrogenase and were calculated relative to the untreated control using the ΔΔCt-method. Data are shown as means ± SEM (control $n = 4$; 5 mM APAP $n = 2$; 10 mM APAP

$n = 3$, 30 mM APAP $n = 1$). Differences between the control and 5 mM or 10 mM APAP were calculated using one-way ANOVA with Dunnett's multiple comparison test and significant changes are indicated in the graphs. Underlying data are available at http://doi.org/10.5281/zenodo.1169306 (qRT_PCR).

3.3. Histological and Immunohistochemical Analysis

Histological investigation and immunofluorescence labeling of hepatic antigens in untreated cultures (control) or those exposed to 10 or 30 mM APAP showed a clear effect of APAP on the tissue organization and distribution pattern of cell-specific markers in bioreactor cultures (Figure 6).

As shown by HE staining (Figure 6A–C), primary human liver cells cultured in control bioreactors were associated in tissue-like cell aggregates between the hollow-fiber capillaries. Cell clusters contained primarily hepatocytes characterized by a large cytoplasm and a round or polygonal shape. The majority of cells appeared morphologically intact. Upon treatment with 10 or 30 mM APAP, cell aggregates and cell–cell connections were partly dissolved, resulting in the occurrence of numerous isolated cells. Most cells displayed a small and condensed cytoplasm and a lack of demarcation of cell nuclei, indicating necrotic and/or apoptotic processes.

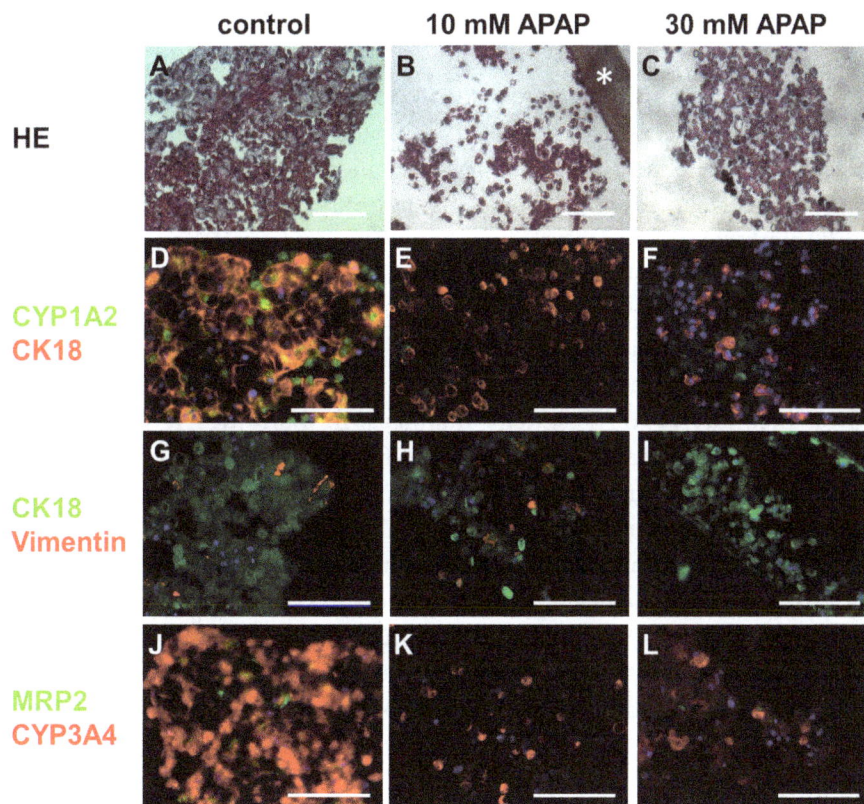

Figure 6. Histological and immunohistochemical staining of primary human liver cells after culture in untreated bioreactors (control) or those treated with 10 mM or 30 mM acetaminophen (APAP). The figure shows the hematoxylin-eosin stain (HE, (**A**–**C**)); double staining of cytochrome P450 family 1, subfamily A, polypeptide 2 (CYP1A2) and cytokeratin 18 (CK18) (**D**–**F**), CK18, and vimentin (**G**–**I**); and double staining of multidrug resistance-associated protein 2 (MRP2) and cytochrome P450 family 3, subfamily A, polypeptide 4 (CYP3A4) (**J**–**L**). The asterisk in (**B**) marks a hollow-fiber capillary membrane. Nuclei were counter-stained with Hoechst 33342 (blue). Scale bars correspond to 100 µm. Source pictures are available at http://doi.org/10.5281/zenodo.1169306 (Histology_Immnunofluorescence).

Immunofluorescence staining confirmed the finding of cell damage and partial disintegration of cell aggregates upon APAP application. In control cultures, the hepatocyte-specific markers CK18 and CYP1A2 (Figure 6D–F) mostly showed an evenly distributed staining. The cytoskeletal marker CK18 was primarily expressed at cell margins, forming a network-like staining pattern, while CYP1A2 was mainly localized in the cytoplasm. In cultures treated with 10 or 30 mM APAP sparse and irregular staining of CK18 and CYP1A2 was observed. Staining of nuclei with Hoechst 33,342 revealed a condensed cytoplasm of most cells and furthermore, the occurrence of isolated nuclei devoid of cytoplasm, as an additional indication of cell death. Double-staining of CK18 and vimentin (Figure 6G–I) showed the presence of some non-parenchymal cells (vimentin-positive) between hepatocytes (CK18-positive), both in control bioreactors and those treated with 10 mM APAP. In contrast, no vimentin-positive cells were detected in cultures exposed to 30 mM APAP, indicating a loss of non-parenchymal cells. Expression of CYP3A4 (Figure 6J–L) was observed in the cytoplasm of most cells in control bioreactors and was still expressed in part of the cells after APAP treatment. The biliary transporter MRP2 being localized in plasma membranes of adjacent cells was detected in around 40–50% of the cells in control cultures. In bioreactors subjected to 10 or 30 mM APAP, the fraction of MRP2 positive cells was decreased to less than 10%.

4. Discussion

In order to precisely predict the hepatotoxicity of compounds during drug development, validated human in vitro models are necessary. Models based on a 3D environment have proven to more accurately reflect the human body compared to conventional 2D models [6]. Various microfluidic culture systems were developed to minimize the amounts of cells and culture materials [12,13]. The microscale 3D bioreactor used in this study is based on an existing four-compartment hollow-fiber technology [14–17] and was down-scaled to a cell compartment volume of 100 µL and a cultivated cell number of 10 million primary human liver cells.

To demonstrate the suitability of the microscale bioreactor system for hepatotoxicity studies, APAP was applied as a gold-standard test substance in concentrations of 5, 10, or 30 mM over a time period of three days. APAP toxicity manifests in many different ways as reviewed by Hinson et al. (2010), such as glutathione depletion, formation of toxic protein adducts, enzyme and cytokine release, and histological alterations [22]. In this study, the focus was on clinical chemistry parameters allowing regular evaluation of APAP toxicity during culture, followed by end-point analyses allowing the judgement of alterations in tissue integrity as well as protein and gene expression.

The time course of clinical chemistry parameters measured in the culture perfusate was generally characterized by an initial peak on the first day of culture. This increase can be ascribed to the cell isolation process, which leads to high levels of cell stress causing the release of intracellular enzymes and metabolites. In addition, disruption of cell–cell and cell–matrix contacts and consequently the loss of cell polarization were described [23]. Liver cells may also become pre-activated due to reperfusion injury associated with oxidative stress [24]. Both disruption of tissue integrity and activation of inflammatory signaling, are associated with dedifferentiation processes starting already during isolation [5].

In concordance with other studies [7,25] we observed dose-dependent toxic effects of APAP on primary human liver cells cultured in the device. Glucose and lactate production rates measured as parameters for energy metabolism showed a dose-dependent decrease from day 3 (beginning of APAP dosing) onwards indicating an impaired cell viability and functionality. The suitability of glucose consumption and lactate production to detect drug-induced changes in cell viability of hepatic cell cultures was previously shown [26]. In contrast to assays based on substrate conversion, for example, MTT or cell titer blue assay, measurement of glucose and lactate allows a regular monitoring of cell activity over time without intervening into the cell metabolism.

Additionally, the enzyme release was measured as an indicator for necrosis or for secondary necrosis following apoptosis [27,28]. Serum levels of the cytoplasmic enzyme LDH are routinely measured in clinical settings in order to assess cell damage within pathological processes, including liver injury [29,30]. AST is distributed both in the cytoplasm and mitochondria of hepatocytes [31]; mild cell injury causes the release of cytosolic enzymes, whereas severe liver damage leads to the release of both, cytoplasmic and mitochondrial enzymes. All experimental groups showed an initial peak of enzyme values on day 1 in consistence with other studies [32]. The absence of an increase in enzyme release during the APAP treatment period is in accordance with previous results investigating diclofenac toxicity in 3D bioreactors [15]. Similarly, periodic treatment of hepatocyte cultures with 18.6 mM APAP for 20 days had no significant effect on LDH release [33]. The absence of response in hepatic enzyme release may be explained by the exhaustion of cytosolic enzyme stores in the initial culture phase, which can be ascribed to cell stress during cell isolation. In a study investigating liver enzyme concentrations in the cytosol and in the supernatant of primary human hepatocytes exposed to APAP, the amount of secreted enzyme relative to the total enzyme content showed a significant increase upon drug application [25]. Hence, the determination of both extracellular and intracellular LDH activity could provide more conclusive results on the actual effect of APAP on enzyme release. However, this would require daily lysis of cell samples, which is difficult to realize in complex 3D culture systems.

Ammonia, a product of amino acid metabolism, is toxic in high concentrations and is therefore converted to urea by hepatocytes. Hence, through the analysis of urea synthesis and ammonia depletion, conclusions can be drawn regarding the functionality of cultured hepatocytes [34]. The observed initial increase in urea and ammonia production rates can be related to cell stress due to cell isolation, in accordance with the observed peaks in enzyme release. The finding of significantly increased ammonia release rates after dosing with 10 or 30 mM APAP from day 3 onwards indicate a dose-dependent influence of APAP on the nitrogen metabolism of the cells. This finding is supported by the fact that APAP toxicity causes mitochondrial dysfunction [25], since part of the enzymes involved in the urea cycle are located in the mitochondria. Moreover, these observations emphasize the suitability of ammonia as a sensitive parameter for hepatocyte functionality and hepatic drug toxicity.

The inflammatory factors PGE2 and IL-6 showed a different course upon drug application. A dose-dependent increase from day 3 (beginning of APAP dosing) onwards was detected for PGE2, a pro-inflammatory factor with immunosuppressive activity, indicating an accumulation of APAP-induced inflammation. Since PGE2 is typically produced by cell types such as endothelial cells and cells of the immune system [35], the finding of an increased PGE2 release confirms the presence of non-parenchymal cells in the liver bioreactor. In contrast, the pro-inflammatory cytokine IL-6, which is secreted by different liver cell types, including hepatocytes [36], showed no significant response to APAP exposure. PGE2 inhibits the production of IL-6 as a mechanism of limiting excessive immune reactions [37], which may be the reason for the lack of IL-6 response to APAP treatment.

Since production rates of metabolic parameters were normalized to the initial cell number, the results from the medium analysis represent the combined effect of cell number and functionality. Thus, values do not allow a distinction between a change either of cell number or of cell activity in the applied culture system.

In the liver, APAP is metabolized mainly by *CYP2E1* to the toxic N-acetyl-*p*-benzochinonimin (NAPQI) [38,39]. Gene expression analysis of the cells subsequent to APAP exposure in comparison to untreated control bioreactors revealed a dose-dependent increase in *CYP2E1* expression especially for 30 mM APAP. This result indicates that high concentrations of APAP lead to an immediate upregulation of *CYP2E1*, which can be seen as a mechanism to accelerate APAP metabolism. In addition, we observed a decrease in gene expression of *CYP1A2* upon APAP treatment compared to the control bioreactor, though less pronounced at increasing APAP concentrations. The contribution of CYP1A2 to APAP metabolism is controversially discussed. While studies considering rat and human liver microsomes [40,41] or recombinant human CYP P450 enzymes [42] showed that CYP1A2 is involved in the metabolism of APAP, other publications, performed in adult human volunteers, reported no direct association of CYP1A2 with APAP depletion [38,43]. These contradictory findings may be caused by different conditions in in vitro experiments as compared with the in vivo situation.

The glutathione S-transferases (GSTs) are enzymes catalyzing the neutralization of free radicals and active drug components using glutathione as reducing agent, which is the main step in phase 2 detoxification [44]. In APAP metabolism, GSTs catalyze the formation of a NAPQI-glutathione adduct, which is then primarily secreted into bile by passing the apical membrane transporter protein MRP2 [45]. Treatment of hepatocytes with APAP resulted in a decrease of *GSTO2* gene expression as compared to the control. Similar findings were reported by Wang et al. (2017), who observed a reduction of GST activities in a mouse model of APAP-induced liver injury [46]. The observed decrease in gene expression of *CPS-1*, an enzyme involved in the production of urea, is in accordance to our observations of increased ammonia release rates in the bioreactors treated with APAP, and further supports the assumption that APAP effects the nitrogen metabolism.

In contrast, the apoptosis-associated gene *CASP3* showed an increase in expression for the bioreactor treated with 5 mM APAP in comparison to the untreated controls, while no change was observed at 10 mM APAP and a decrease was detected after exposure to 30 mM APAP. This observation might be explained by a shift to necrotic cell death, in accordance with findings by Au and colleagues, who observed a transition from apoptosis to necrosis between 10 and 20 mM APAP exposure using HepG2 cells [47]. However, another apoptosis-associated factor, *AIFM1*, revealed reduced expression values for all drug-treated bioreactors as compared to the control. The pro-apoptotic function of AIFM1 is based on the activation of a caspase-independent pathway upon apoptotic stimuli, whereas its anti-apoptotic function is part of the regular mitochondria metabolism via NADH oxidoreduction [48]. As APAP treatment results in mitochondrial dysfunction, it might consequently also lead to a reduced gene expression of *AIFM1*. Since APAP is known to induce both necrosis and apoptosis [49], a more detailed characterization of the type of cell death would be of interest in future studies to differentiate between necrosis and apoptosis.

Histological and immunohistochemical analyses revealed that the cells cultured in control bioreactors were reorganized in tissue-like formations, which resembled those observed in larger scale bioreactors with higher initial cell amounts [14,16]. Typical structural and functional markers of hepatocytes, including CK18, CYP1A1 and CYP3A4 were regularly detected and showed an in vivo-like distribution pattern. Staining of MRP2, which is the main transporter for biliary excretion of acetaminophen sulfate [50,51], indicates cell polarization with formation of bile canaliculi. The characterization of the cells by means of cell-specific markers showed that in addition to hepatocytes, the aggregates also comprised non-parenchymal cells identified by vimentin staining. Since the non-parenchymal cells of the liver play a major role in drug-induced liver injury [52], the presence of these cells in liver models is critical to assess complex drug effects mediated by different liver cell populations.

Exposure to APAP at a concentration of 10 or 30 mM resulted in partial disintegration of cell aggregates and loss of cell integrity. In particular, MRP2 was rarely detectable in APAP-treated bioreactors indicating a depolarization of hepatocytes. This is in line with findings by Bhise and colleagues, who showed a massive reduction of MRP2 immunostaining after treatment of hepatic spheroids composed of hepatoma cells with 15 mM APAP [53]. The reduction of membrane transporters may lead to impaired excretion of APAP metabolites and therefore cause accumulation of toxic products in the cells, if glutathione is not sufficiently available.

In summary, we were able to detect dose-dependent hepatotoxic effects of 5 to 30 mM APAP on primary human hepatocytes cultured in the microscale 3D bioreactor. Au and colleagues identified hepatotoxic influences of APAP at 10 mM using spheroids comprising HepG2 cells and fibroblasts cultured on a microfluidic platform [47]. Other microfluidic studies using HepG2 cells found a reduction of cell viability by more than 50% upon treatment with 15 mM APAP [53] resp. 20 mM APAP [54]. However, in the human body, plasma concentrations of 0.5 to 3 mM APAP were observed in overdose scenarios [55]. A potential reason for the discrepancy between toxic APAP concentrations in vivo and in vitro can be seen in the contribution of systemic influences, such as nutritional status [56] and blood cells [57], to APAP toxicity. In addition, the non-parenchymal liver cells, such as sinusoidal endothelial cells [58] and Kupffer cells [59], have been shown to play a role in APAP toxicity. Although in the present study the obtained mixture of primary liver cells after enzymatic digestion of the organ was used without further purification of hepatocytes, the number of non-parenchymal cells might not have been sufficient to correctly imitate the human in vivo liver.

Hence, future studies would be of interest to further investigate the role of individual liver cell populations in APAP metabolism and toxicity. The addition of non-parenchymal liver cells to the microscale 3D liver culture in ratios comparable to the in vivo situation could increase its sensitivity for APAP toxicity. Methods for isolation of endothelial cells, Kupffer cells and stellate cells from human liver tissue were recently described [18] and can be used to provide defined amounts of these cells for human liver cell models. Other microfluidic systems attempt to recapitulate the microarchitecture of the liver sinusoid by providing several compartments for the different liver cell types [60,61]. A further important factor in drug susceptibility might be the oxygen concentration, as indicated by results from co-cultures of primary rat hepatocytes and fibroblasts, which proved to be more sensitive to APAP exposure in low-oxygen regions [62]. Thus, the creation of physiological oxygen gradients might increase the predictive power of in vitro drug effects. To assess systemic effects mediated by other organs, integration of the microscale 3D liver bioreactor into a multi-organ platform would be an attractive approach. In this context, the structure of the bioreactor provides suitable conditions for realization of microscale systems integrating various organ constructs, such as kidney, heart, and lung.

A general limitation of microscale systems can be seen in the small amount of cell material available for end-point analyses such as immunohistochemistry, qPCR, or Western blots. Thus, studies investigating microscale liver tissues often show results from only one end-point analysis, mostly immunohistochemistry or Western blotting [32,47,53,54,62]. Hence, the implementation of analytic methods allowing analyses from minimal cell numbers is required to generate a larger range of data from microscale systems.

5. Conclusions

In conclusion, the results from APAP application in this study demonstrate that the microscale 3D liver bioreactor provides a useful tool to detect hepatotoxic effects of drugs in a perfused human in vitro culture environment. Our observations emphasize the potential of clinical chemistry parameters, such as lactate production and ammonia release, as sensitive parameters for monitoring dose-dependent hepatotoxic effects throughout the culture period. The analysis of inflammatory factors showed that mainly PGE2 was affected by APAP exposure in the model. The toxic effect of APAP in the in vitro model was confirmed by end-point analyses, including histological and immunohistochemical evaluation and PCR analysis. In order to increase the sensitivity of the present microscale 3D liver bioreactor for toxicity studies at physiologically relevant drug concentrations, co-cultures supplemented with different non-parenchymal cell types at physiological ratios, and also creation of defined oxygen gradients could be applied in future studies.

Acknowledgments: The work for the study was supported, in part, by the Defense Threat Reduction Agency (program: Integration of Novel Technologies for Organ Development and Rapid Assessment of Medical Countermeasures, Interagency Agreement DTRA100271A5196), USA, and in part by the German Ministry for Education and Research (BMBF) within the funding network "Virtual Liver" (FKZ 0315741).

Author Contributions: Nora Freyer and Selina Greuel analyzed the data and wrote the manuscript; Fanny Knöspel conceived and designed the experiments and evaluated the data; Florian Gerstmann and Lisa Storch performed the experiments and contributed to data analysis; Georg Damm and Daniel Seehofer were responsible for isolation, processing, and quality analysis of fresh primary human liver cells; Jennifer Foster Harris and Rashi Iyer contributed to the conception and design of experiments and writing of the manuscript; Frank Schubert was responsible for technical developments and design of bioreactor prototypes; and Katrin Zeilinger contributed to the design of experiments, evaluation of results and writing of the manuscript.

Conflicts of Interest: The authors declare no conflict of interest.

References

1. Bhakuni, G.S.; Bedi, O.; Bariwal, J.; Deshmukh, R.; Kumar, P. Animal models of hepatotoxicity. *Inflamm. Res.* **2016**, *65*, 13–24. [CrossRef] [PubMed]
2. Maes, M.; Vinken, M.; Jaeschke, H. Experimental models of hepatotoxicity related to acute liver failure. *Toxicol. Appl. Pharmacol.* **2016**, *290*, 86–97. [CrossRef] [PubMed]
3. Sharma, V.; McNeill, J.H. To scale or not to scale: The principles of dose extrapolation. *Br. J. Pharmacol.* **2009**, *157*, 907–921. [CrossRef] [PubMed]
4. Olson, H.; Betton, G.; Robinson, D.; Thomas, K.; Monro, A.; Kolaja, G.; Lilly, P.; Sanders, J.; Sipes, G.; Bracken, W.; et al. Concordance of the toxicity of pharmaceuticals in humans and in animals. *Regul. Toxicol. Pharmacol.* **2000**, *32*, 56–67. [CrossRef] [PubMed]
5. Elaut, G.; Henkens, T.; Papeleu, P.; Snykers, S.; Vinken, M.; Vanhaecke, T.; Rogiers, V. Molecular mechanisms underlying the dedifferentiation process of isolated hepatocytes and their cultures. *Curr. Drug Metab.* **2006**, *7*, 629–660. [CrossRef] [PubMed]
6. Elliott, N.T.; Yuan, F. A review of three-dimensional in vitro tissue models for drug discovery and transport studies. *J. Pharm. Sci.* **2011**, *100*, 59–74. [CrossRef] [PubMed]
7. Schyschka, L.; Sánchez, J.J.; Wang, Z.; Burkhardt, B.; Müller-Vieira, U.; Zeilinger, K.; Bachmann, A.; Nadalin, S.; Damm, G.; Nussler, A.K. Hepatic 3D cultures but not 2D cultures preserve specific transporter activity for acetaminophen-induced hepatotoxicity. *Arch. Toxicol.* **2013**, *87*, 1581–1593. [CrossRef] [PubMed]
8. Bell, C.C.; Hendriks, D.F.; Moro, S.M.; Ellis, E.; Walsh, J.; Renblom, A.; Fredriksson Puigvert, L.; Dankers, A.C.; Jacobs, F.; Snoeys, J.; et al. Characterization of primary human hepatocyte spheroids as a model system for drug-induced liver injury, liver function and disease. *Sci. Rep.* **2016**, *6*, 25187. [CrossRef] [PubMed]
9. Ohkura, T.; Ohta, K.; Nagao, T.; Kusumoto, K.; Koeda, A.; Ueda, T.; Jomura, T.; Ikeya, T.; Ozeki, E.; Wada, K.; et al. Evaluation of human hepatocytes cultured by three-dimensional spheroid systems for drug metabolism. *Drug Metab. Pharmacokinet.* **2014**, *29*, 373–378. [CrossRef] [PubMed]
10. Dash, A.; Simmers, M.B.; Deering, T.G.; Berry, D.J.; Feaver, R.E.; Hastings, N.E.; Pruett, T.L.; LeCluyse, E.L.; Blackman, B.R.; Wamhoff, B.R. Hemodynamic flow improves rat hepatocyte morphology, function, and metabolic activity in vitro. *Am. J. Physiol. Cell Physiol.* **2013**, *304*, C1053–C1063. [CrossRef] [PubMed]
11. De Bartolo, L.; Salerno, S.; Curcio, E.; Piscioneri, A.; Rende, M.; Morelli, S.; Tasselli, F.; Bader, A.; Drioli, E. Human hepatocyte functions in a crossed hollow fiber membrane bioreactor. *Biomaterials* **2009**, *30*, 2531–2543. [CrossRef] [PubMed]
12. Aziz, A.U.R.; Geng, C.; Fu, M.; Yu, X.; Qin, K.; Liu, B. The role of microfluidics for organ on chip simulations. *Bioengineering* **2017**, *4*, 39. [CrossRef] [PubMed]
13. Gupta, N.; Liu, J.R.; Patel, B.; Solomon, D.E.; Vaidya, B.; Gupta, V. Microfluidics-based 3D cell culture models: Utility in novel drug discovery and delivery research. *Bioeng. Transl. Med.* **2016**, *1*, 63–81. [CrossRef] [PubMed]
14. Zeilinger, K.; Schreiter, T.; Darnell, M.; Söderdahl, T.; Lübberstedt, M.; Dillner, B.; Knobeloch, D.; Nüssler, A.K.; Gerlach, J.C.; Andersson, T.B. Scaling down of a clinical three-dimensional perfusion multicompartment hollow fiber liver bioreactor developed for extracorporeal liver support to an analytical scale device useful for hepatic pharmacological in vitro studies. *Tissue Eng. Part C* **2011**, *17*, 549–556. [CrossRef] [PubMed]
15. Knöspel, F.; Jacobs, F.; Freyer, N.; Damm, G.; De Bondt, A.; van den Wyngaert, I.; Snoeys, J.; Monshouwer, M.; Richter, M.; Strahl, N.; et al. In vitro model for hepatotoxicity studies based on primary human hepatocyte cultivation in a perfused 3D bioreactor system. *Int. J. Mol. Sci.* **2016**, *17*, 584. [CrossRef] [PubMed]
16. Lübberstedt, M.; Müller-Vieira, U.; Biemel, K.M.; Darnell, M.; Hoffmann, S.A.; Knöspel, F.; Wönne, E.C.; Knobeloch, D.; Nüssler, A.K.; Gerlach, J.C.; et al. Serum-free culture of primary human hepatocytes in a miniaturized hollow-fibre membrane bioreactor for pharmacological in vitro studies. *J. Tissue Eng. Regen. Med.* **2015**, *9*, 1017–1026. [CrossRef] [PubMed]
17. Hoffmann, S.A.; Müller-Vieira, U.; Biemel, K.; Knobeloch, D.; Heydel, S.; Lübberstedt, M.; Nüssler, A.K.; Andersson, T.B.; Gerlach, J.C.; Zeilinger, K. Analysis of drug metabolism activities in a miniaturized liver cell bioreactor for use in pharmacological studies. *Biotechnol. Bioeng.* **2012**, *109*, 3172–3181. [CrossRef] [PubMed]

18. Kegel, V.; Deharde, D.; Pfeiffer, E.; Zeilinger, K.; Seehofer, D.; Damm, G. Protocol for isolation of primary human hepatocytes and corresponding major populations of non-parenchymal liver cells. *J. Vis. Exp.* **2016**, *109*, e53069. [CrossRef] [PubMed]

19. Pfeiffer, E.; Kegel, V.; Zeilinger, K.; Hengstler, J.G.; Nüssler, A.K.; Seehofer, D.; Damm, G. Featured Article: Isolation, characterization, and cultivation of human hepatocytes and non-parenchymal liver cells. *Exp. Biol. Med.* **2015**, *240*, 645–656. [CrossRef] [PubMed]

20. Freyer, N.; Knöspel, F.; Strahl, N.; Amini, L.; Schrade, P.; Bachmann, S.; Damm, G.; Seehofer, D.; Jacobs, F.; Monshouwer, M.; et al. Hepatic differentiation of human induced pluripotent stem cells in a perfused three-dimensional multicompartment bioreactor. *BioRes. Open Access* **2016**, *5*, 235–248. [CrossRef] [PubMed]

21. Livak, K.J.; Schmittgen, T.D. Analysis of relative gene expression data using real-time quantitative PCR and the $2^{-\Delta\Delta Ct}$ Method. *Methods* **2001**, *25*, 402–408. [CrossRef] [PubMed]

22. Hinson, J.A.; Roberts, D.W.; James, L.P. Mechanisms of acetaminophen-induced liver necrosis. *Handb. Exp. Pharmacol.* **2010**, *196*, 369–405. [CrossRef]

23. Treyer, A.; Müsch, A. Hepatocyte polarity. *Compr. Physiol.* **2013**, *3*, 243–287. [CrossRef] [PubMed]

24. Tormos, A.M.; Taléns-Visconti, R.; Bonora-Centelles, A.; Pérez, S.; Sastre, J. Oxidative stress triggers cytokinesis failure in hepatocytes upon isolation. *Free Radic. Res.* **2015**, *49*, 927–934. [CrossRef] [PubMed]

25. Xie, Y.; McGill, M.R.; Dorko, K.; Kumer, S.C.; Schmitt, T.M.; Forster, J.; Jaeschke, H. Mechanisms of acetaminophen-induced cell death in primary human hepatocytes. *Toxicol. Appl. Pharmacol.* **2014**, *279*, 266–274. [CrossRef] [PubMed]

26. Prill, S.; Jaeger, M.S.; Duschl, C. Long-term microfluidic glucose and lactate monitoring in hepatic cell culture. *Biomicrofluidics* **2014**, *8*, 034102. [CrossRef] [PubMed]

27. Cummings, B.S.; Wills, L.P.; Schnellmann, R.G. Measurement of cell death in mammalian cells. *Curr. Protoc. Pharmacol.* **2012**. [CrossRef]

28. Silva, M.T. Secondary necrosis: The natural outcome of the complete apoptotic program. *FEBS Lett.* **2010**, *584*, 4491–4499. [CrossRef] [PubMed]

29. Jialal, I.; Sokoll, L.J. Clinical utility of lactate dehydrogenase: A historical perspective. *Am. J. Clin. Pathol.* **2015**, *143*, 158–159. [CrossRef] [PubMed]

30. Cassidy, W.M.; Reynolds, T.B. Serum lactic dehydrogenase in the differential diagnosis of acute hepatocellular injury. *J. Clin. Gastroenterol.* **1994**, *19*, 118–121. [CrossRef] [PubMed]

31. Wolf, P.L. Biochemical diagnosis of liver disease. *Indian J. Clin. Biochem.* **1999**, *14*, 59–90. [CrossRef] [PubMed]

32. Choi, K.; Pfund, W.P.; Andersen, M.E.; Thomas, R.S.; Clewell, H.J.; LeCluyse, E.L. Development of 3D dynamic flow model of human liver and its application to prediction of metabolic clearance of 7-ethoxycoumarin. *Tissue Eng. Part C Methods* **2014**, *20*, 641–651. [CrossRef] [PubMed]

33. Ullrich, A.; Stolz, D.B.; Ellis, E.C.; Strom, S.C.; Michalopoulos, G.K.; Hengstler, J.G.; Runge, D. Long term cultures of primary human hepatocytes as an alternative to drugtesting in animals. *ALTEX* **2009**, *26*, 295–302. [CrossRef] [PubMed]

34. Bolleyn, J.; Rogiers, V.; Vanhaecke, T. Functionality testing of primary hepatocytes in culture by measuring urea synthesis. *Methods Mol. Biol.* **2015**, *1250*, 317–321. [CrossRef] [PubMed]

35. Simmons, D.L.; Botting, R.M.; Hla, T. Cyclooxygenase isozymes: The biology of prostaglandin synthesis and inhibition. *Pharmacol. Rev.* **2004**, *56*, 387–437. [CrossRef] [PubMed]

36. Norris, C.A.; He, M.; Kang, L.I.; Ding, M.Q.; Radder, J.E.; Haynes, M.M.; Yang, Y.; Paranjpe, S.; Bowen, W.C.; Orr, A.; et al. Synthesis of IL-6 by hepatocytes is a normal response to common hepatic stimuli. *PLoS ONE* **2014**, *9*, e96053. [CrossRef] [PubMed]

37. Kalinski, P. Regulation of immune responses by prostaglandin E2. *J. Immunol.* **2012**, *188*, 21–28. [CrossRef] [PubMed]

38. Manyike, P.T.; Kharasch, E.D.; Kalhorn, T.F.; Slattery, J.T. Contribution of CYP2E1 and CYP3A to acetaminophen reactive metabolite formation. *Clin. Pharmacol. Ther.* **2000**, *67*, 275–282. [CrossRef] [PubMed]

39. Lee, S.S.; Buters, J.T.; Pineau, T.; Fernandez-Salguero, P.; Gonzalez, F.J. Role of CYP2E1 in the hepatotoxicity of acetaminophen. *J. Biol. Chem.* **1996**, *271*, 12063–12067. [CrossRef] [PubMed]

40. Patten, C.J.; Thomas, P.E.; Guy, R.L.; Lee, M.; Gonzalez, F.J.; Guengerich, F.P.; Yang, C.S. Cytochrome P450 enzymes involved in acetaminophen activation by rat and human liver microsomes and their kinetics. *Chem. Res. Toxicol.* **1993**, *6*, 511–518. [CrossRef] [PubMed]

41. Raucy, J.L.; Lasker, J.M.; Lieber, C.S.; Black, M. Acetaminophen activation by human liver cytochromes P450IIE1 and P450IA2. *Arch. Biochem. Biophys.* **1989**, *271*, 270–283. [CrossRef]

42. Laine, J.E.; Auriola, S.; Pasanen, M.; Juvonen, R.O. Acetaminophen bioactivation by human cytochrome P450 enzymes and animal microsomes. *Xenobiotica* **2009**, *39*, 11–21. [CrossRef] [PubMed]

43. Sarich, T.; Kalhorn, T.; Magee, S.; Al-Sayegh, F.; Adams, S.; Slattery, J.; Goldstein, J.; Nelson, S.; Wright, J. The effect of omeprazole pretreatment on acetaminophen metabolism in rapid and slow metabolizers of S-mephenytoin. *Clin. Pharmacol. Ther.* **1997**, *62*, 21–28. [CrossRef]

44. Khosravi, M.; Saadat, I.; Karimi, M.H.; Geramizadeh, B.; Saadat, M. Glutathione S-transferase omega 2 genetic polymorphism and risk of hepatic failure that lead to liver transplantation in iranian population. *Int. J. Organ Transplant. Med.* **2013**, *4*, 16–20. [PubMed]

45. Chen, C.; Hennig, G.E.; Manautou, J.E. Hepatobiliary excretion of acetaminophen glutathione conjugate and its derivatives in transport-deficient (TR-) hyperbilirubinemic rats. *Drug Metab. Dispos.* **2003**, *31*, 798–804. [CrossRef] [PubMed]

46. Wang, J.X.; Zhang, C.; Fu, L.; Zhang, D.G.; Wang, B.W.; Zhang, Z.H.; Chen, Y.H.; Lu, Y.; Chen, X.; Xu, D.X. Protective effect of rosiglitazone against acetaminophen-induced acute liver injury is associated with down-regulation of hepatic NADPH oxidases. *Toxicol. Lett.* **2017**, *265*, 38–46. [CrossRef] [PubMed]

47. Au, S.H.; Chamberlain, M.D.; Mahesh, S.; Sefton, M.V.; Wheeler, A.R. Hepatic organoids for microfluidic drug screening. *Lab Chip* **2014**, *14*, 3290–3299. [CrossRef] [PubMed]

48. Rinaldi, C.; Grunseich, C.; Sevrioukova, I.F.; Schindler, A.; Horkayne-Szakaly, I.; Lamperti, C.; Landouré, G.; Kennerson, M.L.; Burnett, B.G.; Bönnemann, C.; et al. Cowchock syndrome is associated with a mutation in apoptosis-inducing factor. *Am. J. Hum. Genet.* **2012**, *91*, 1095–1102. [CrossRef] [PubMed]

49. Guicciardi, M.E.; Malhi, H.; Mott, J.L.; Gores, G.J. Apoptosis and necrosis in the liver. *Compr. Physiol.* **2013**, *3*, 977–1010. [CrossRef] [PubMed]

50. Jaeschke, H.; Bajt, M.L. Intracellular signaling mechanisms of acetaminophen-induced liver cell death. *Toxicol. Sci.* **2006**, *89*, 31–41. [CrossRef] [PubMed]

51. Zamek-Gliszczynski, M.J.; Hoffmaster, K.A.; Tian, X.; Zhao, R.; Polli, J.W.; Humphreys, J.E.; Webster, L.O.; Bridges, A.S.; Kalvass, J.C.; Brouwer, K.L. Multiple mechanisms are involved in the biliary excretion of acetaminophen sulfate in the rat: Role of MRP2 and BCRP1. *Drug Metab. Dispos.* **2005**, *33*, 1158–1165. [CrossRef] [PubMed]

52. Godoy, P.; Hewitt, N.J.; Albrecht, U.; Andersen, M.E.; Ansari, N.; Bhattacharya, S.; Bode, J.G.; Bolleyn, J.; Borner, C.; Böttger, J.; et al. Recent advances in 2D and 3D in vitro systems using primary hepatocytes, alternative hepatocyte sources and non-parenchymal liver cells and their use in investigating mechanisms of hepatotoxicity, cell signaling and ADME. *Arch. Toxicol.* **2013**, *87*, 1315–1530. [CrossRef] [PubMed]

53. Bhise, N.S.; Manoharan, V.; Massa, S.; Tamayol, A.; Ghaderi, M.; Miscuglio, M.; Lang, Q.; Shrike Zhang, Y.; Shin, S.R.; Calzone, G.; et al. A liver-on-a-chip platform with bioprinted hepatic spheroids. *Biofabrication* **2016**, *8*, 014101. [CrossRef] [PubMed]

54. Ma, C.; Zhao, L.; Zhou, E.M.; Xu, J.; Shen, S.; Wang, J. On-chip construction of liver lobule-like microtissue and its application for adverse drug reaction assay. *Anal. Chem.* **2016**, *88*, 1719–1727. [CrossRef] [PubMed]

55. Cairney, D.G.; Beckwith, H.K.; Al-Hourani, K.; Eddleston, M.; Bateman, D.N.; Dear, J.W. Plasma paracetamol concentration at hospital presentation has a dose-dependent relationship with liver injury despite prompt treatment with intravenous acetylcysteine. *Clin. Toxicol.* **2016**, *54*, 405–410. [CrossRef] [PubMed]

56. Whitcomb, D.C.; Block, G.D. Association of acetaminophen hepatotoxicity with fasting and ethanol use. *JAMA* **1994**, *272*, 1845–1850. [CrossRef] [PubMed]

57. Jaeschke, H.; Williams, C.D.; Ramachandran, A.; Bajt, M.L. Acetaminophen hepatotoxicity and repair: The role of sterile inflammation and innate immunity. *Liver Int.* **2012**, *32*, 8–20. [CrossRef] [PubMed]

58. McCuskey, R.S. Sinusoidal endothelial cells as an early target for hepatic toxicants. *Clin. Hemorheol. Microcirc.* **2006**, *34*, 5–10. [PubMed]

59. Kegel, V.; Pfeiffer, E.; Burkhardt, B.; Liu, J.L.; Zeilinger, K.; Nüssler, A.K.; Seehofer, D.; Damm, G. Subtoxic concentrations of hepatotoxic drugs lead to Kupffer cell activation in a human in vitro liver model: An approach to study DILI. *Mediators Inflamm.* **2015**, *2015*, 640631. [CrossRef] [PubMed]

60. Prodanov, L.; Jindal, R.; Bale, S.S.; Hegde, M.; McCarty, W.J.; Golberg, I.; Bhushan, A.; Yarmush, M.L.; Usta, O.B. Long-term maintenance of a microfluidic 3D human liver sinusoid. *Biotechnol. Bioeng.* **2016**, *113*, 241–246. [CrossRef] [PubMed]

61. Rennert, K.; Steinborn, S.; Gröger, M.; Ungerböck, B.; Jank, A.M.; Ehgartner, J.; Nietzsche, S.; Dinger, J.; Kiehntopf, M.; Funke, H.; et al. A microfluidically perfused three dimensional human liver model. *Biomaterials* **2015**, *71*, 119–131. [CrossRef] [PubMed]

62. Allen, J.W.; Khetani, S.R.; Bhatia, S.N. In vitro zonation and toxicity in a hepatocyte bioreactor. *Toxicol. Sci.* **2005**, *84*, 110–119. [CrossRef] [PubMed]

6

Mesenchymal Stem Cells Derived from Healthy and Diseased Human Gingiva Support Osteogenesis on Electrospun Polycaprolactone Scaffolds

Catherine Jauregui [1,†], Suyog Yoganarasimha [2] and Parthasarathy Madurantakam [1,3,*]

1   Philips Institute, School of Dentistry, Virginia Commonwealth University, Richmond, VA 23298, USA; cjauregui@augusta.edu

2   Department of Biomedical Engineering, School of Engineering, Virginia Commonwealth University, Richmond, VA 23284, USA; suyog.yoganarasimha@gmail.com

3   Department of General Practice, School of Dentistry, Virginia Commonwealth University, Richmond, VA 23298, USA

*   Correspondence: madurantakap@vcu.edu

†   Currently affiliation: Department of Oral Biology, Dental College of Georgia, Augusta University, Augusta, GA 30912, USA.

Abstract: Periodontitis is a chronic inflammatory disease affecting almost half of the adult US population. Gingiva is an integral part of the periodontium and has recently been identified as a source of adult gingiva-derived mesenchymal stem cells (GMSCs). Given the prevalence of periodontitis, the purpose of this study is to evaluate differences between GMSCs derived from healthy and diseased gingival tissues and explore their potential in bone engineering. Primary clonal cell lines were established from harvested healthy and diseased gingival and characterized for expression of known stem-cell markers and multi-lineage differentiation potential. Finally, they were cultured on electrospun polycaprolactone (PCL) scaffolds and evaluated for attachment, proliferation, and differentiation. Flow cytometry demonstrated cells isolated from healthy and diseased gingiva met the criteria defining mesenchymal stem cells (MSCs). However, GMSCs from diseased tissue showed decreased colony-forming unit efficiency, decreased alkaline phosphatase activity, weaker osteoblast mineralization, and greater propensity to differentiate into adipocytes than their healthy counterparts. When cultured on electrospun PCL scaffolds, GMSCs from both sources showed robust attachment and proliferation over a 7-day period; they exhibited high mineralization as well as strong expression of alkaline phosphatase. Our results show preservation of 'stemness' and osteogenic potential of GMSC even in the presence of disease, opening up the possibility of using routinely discarded, diseased gingival tissue as an alternate source of adult MSCs.

Keywords: gingiva; mesenchymal stem cells; electrospinning; scaffolds; bone tissue engineering; osteogenesis; alkaline phosphatase; alizarin red

1. Introduction

Bone is a unique tissue with a remarkable potential to undergo complete regeneration following injury, even in adult life. However, when the severity of the defect surpasses a critical-size, external invention in some form of grafting is required. With more than two million bone grafting procedures annually worldwide, bone represents the second most transplanted tissue next only to blood products [1]. The most common hard tissue graft is an autograft, where bone is taken from the patient's own body and reimplanted into the defect site. Autologous bone grafts are osteoconductive (provide a scaffold where bone cells can proliferate), osteoinductive (induce proliferation of undifferentiated cells and their differentiation into osteoblasts), and osteogenic (provide a reservoir of skeletal stem and progenitor cells that can form new bone). However, autografts are limited in availability, require additional invasive surgery, and have donor site morbidity [2]. Allografts derived from human donors or cadavers eliminate donor site morbidity, are osteoconductive, and are available in large quantities. However, these are associated with increased costs, laborious processing, decreased mechanical strength, limited osteoinduction, and increased risk of infection. Xenografts are cheap and are readily available but have risks of immunogenicity and slow integration [3]. Disadvantages associated with traditional sources have driven the development of synthetic bone substitutes that includes ceramics, metals, and polymers [4]. Bone tissue engineering is an alternate strategy that integrates the current advances in material science, cell biology, and bioengineering to construct viable 3-dimensional constructs that can restore structure and function to bone lost from injury or disease. The fundamental tenet of tissue engineering is the triad of scaffolds (extracellular matrix mimics), cells (including mesenchymal stem cells), and biological signaling molecules [5]. Electrospinning is a popular scaffold fabrication strategy that generates 3-dimensional, porous, nanofibrous scaffolds with a large surface area to volume ratio in a physical dimension similar to native tissues. Furthermore, electrospinning offers wide choice in the polymer composition, fiber diameter and alignment, as well as controlling scaffold density and degradation [6,7].

Adult mesenchymal stem cells (MSCs) are promising cell types in tissue engineering because of their high proliferation potential, ability to differentiate into multiple mesodermal lineages (cartilage, fat, and bone), and their ability to be manipulated in culture [8]. MSCs have been isolated from many different adult tissues, including bone marrow, adipose tissue, and peripheral blood [9]. The differentiation capabilities of MSCs into cartilage, fat, and bone from these different sources as well as their immunosuppressive properties have been widely documented [10]. However, each source also is associated with limitations: while bone marrow derived stem cells include low yield, difficulties in cell collection (harvest), aging, and limited proliferative capacity [11], adipose-derived MSCs lose genetic stability over time and are prone to tumor formation [12]. Thus, there is a need to find alternate sources for adult MSCs.

Human gingiva is a recently identified stem cell source that has the advantages of abundant tissue availability, ease of access, and scarless regeneration following harvest. MSCs isolated from human gingiva (gingiva-derived mesenchymal stem cells, GMSCs) are an attractive option in stem cell-based therapies because of their relative abundance, ease of isolation, and rapid ex vivo expansion [13]. In addition to their differentiation and self-renewal properties, GMSCs also possess unique immunomodulatory and anti-inflammatory functions [14]. Adult oral mucosal connective tissue cells, and in particular MSCs, can be easily harvested with little morbidity and possess distinct characteristics, including neural crest origin, multipotent differentiation capacity, fetal-like phenotype that may be utilized to promote tissue regeneration, and fast, scar-free wound healing [15]. Most of the abovementioned studies have used healthy gingiva as the source for isolating the stem cells.

In spite of these advantages, human periodontium (of which gingiva is a component) is one of the most commonly diseased tissues. According to the Centers for Disease Control and Prevention (CDC), half of Americans aged 30 or older have some form of periodontitis [16], a chronic infection initiated by bacteria and modulated by host immune response. A common treatment for chronic periodontitis is surgical resection of diseased tissue and thorough debridement of infected root surface. The resected diseased gingival/periodontal tissue is routinely discarded as part of surgery and is considered a medical waste. It would be of huge clinical interest to explore the potential of GMSCs derived from diseased gingiva and compare it against that of healthy tissue in the context of bone tissue engineering. In fact, one study showed that MSCs isolated from diseased gingival tissue are functionally equivalent to MSC-like cells derived from healthy gingival tissue [13].

Polycaprolactone (PCL) is a synthetic polymer on the US Food and Drug Administration's (FDA's) generally regarded as safe (GRAS) list. It exhibits excellent biocompatibility, complete degradation in vivo, and has been approved for drug delivery and medical devices applications. PCL has been successfully used as micro- and nano-spheres in controlled drug delivery systems [17,18] and in tissue-engineering applications [19].

Our study expands on this idea and evaluated the survival and proliferation of GMSCs from healthy and diseased gingival tissues on electrospun polycaprolactone (e-PCL) scaffolds. We began by evaluating differences in GMSCs derived from diseased and healthy human gingiva using well-established methods. Subsequently, we cultured these different populations on previously characterized e-PCL scaffolds in osteogenic and adipogenic environments to assess their ability to support osteogenesis.

2. Materials and Methods

The study protocol and the consent documents were reviewed and approved by the Institutional Review Board at Virginia Commonwealth University (Approval # HM 14826). All tissue culture reagents were from Invitrogen and culture vessels were from Corning, unless otherwise noted.

2.1. Patient and Sample Identification

We identified three healthy adults (ages 18–24) who presented to the surgery clinic for extraction of complete bony impacted third molars with healthy overlying gingiva. Patients with soft tissue impactions and pericoronitis were excluded. Gingival tissues for harvest had no clinical signs of infection/inflammation. Diseased gingival samples were tissues that were resected from patients with chronic periodontitis undergoing flap and osseous surgery ($n = 9$; ages 32–55). Soft, friable gingival tissue directly overlying the deepest periodontal pocket was identified as the diseased sample and used for this study.

2.2. Sample Collection and Establishment of Primary Clonal Cell Lines

Harvested gingival tissues were collected in cold, sterile saline (4 °C) and were transported to the laboratory within 30 min. Following a brief dip in 70% ethanol, tissues were washed three times in sterile phosphate buffered saline (PBS). The tissues were then finely chopped using dissecting scissors into 1 mm × 1 mm size pieces and then treated with 1 mg/mL dispase in α minimum essential medium (αMEM) for 30 min under gentle agitation at 37 °C. After brief centrifugation, the supernatant was removed and replaced with 0.66 mg/mL collagenase for 1 h at 37 °C. After centrifugation at 800 g × 5 min, the pellet was re-suspended in fresh αMEM containing 10% fetal bovine serum (FBS, Atlanta Biologicals, Flowery Branch, GA, USA) and 1% Antibiotic-Antimycotic (GIBCO) and passed through a 70 μM cell strainer, prior to seeding in 75 cm^2 culture flasks. Flasks were then incubated undisturbed under standard culture conditions (5% CO_2, 100% humidity, and 37 °C) for 7–10 days until confluent. Adherent cells were then isolated by trypsinization and frozen stocks prepared in Bambanker (Wako Chemicals, Richmond, VA, USA) and stored at −80 °C.

Mesenchymal stem cells derived from healthy human gingiva (hGMSCs) are referred to as Healthy samples, #A and #C. MSCs derived from diseased gingival tissues (dGMSCs) are referred to as Diseased samples, #8 and #9.

2.3. Routine Cell Culture

Cells were maintained in complete culture media (CCM) containing AdvanceStem™ Cell Culture Medium (HyClone, Logan, UT, USA) which contains antibiotics under conventional conditions. Media was changed every 3 days until 70–80% confluent, at which time the cells were passaged into fresh flasks, or plates/dishes as needed for assays described below. Cells between p3 and p7 were utilized for all experiments in this study. All experiments for characterizing GMSCs were done in triplicate and repeated for reproducibility.

2.4. Colony Forming Unit (CFU) Assay

Cells were seeded at a density of 1.0×10^2 cells in 10-cm dishes and cultured under conventional conditions ($n = 3$). Non-adherent cells were removed after 2–3 days, and cells were subsequently fed every 3 days for 14 days. Colonies were then washed twice with PBS, incubated for 30 min in 0.5% crystal violet in (100% methanol), and counted.

2.5. Flow Cytometric Analysis

Cells were seeded at a density of 5×10^5 cells in 75 cm^2 flasks and cultured under conventional conditions for 72 h. Subsequently, cells were harvested using 0.05% trypsin- ethylenediaminetetraacetic acid and cell pellets re-suspended in PBS prior to cytometric analysis by a Human MSC analysis kit (BD Biosciences, San Jose, CA, USA). Briefly, cells were incubated with fluorescein (FITC) mouse anti-human CD90, adenomatous polyposis coli (APC) mouse anti-human CD73, PerCP-Cy5.5 mouse anti-human CD105, and a PE-conjugated negative cocktail (anti-human CD34, CD11b, CD19, CD45, and HLA-DR) on ice for 30 min. Cells were then washed in PBS and analyzed using a BD Biosciences Aria II flow cytometer.

2.6. Differentiation Assays

Cells were seeded in 6-well plates at a density of 5×10^4 cells per well and grown to confluence under standard culture conditions. Cells were then maintained in osteogenic medium, adipogenic medium, or control CCM (HyClone), and the media were changed every three days Following 21 days incubation, wells were washed in PBS and cells fixed in 10% neutral-buffered formalin for 1 h at room temperature. Cells were then stained with either 2% Alizarin Red (Sigma-Aldrich, St. Louis, MO, USA) or 0.5% Oil Red O (Sigma-Aldrich) for 20 min at room temperature, and subsequently washed 4 times in PBS prior to microscopy. Controls for negative differentiation (cells grown in CCM stained with either stain) and negative staining (cells grown in osteogenic medium were stained with Oil Red O and cells maintained in adipogenic medium were stained with Alizarin Red) were employed.

2.6.1. Scale Used to Assess Osteogenesis

0 = no mineralization over total well surface; 1 = very few cells producing mineralized nodules visible over total well surface; 2 = very few cells producing mineralized nodules but visible in several microscope fields; 3 = very few cells producing mineralized nodules but visible in majority of microscope fields; 4 = high number of cells producing mineralized nodules; 5 = positive staining of entire well surface; virtually all cells producing mineralized nodules.

2.6.2. Scale Used to Assess Adipogenesis

0 = no adipocytes over total well surface; 1 = very few adipocytes visible over total well surface; 2 = very few adipocytes but visible in several microscope fields; 3 = very few adipocytes but visible in majority of microscope fields; 4 = high number of cells are adipocytes; 5 = positive staining of entire well surface, virtually all cells are adipocytes.

2.7. Cell Proliferation

Cells were seeded at a density of 2.5×10^4 cells/well in 24-well plates and cultured until 70% confluent. Cells were starved in serum-free medium for 24 h prior to the assay and then incubated in CCM for the required times: 72 h, 1 week, 2 weeks, or 3 weeks. At each time point, the cell proliferation assay (MTS:PMS assay) was performed according to manufacturers' instructions (Promega, Madison, WI, USA). Briefly, wells were washed twice in PBS, 100 μL of MTS:PMS reagent was dispensed in each well and was incubated for 1 h. Samples (100 μL) were subsequently taken from each well and absorbance at 490 nm read using a multi-plate reader. Dehydrogenases in metabolically active cells convert MTS into aqueous, soluble formazan, and this can be used to estimate the number of living cells in culture. A standard curve was generated for each cell line and used to normalize the data from alkaline phosphatase assays.

2.8. Measurement of Alkaline Phosphatase (ALP) Activity

The treatment of cells to quantify ALP activity was similar to that previously described [20]. Briefly, cellular monolayers were lysed with 100 μL of 25 mM sodium carbonate (pH 10.3), 0.1% (v/v)

Triton X-100 (Lysis buffer), and treated with 200 μL of 15 mM p-nitrophenyl phosphate (di-tris salt, Sigma, Cream Ridge, NJ, USA) in 250 mM sodium carbonate (pH 10.3) and 1.5 mM $MgCl_2$ (substrate buffer). Lysates were then left under conventional cell culturing conditions for 2 h. Samples were subsequently taken from each well (100 μL), dispensed into a 96-well plate, and the absorbance at 405 nm read using a multiplate reader. A blank of lysis buffer and p-nitrophenyl phosphate was used to normalize data. A series of p-nitrophenol (25–500 μM) prepared in the substrate buffer enabled quantification of product formation.

2.9. Electrospinning PCL Scaffolds

PCL (MW 80,000, Sigma) was dissolved in a binary solvent system of formic acid:acetic acid (1:3) at a concentration of 100 mg/mL. An electrospinning apparatus (EC-DIG, IME Technologies, Geldrop, The Netherlands) was used at optimized process conditions to generate continuous, non-woven fibers that were collected onto 18 mm diameter circular glass coverslips attached to a cylindrical drum mandrel (100 mm diameter) rotating at 100 rpm. After electrospinning, coverslips were removed from the mandrel, dried in a fume hood overnight, and stored in an airtight desiccator until use.

2.10. Scanning Electron Microscopy

Electrospun scaffolds were mounted on aluminum stubs using standard double-sided tape, sputter coated with platinum, and examined at an accelerating voltage of 20 KV using JEOL JSM 5610LV (JEOL, Peabody, MA, USA) scanning electron microscope. Average fiber diameter was calculated from a total of 50 randomly selected fibers from SEM images using Image J (NIH).

2.11. Scaffold Disinfection and Cell Seeding

The scaffolds on coverslips were disinfected with 70% ethanol for 30 min. Following disinfection, scaffolds were washed thrice with PBS (10 min each) and incubated in cell culture media overnight prior to cell seeding. Primary cell lines (2 hGMSC and 2 dGMSC) used between p3 and p7 for all assays. New batches of electrospun PCL scaffolds were generated for each experiment (cell survival, proliferation, and differentiation) and disinfected using the above-mentioned protocol. All experiments were carried out in a 12-well plate format and 100 μL of cell suspension (20,000 cells) was placed in the center of the scaffold within a 10mm diameter sterile, glass-cloning ring (Corning). Cells were allowed to attach to the scaffold for the first 24 h, before more media was added. Media was changed every 3 days for the duration of the experiments. All experiments were performed in triplicate and repeated to ensure reproducibility of results.

2.12. Cell Survival Using Live/Dead Assay

Live/Dead assay (Life Technologies, Carlsbad, CA, USA) was performed to evaluate scaffold cytocompatibility and early survival of GMSCs seeded onto electrospun scaffolds. At 24 h and 7 days, constructs were washed with PBS and Live/Dead stain was added at 5× concentration directly to the cell-scaffold constructs and imaged using a Nikon fluorescence microscope (10×. All images are overlaid with green and red channels indicating live and dead cells, respectively.

2.13. Cell Proliferation Using MTS Assay

Cell proliferation was evaluated using MTS (Promega, Madison, WI, USA) assay. Cells from the four experimental groups were seeded onto scaffolds and a modified MTS assay was performed on days 1 and 7. Scaffolds without cells served as controls. Briefly, cell seeded scaffolds were washed with PBS at designated time points and incubated with MTS reagent for 2 h. The absorbance of the supernatant was read at 490 nm using a BioTek Synergy 2 microplate reader. Experiments were performed in triplicates and repeated to confirm the results.

2.14. GMSC Differentiation on Scaffolds

A total of 20,000 cells were placed in the center of the scaffolds in 100 μL of culture media for 24 h to facilitate cell attachment to scaffolds. At 24 h, the media was replaced with specialized osteogenic or adipogenic media and replaced every 3 days for a total of 3 weeks. Control wells had cells-scaffolds incubate in maintenance media. Osteogenicity was evaluated with staining by Alizarin Red and Alkaline Phosphatase, while Oil Red O verified adipogenicity. The hGMSCs and dGMSCs cultured in maintenance media served as controls. These were also stained for Alizarin red, Oil Red O, and alkaline phosphastase. The layout for the differentiation assay was modified from published protocol [21] and is given in Figure 1.

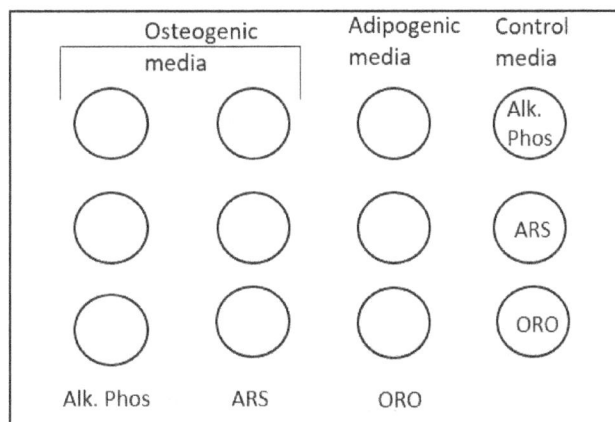

Figure 1. The setup for differentiation assay for GMSCs on electrospun scaffolds on a 12-well plate. Each well had 20,000 GMSCs overlaid on top of identical e-PCL scaffolds. Cells in the first two columns were incubated in osteogenic media; third column had adipogenic media and the fourth column had cells in maintenance (control) media. After three weeks, the cell-scaffold constructs were stained for Alk. Phos (alkaline phosphatase); ARS (alizarin red S) and ORO (Oil Red O). Control wells were also stained as specified. This setup allows for controls of media and stain specificity.

2.15. Statistical Analyses

Values were presented as means and standard deviation, where appropriate. Analyses that compared healthy and diseased tissue with respect to CFU efficiency, alkaline phosphatase production used analysis of variance, and significant differences were described using Tukey's honest significant difference (HSD). All analyses were performed using SAS software (JMP version 10; SAS Institute, Inc., Cary, NC, USA). Cell proliferation assay on scaffolds were tested by a paired t-test for significant difference. Significance set a-priori at 0.05.

3. Results

3.1. Adherent Cells Isolated from Healthy and Diseased Gingiva Showed Characteristics of MSC

The Mesenchymal and Tissue Stem Cell Committee of the International Society for Cellular Therapy proposed these criteria [22] when defining human MSCs:

a.   Must be plastic-adherent under standard culture conditions
b.   Must express CD105, CD73, and CD90. MSCs should not express CD45, CD34, CD14, or CD11b, CD79alpha, or HLA-DR surface molecules and,
c.   Must differentiate into multiple lineages in vitro

3.1.1. Adherent Cells from Both Diseased and Healthy Gingiva Exhibit CFU Activity, Although to Different Degree

Cultures from both healthy and diseased gingival tissue formed adherent colony-forming units on plastic after 14 days incubation (Figure 2A). Qualitatively, the CFUs were less defined and fewer in diseased tissue compared to healthy gingiva. Counting of the individual colonies showed a significant difference in the CFU forming ability between the two types of tissues (Figure 2B). While cells from healthy gingiva patients exhibited an average CFU efficiency of 80.5%, cells from diseased gingiva had a CFU efficiency of 45% (t-test two-tailed p-value: <0.0001).

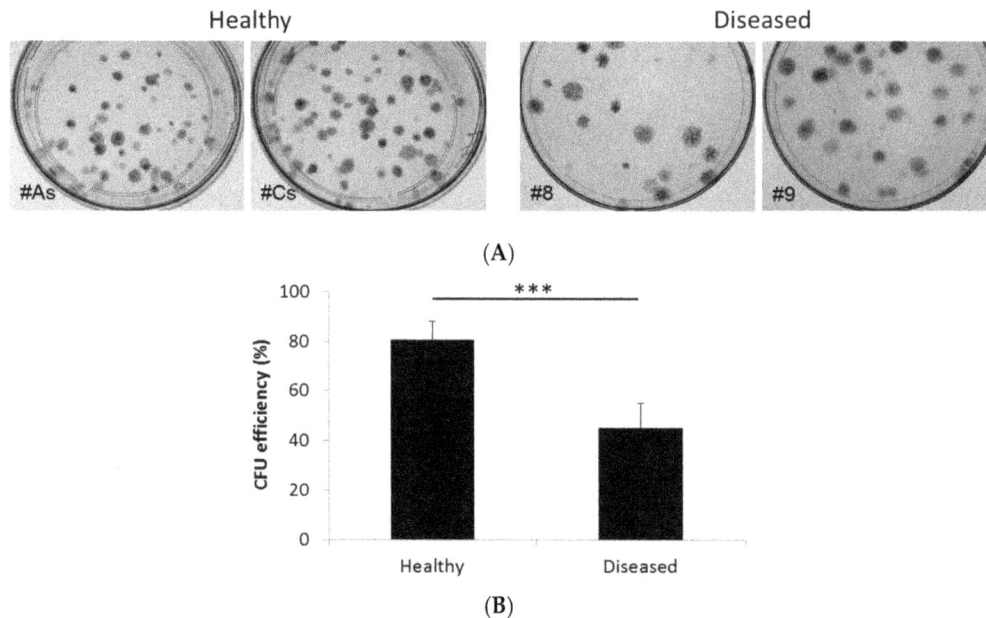

Figure 2. CFU efficiency adherent cells derived from healthy and diseased gingiva after 14 days of culture. (**A**) Representative photographs of the plates after staining with crystal violet; (**B**) Quantification of CFU efficiency shows a significant difference between the two types. *** t-test, two-tailed p-value <0.0001.

3.1.2. Flow Cytometry: Adherent Cells from Both Tissues Express Cell Surface Markers for Adult MSC

Both populations of GMSCs analyzed were positive for the cell surface markers CD90 (Figure 3. top panel, y-axis), CD73 (Figure 3. bottom panel, y-axis), and CD105 (Figure 3. bottom panel, x-axis). Healthy and diseased GMSCs were negative for CD11b, CD19, CD34, CD45, HLA-DR (negative cocktail) (Figure 3. top panel, x-axis). Flow cytometric analysis showed no apparent phenotypic differences between healthy and diseased groups.

3.1.3. Adherent Cells from Healthy Gingiva Showed Higher Osteogenicity, while Cells from Diseased Gingiva Showed Increased Adipogenesis

Both populations of adherent cells underwent osteogenesis (Figure 4. Top panel); however, the cells derived from healthy gingiva had very high levels of mineralization, with a score of 5 (positive staining of entire well surface with virtually all cells producing mineralized nodules) while cells from diseased gingiva showed less apparent mineralization, with an average score of 4 (a high number of cells producing mineralized nodules).

While cells derived from healthy gingiva exhibited very few adipocytes in a few fields (score 1), cells derived from diseased gingiva showed a higher propensity for adipogenesis (score 4) with well-defined lipid droplets (Figure 4. middle panel).

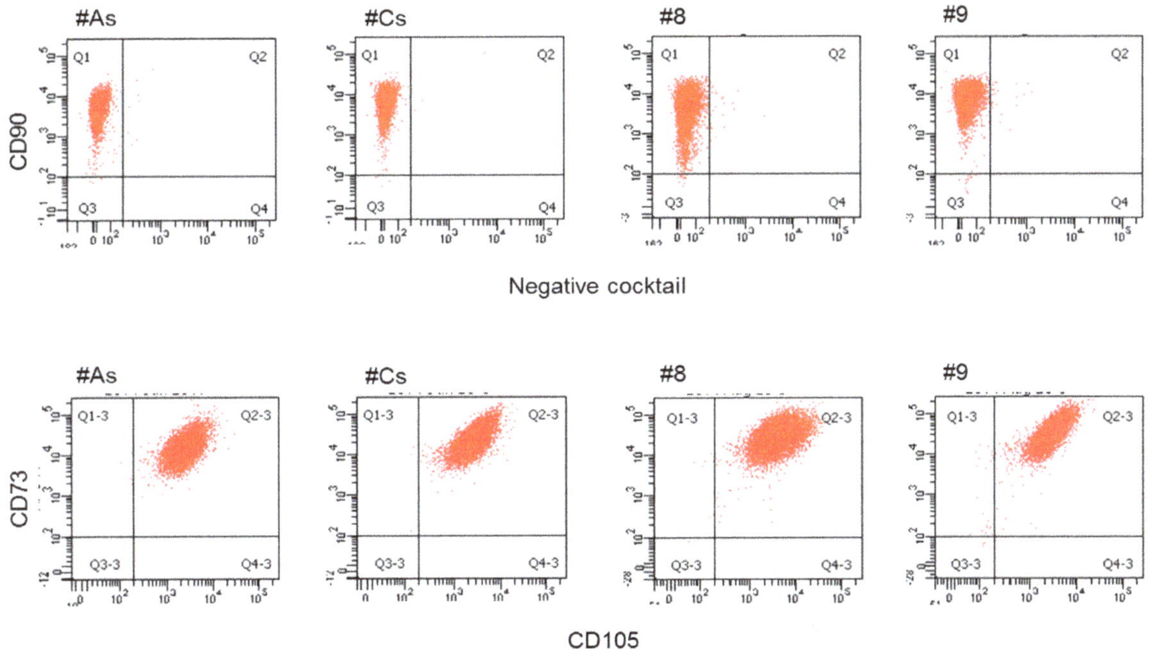

Figure 3. Flow cytometry data of cells derived from healthy (**left panel**) and diseased (**right panel**) gingiva both show identical expression of cell-surface markers characteristic of MSC. Cells were labeled with fluorescent conjugated antibodies and characterized using flow-cytometry.

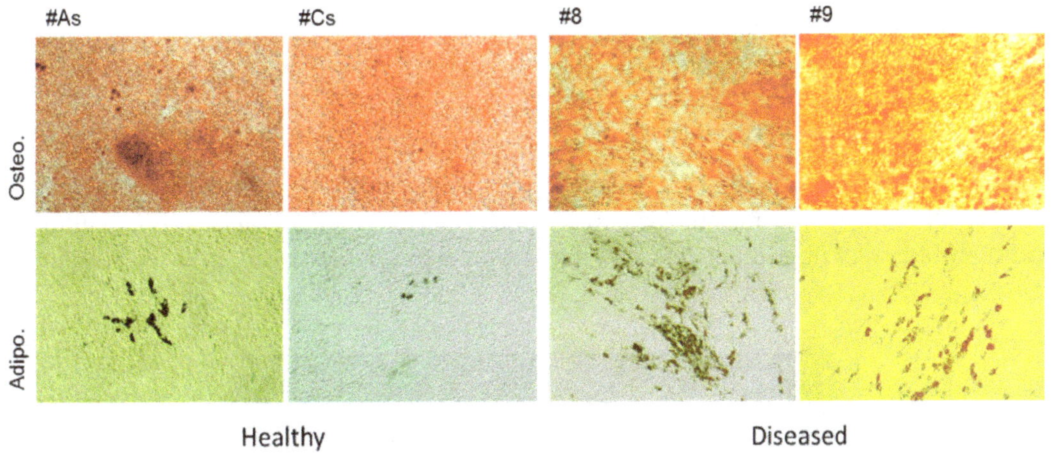

Figure 4. Multi-lineage differentiation of cells derived from healthy (**left**) and diseased (**right**) gingiva over 21 days. The top panel represents cells cultured in osteogenic media and stained with Alizarin red; the middle panel represents MSC cultured in adipogenic media, stained with Oil Red O, while the lower panel represents culture in maintenance media. All photographs were taken at the same magnification to illustrate sparse adipocytic differentiation compared to osteogenic differentiation.

Based on these results, we can safely conclude that the adherent cells derived from healthy and diseased gingiva can be recognized as MSCs. Henceforth, we will call these cell populations as gingiva-derived MSCs (GMSCs), specifically hGMSCs and dGMSCs to refer to healthy or diseased status of the donor site, respectively. It is important to realize that the cell populations differ in their propensity towards osteo- and adipo-genesis, at least on tissue culture plastic.

3.2. Alkaline Phosphatase Is Produced in Higher Levels in Healthy GMSCs (hGMSCs) Compared to Diseased GMSCs (dGMSCs)

Levels of ALP, a marker of osteoblastic differentiation, were assessed by colorimetric assay (Figure 5). In both healthy and diseased groups, levels of ALP increased over time, appearing to level out by 3 weeks. There was significantly less ALP in diseased cells compared with that of healthy cells at 1, 2, and 3 weeks.

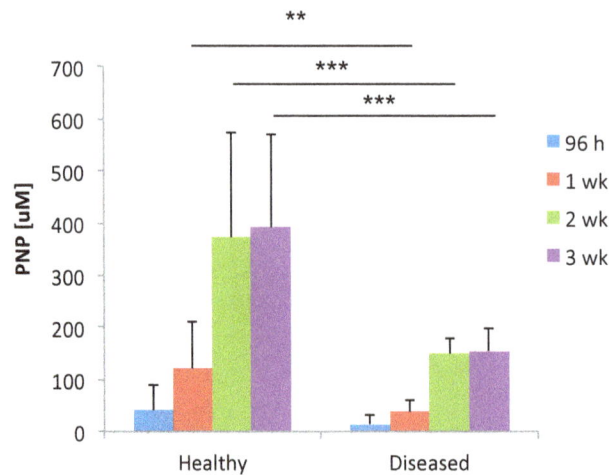

Figure 5. Alkaline phosphatase production in hGMSCs and dGMSCs. Cell number, assessed by MTS assay, was used to normalize the colorimetric signal prior to analysis. Assays were done at 72 h, 1 week, 2 weeks, or 3 weeks. In both groups, levels of ALP increased over time. There was significantly less ALP in the dGMSCs compared to that of the hGMSCs at 1 week, 2 weeks, and 3 weeks. (*t*-test, two-tailed *p*-values: ** <0.01, *** <0.001).

3.3. Electrospun Scaffold Characterization by SEM

Porous, nanofibrous scaffolds were generated after the optimization of electrospinning conditions (voltage: 25 kV; Flow rate: 2 mL/h; Air gap distance: 12.5 cm). SEM analyses (Figure 6) revealed a broad distribution of fiber diameters ranging from 136 to 2200 nm with an average diameter of 0.82 ± 0.52 µm.

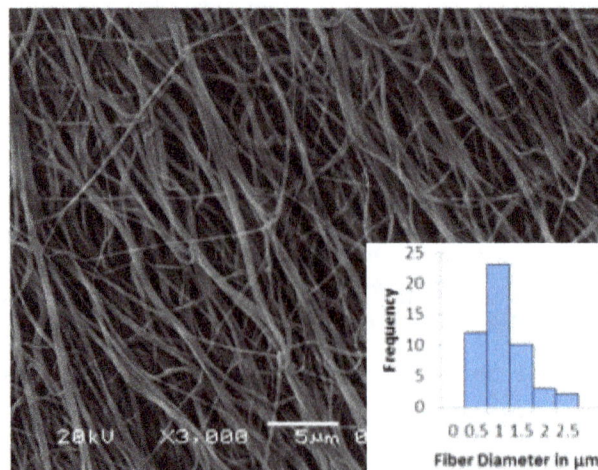

Figure 6. SEM of e-PCL fibers showing porous, fibrous morphology. Inset is the histogram of the fiber diameters of 50 randomly chosen fibers, showing the distribution. The average fiber diameter was 0.82 µm.

3.4. 24-Hours Cell Survival and 7-Days Proliferation on Electrospun Scaffolds

The GMSC from both the healthy and diseased gingiva showed a strong initial attachment and survival at 24 h as well as at 7 days evidenced by the Live/Dead Assay (Figure 7). This was consistent across different primary cell lines. We observed that the cells tended to align along the direction of the fiber, as has been confirmed in previous studies. When analyzing the cell proliferation over 7 days, we were able to see a significant increase in cell numbers compared to that at baseline (initial cell seeding density) irrespective of the cell line or the health status of the gingival tissue (Figure 8). There were, however, differences between individual cell populations due to intrinsic differences in primary cell lines. We did not compare the cell proliferation across different cell populations for the same reason.

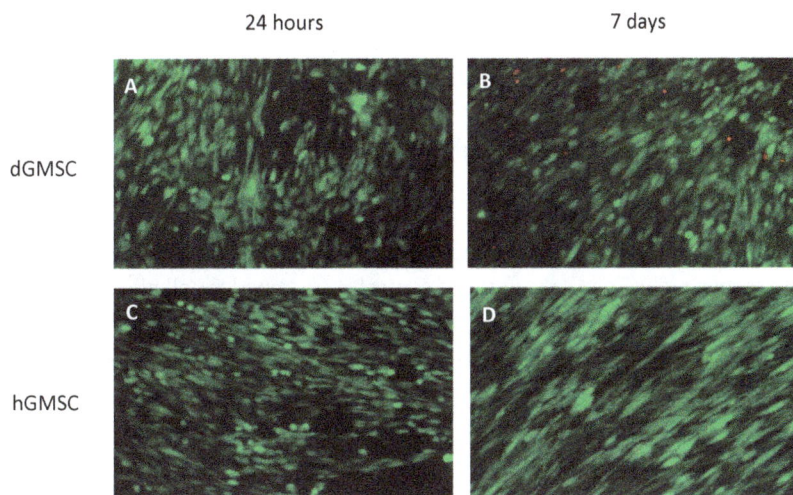

Figure 7. Live/Dead Assay of SEM of GMSC derived from diseased (dGMSC) and healthy (hGMSC) human gingiva at 24 h of culture. (**A,C**) represent the images at 24 h while (**B,D**) are taken at 7 days of culture.

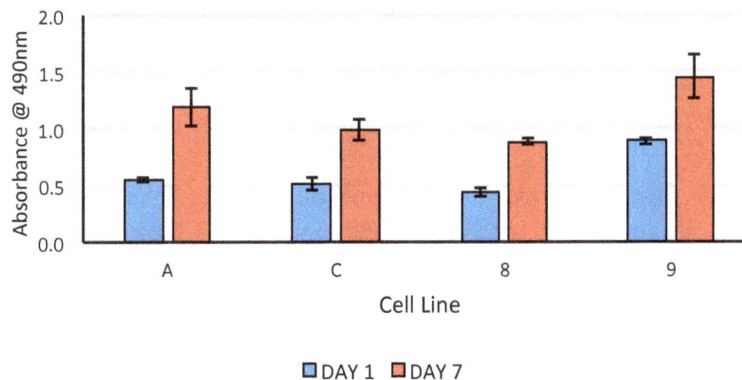

Figure 8. Cell proliferation as assessed by MTS in GMSC cultured on electrospun PCL scaffolds in 24 h and 7 days. There is clear evidence of cell proliferation over time irrespective of the health of donor gingival tissue even though there are differences in the degree of proliferation rates.

3.5. GMSC Differentiation on Electrospun PCL Scaffolds

The behavior of the GMSC on electrospun scaffolds was interesting. Both the hGMSCs and dGMSCs showed a strong tendency for osteogenesis as reflected by the high intensity staining for Alkaline Phosphatase and Alizarin Red (Figure 9). The setup of the experiment allowed us to rule out non-specific staining of cell-laden scaffolds. Control wells (GMSCs cultured in maintenance media) free of staining indicate that electrospun PCL scaffolds supported osteogenic differentiation. This was true of both h- and d-GMSCs.

When the cells were cultured on adipogenic media, the results were not very clear. This was because of non-specific binding of Oil Red O to the cell-scaffold constructs in the control wells.

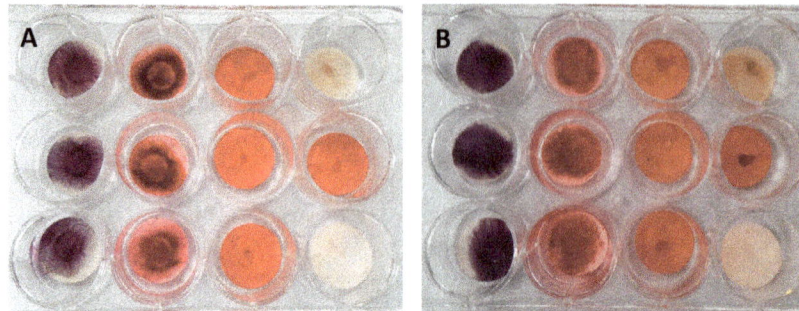

Figure 9. Results of 21-day differentiation assay of GMSCs; **Panel A**—dGMSCs and **Panel B**—hGMSCs. The setup outlined in Figure 1 was followed in these experiments. The first two columns represent cells cultured in osteogenic media while the third was cultured in adipogenic media and the fourth column represents control. Columns 1, 2s and 3 were stained for alkaline phosphatase, Alizarin red, and Oil Red O, respectively. The top well in the control was stained for alkaline phosphatase, middle well for Oil Red O, and bottom well for Alizarin red. As can be seen, **panel B** stains more strongly for osteogenic markers compared to **panel A**, suggesting increased osteogenic potential for hGMSCs. Oil Red O demonstrates non-specific binding of the stain to the cell-scaffold construct.

4. Discussion

Bone grafting is the standard of care for treating larger than critical-sized defects, common after infections, surgery, or trauma. Options include autografts and allografts; the former is associated with additional surgery and increased morbidity while the latter is associated with risk of disease transmission. Given these limitations associated with bone substitutes, tissue engineering has emerged as an alternative approach. Tissue engineering draws on principles of cell biology, developmental biology, and biomaterials to fabricate new structures to replace lost or damaged tissues [23]. The goal of bone engineering is to restore the structure and function of the missing tissue using a combination of scaffolds, cells, and signaling molecules. Upon implantation, these cell-scaffold constructs are expected to induce host cells to lay down new bone and eventually turn over and integrate with the host bone [7].

MSCs are attractive in scaffold-based tissue engineering because of their unique properties: clonal expansion and multi-lineage differentiation. These properties allow for harvesting of relatively small numbers of MSCs, expanding them in the lab, and directing differentiation by exposing them to specific biochemical signals [24–26]. MSC-derived cells can be seeded in biocompatible scaffolds, which can be shaped into the anatomical structure and then surgically implanted to heal the defect [27].

Even though bone marrow and adipose tissue have been the preferred sources for MSCs, because of ease of harvest and potential for autologous use, they have several disadvantages. First, bone marrow-derived stem cells BMSCs are associated with extremely low yields (0.001% to 0.01%) that require large quantities of bone marrow aspiration with attendant donor site morbidity [28]; both BMSCs and adipose-derived stem cells ADSCs are present as mixed cell populations and require several passages in culture to purify and enrich the MSCs [29,30]. Another limitation of BMSC is the functional heterogeneity amongst seemingly pure preparations as determined by clonal expansion assays [31].

Gingiva refers to specific oral tissue that forms a soft tissue cuff around the tooth and forms a first line of defense against physical, chemical, and biological assault. It is a part of the oral mucosal immunity. The intrinsic wound healing properties of gingiva—reduced inflammation, rapid re-epithelialization, and fetal-like scarless healing—are driven primarily by the mesenchymal stem cells that reside in the lamina propria of the gingiva.

GMSCs lend themselves to rapid in vitro expansion owing to shorter population doubling and relative homogeneous cell population. It has been shown that adherent cells isolated from a small piece of gingival tissue (~2 × 2 mm^2) usually reached confluence (~1–2 × 10^6 cells) after culture for 10~14 days. GMSCs were shown to be similar to BMSCs in terms of CFU efficiency (4–6%), population doublings, and ability to maintain early phenotypes at passage six. GMSCs were found to be superior to BMSCs in terms of proliferation rate (40 h vs. 80 h), maintenance of normal karyotype and telomerase activity in long-term cultures, and not being tumorigenic [32,33].

Within our study, we found differences in both of these aspects between GMSCs derived from healthy and diseased gingiva: hGMSCs had higher CFU and more osteogenic tendency compared to those of dGMSCs (which demonstrated more adipogenicity). It is interesting to explore the reasons for these observed differences. While this could be attributed to primary cell lines derived from innately different donors, role of disease status should merit consideration. This is because of increased understanding of the role of epigenetics in the development and progression of periodontal disease [34–36]. Recently, Xu et al. reported that the MSC fate was also affected by epigenetic factors [37]. In a comparative study of MSCs derived from bone marrow and adipose tissue from the same patient, they found hypomethylation of Runx2 promoter (an osteogenesis specific transcription factor) in BMSCs as well as of the PPARγ promoter (an adipogenesis specific transcription factor) in ADSC. Since the degree of methylation is inversely related to gene expression, these epigenetic differences accounted for differential behavior of these cell lines. We hypothesize that differences in behavior between GMSCs derived from diseased and healthy gingiva could be attributed to epigenetic changes associated with chronic periodontitis. If DNA methylation studies verify these changes, it could permit resetting of the epigenetic memory of GMSCs derived from diseased gingiva towards osteoblastic lineage resulting in favorable bone regeneration. Since periodontitis can be localized to one to few teeth, it will be interesting to compare the behavior of GMSCs harvested from healthy and diseased sites within the same donor.

In this study, we demonstrate that GMSCs can be isolated from healthy and periodontally diseased tissues. MSCs from each of these sources share key characteristics such as fibroblast-like appearance, adherence to plastic, expression of certain cell surface antigens (CD73, CD90, CD105), and lack of expression of others (CD11b, CD14, CD19, CD34, CD45, CD78) [9]. Our results indicated no difference in the expression of cell surface markers between diseased and healthy cells, as confirmed by prior studies [13,38]. Whilst the above criteria fit well for cells isolated from gingival tissues, it is important to realize that satisfying the above mentioned requirements cannot reliably confirm MSCs from other cell types such as fibroblasts [15]. The true test of MSCs is in fact their multi-lineage differentiation potential under appropriate conditions, whereas fibroblasts do not differentiate into osteoblasts and adipocytes [39].

The potential of mesenchymal stem cells in therapeutic interventions is already being explored in a variety of clinical scenarios that have limited treatment options. Preclinical and clinical data support the use of these cells in plastic surgery, orthopedics, myocardial infarction, graft vs. host disease, and autoimmune diseases [40]. It is important to realize that most of the clinical data comes from BMSCs or ADSCs. However, with the potential of MSCs beginning to be understood, it is important to explore alternate sources of adult MSCs. In this sense, GMSCs represents a readily available cell source that can be potentially used in alveolar bone regeneration, the most challenging situation in chronic periodontitis. Recent studies showing GMSCs can be successful in generating functional constructs with alginate microspheres [41–43]. By demonstrating the viability of 3D tissue constructs using GMSCs and e-PCL scaffolds, as well as their ability to support osteogenesis, our study expands the application of this novel cell source.

5. Strengths, Limitations, and Future Research Directions

a. Generalizability of results: One of the strengths of the study is that the GMSCs evaluated are primary cell sources that are not modified (transformed) in any way. Using finite lines for MSCs is challenging because they are slow and hard to establish, impose limits on the passage numbers that can be used for differentiation, and are expensive to maintain (with specialized media). However, these are the biologically more appropriate cells to study because of their relevance in clinical translation. The results of this study should be interpreted with caution because the data is derived from four primary cell sources that have inherent variability. Future efforts could involve establishing a gingival tissue repository and its distribution to laboratories for evaluating the behavior of these cells. As mentioned above, another option would be to study MSCs derived from healthy and diseased gingiva from the same patient.

b. Data quality: The current study used qualitative measures (staining for Alizarin red, Oil Red O, Alkaline phosphatase) to study GMSC differentiation. We adopted this approach because staining can provide early proof-of-concept information about the behavior of GSMCs on electrospun scaffolds. Careful experimental setup with proper controls allowed us to visually verify the differentiation of GMSCs. However, lack of quantitative measures is a limitation and future studies with robust quantifiable data will improve the strength of conclusions. More detailed experiments involving RT-PCR and Western blotting can provide insights into the mechanisms controlled the fate of MSCs.

c. A minor limitation of this study could be not validating the GMSCs for their chondrogenic lineage. Detailed characterization of GMSCs by previous research groups has established that GMSCs do possess multi-lineage potential and, hence, these tests were not repeated in our study. Since our primary goal was to evaluate effectiveness of GMSCs in bone engineering, we felt that doing chondrogenic assays would not add significantly to the scientific merit of the study.

6. Conclusions

Bone tissue engineering involves the triad of scaffolds, cells, and signaling molecules. While electrospun PCL has remained a popular choice in bone engineering, gingiva has been recently identified as a tissue source for MSCs. This study extends the scope of GMSCs in bone engineering by investigating the potential of MSCs derived from healthy and diseased gingival tissues on electrospun matrices. Results from 2D culture clearly indicate superior osteogenicity of hGMSCs while the 3D culture on PCL scaffolds showed comparable osteogenicity in dGMSCs and hGMSCs. Future studies should explore the role of electrospun matrices in directing differentiation, compared to tissue culture plastic substrate, particularly in the context of cell signaling. Overall, the study results indicate that diseased gingiva could serve as a viable source of GMSCs given the fact that these tissues are routinely discarded during surgery. Evaluation of osteogenicity of GMSC-laden electrospun scaffolds in appropriate animal models can provide critical insights into their potential for bone regeneration. Such information would be important prior to clinical application in humans.

Acknowledgments: The project was supported by CTSA (UL1TR000058) from the National Center for Advancing Translational Sciences and the CCTR Endowment Fund of Virginia Commonwealth University. The contents are solely the responsibility of the authors and do not necessarily represent official views of the National Center for Advancing Translational Sciences or the National Institutes of Health. The authors acknowledge Ramez Fahmy for his help with the manuscript.

Author Contributions: Catherine Jauregui and Parthasarathy Madurantakam conceived and designed the experiments. Catherine Jauregui was primarily responsible for establishing the primary cell lines for GMSCs as well as their detailed characterization. Suyog Yoganarasimha generated electrospun scaffolds and performed all the GMSC experiments on scaffolds. Parthasarathy Madurantakam wrote the paper.

Conflicts of Interest: The authors declare no conflict of interest. The funding sponsors had no role in the design of the study; in the collection, analyses, or interpretation of data; in the writing of the manuscript, and in the decision to publish the results.

References

1. Campana, V.; Milano, G.; Pagano, E.; Barba, M.; Cicione, C.; Salonna, G.; Lattanzi, W.; Logroscino, G. Bone substitutes in orthopaedic surgery: From basic science to clinical practice. *J. Mater. Sci. Mater. Med.* **2014**, *25*, 2445–2461. [CrossRef] [PubMed]

2. Logeart-Avramoglou, D.; Anagnostou, F.; Bizios, R.; Petite, H. Engineering bone: Challenges and obstacles. *J. Cell. Mol. Med.* **2005**, *9*, 72–84. [CrossRef] [PubMed]

3. Giannoudis, P.V.; Dinopoulos, H.; Tsiridis, E. Bone substitutes: An update. *Injury* **2005**, *36*, S20–S27. [CrossRef] [PubMed]

4. Laurencin, C.; Khan, Y.; El-Amin, S.F. Bone graft substitutes. *Expert Rev. Med. Dev.* **2006**, *3*, 49–57. [CrossRef] [PubMed]

5. Yousefi, A.; James, P.F.; Akbarzadeh, R.; Subramanian, A.; Flavin, C.; Oudadesse, H. Prospect of stem cells in bone tissue engineering: A review. *Stem Cells Int.* **2016**, *2016*, 6180487. [CrossRef] [PubMed]

6. Madurantakam, P.A.; Cost, C.P.; Simpson, D.G.; Bowlin, G.L. Science of nanofibrous scaffold fabrication: Strategies for next generation tissue-engineering scaffolds. *Nanomedicine* **2009**, *4*, 193–206. [CrossRef] [PubMed]

7. Sell, S.; Barnes, C.; Smith, M.; McClure, M.; Madurantakam, P.; Grant, J.; McManus, M.; Bowlin, G. Extracellular matrix regenerated: Tissue engineering via electrospun biomimetic nanofibers. *Polym. Int.* **2007**, *56*, 1349–1360. [CrossRef]

8. Mitrano, T.I.; Grob, M.S.; Carrión, F.; Nova-Lamperti, E.; Luz, P.A.; Fierro, F.S.; Quintero, A.; Chaparro, A.; Sanz, A. Culture and characterization of mesenchymal stem cells from human gingival tissue. *J. Periodontol.* **2010**, *81*, 917–925. [CrossRef] [PubMed]

9. Elahi, K.C.; Klein, G.; Avci-Adali, M.; Sievert, K.D.; Macneil, S.; Aicher, W.K. Human mesenchymal stromal cells from different sources diverge in their expression of cell surface proteins and display distinct differentiation patterns. *Stem Cells Int.* **2016**, *2016*. [CrossRef] [PubMed]

10. Trento, C.; Dazzi, F. Mesenchymal stem cells and innate tolerance: Biology and clinical applications. *Swiss Med. Wkly.* **2010**, *140*, w13121. [CrossRef] [PubMed]

11. Wang, F.; Yu, M.; Yan, X.; Wen, Y.; Zeng, Q.; Yue, W.; Yang, P.; Pei, X. Gingiva-Derived Mesenchymal Stem Cell-Mediated Therapeutic Approach for Bone Tissue Regeneration. *Stem Cells Dev.* **2011**, *20*, 2093–2102. [CrossRef] [PubMed]

12. Huang, G.T.-J.; Gronthos, S.; Shi, S. Mesenchymal stem cells derived from dental tissues vs. those from other sources: Their biology and role in regenerative medicine. *J. Dent. Res.* **2009**, *88*, 792–806. [CrossRef] [PubMed]

13. Ge, S.; Mrozik, K.M.; Menicanin, D.; Gronthos, S.; Bartold, P.M. Isolation and characterization of mesenchymal stem cell-like cells from healthy and inflamed gingival tissue: Potential use for clinical therapy. *Regen. Med.* **2012**, *7*, 819–832. [CrossRef] [PubMed]

14. Zhang, Q.Z.; Su, W.R.; Shi, S.H.; Wilder-Smith, P.; Xiang, A.P.; Wong, A.; Nguyen, A.L.; Kwon, C.W.; Le, A.D. Human gingiva-derived mesenchymal stem cells elicit polarization of M2 macrophages and enhance cutaneous wound healing. *Stem Cells* **2010**, *28*, 1856–1868. [CrossRef] [PubMed]

15. Fournier, B.P.J.; Larjava, H.; Häkkinen, L. Gingiva as a source of stem cells with therapeutic potential. *Stem Cells Dev.* **2013**, *22*, 3157–3177. [CrossRef] [PubMed]

16. Eke, P.I.; Dye, B.A.; Wei, L.; Thornton-Evans, G.O.; Genco, R.J. Prevalence of periodontitis in adults in the United States: 2009 and 2010. *J. Dent. Res.* **2012**, *91*, 914–920. [CrossRef] [PubMed]

17. Lemmouchi, Y.; Schacht, E.; Kageruka, P.; De Deken, R.; Diarra, B.; Diall, O.; Geerts, S. Biodegradable polyesters for controlled release of trypanocidal drugs: In vitro and in vivo studies. *Biomaterials* **1998**, *19*, 1827–1837. [CrossRef]

18. Zhang, S.; Uludağ, H. Nanoparticulate systems for growth factor delivery. *Pharm. Res.* **2009**, *26*, 1561–1580. [CrossRef] [PubMed]

19. Dash, T.K.; Konkimalla, V.B. Poly-ε-caprolactone based formulations for drug delivery and tissue engineering: A review. *J. Control. Release* **2011**, *158*, 15–33. [CrossRef] [PubMed]

20. Coss, D.; Yang, L.; Kuo, C.B.; Xu, X.; Luben, R.A.; Walker, A.M. Effects of prolactin on osteoblast alkaline phosphatase and bone formation in the developing rat. *Am. J. Physiol. Endocrinol. Metab.* **2000**, *279*, E1216–E1225. [CrossRef] [PubMed]

21. Reger, R.; Tucker, A.; Wolfe, M. *Mesenchymal Stem Cells*; Prockop, D.J., Phinney, D.G., Bunnell, B.A., Eds.; Humana Press: New York, NY, USA, 2008; pp. 93–109.

22. Dominici, M.; Le Blanc, K.; Mueller, I.; Slaper-Cortenbach, I.; Marini, F.C.; Krause, D.S.; Deans, R.J.; Keating, A.; Prockop, D.J.; Horwitz, E.M. Minimal criteria for defining multipotent mesenchymal stromal cells. The International Society for Cellular Therapy position statement. *Cytotherapy* **2006**, *8*, 315–317. [CrossRef] [PubMed]

23. Han, J.; Menicanin, D.; Gronthos, S.; Bartold, P.M. Stem cells, tissue engineering and periodontal regeneration. *Aust. Dent. J.* **2014**, *59* (Suppl. S1), 117–130. [CrossRef] [PubMed]

24. Marion, N.W.; Mao, J.J. Mesenchymal stem cells and tissue engineering. *Methods Enzymol.* **2006**, *420*, 339–361. [PubMed]

25. Tuan, R.S.; Boland, G.; Tuli, R. Adult mesenchymal stem cells and cell-based tissue engineering. *Arthritis Res. Ther.* **2003**, *5*, 32–45. [CrossRef] [PubMed]

26. Tae, S.-K.; Lee, S.-H.; Park, J.-S.; Im, G.-I. Mesenchymal stem cells for tissue engineering and regenerative medicine. *Biomed. Mater.* **2006**, *1*, 63–71. [CrossRef] [PubMed]

27. Parekkadan, B.; Milwid, J.M. Mesenchymal stem cells as therapeutics. *Annu. Rev. Biomed. Eng.* **2010**, *12*, 87–117. [CrossRef] [PubMed]

28. Bonab, M.M.; Alimoghaddam, K.; Talebian, F.; Ghaffari, S.H.; Ghavamzadeh, A.; Nikbin, B. Aging of mesenchymal stem cell in vitro. *BMC Cell Biol.* **2006**, *7*, 14. [CrossRef] [PubMed]

29. Izadpanah, R.; Trygg, C.; Patel, B.; Kriedt, C.; Dufour, J.; Gimble, J.M.; Bunnell, B.A. Biologic properties of mesenchymal stem cells derived from bone marrow and adipose tissue. *J. Cell. Biochem.* **2006**, *99*, 1285–1297. [CrossRef] [PubMed]

30. Requicha, J.F.; Viegas, C.A.; Albuquerque, C.M.; Azevedo, J.M.; Reis, R.L.; Gomes, M.E. Effect of anatomical origin and cell passage number on the stemness and osteogenic differentiation potential of canine adipose-derived stem cells. *Stem Cell Rev. Rep.* **2012**, *8*, 1211–1222. [CrossRef] [PubMed]

31. Russell, K.C.; Lacey, M.R.; Gilliam, J.K.; Tucker, H.A.; Phinney, D.G.; O'Connor, K.C. Clonal analysis of the proliferation potential of human bone marrow mesenchymal stem cells as a function of potency. *Biotechnol. Bioeng.* **2011**, *108*, 2716–2726. [CrossRef] [PubMed]

32. Zhang, Q.; Shi, S.; Liu, Y.; Uyanne, J.; Shi, Y.; Shi, S.; Le, A.D. Mesenchymal stem cells derived from human gingiva are capable of immunomodulatory functions and ameliorate inflammation-related tissue destruction in experimental colitis. *J. Immunol.* **2009**, *183*, 7787–7798. [CrossRef] [PubMed]

33. Tomar, G.B.; Srivastava, R.K.; Gupta, N.; Barhanpurkar, A.P.; Pote, S.T.; Jhaveri, H.M.; Mishra, G.C.; Wani, M.R. Human gingiva-derived mesenchymal stem cells are superior to bone marrow-derived mesenchymal stem cells for cell therapy in regenerative medicine. *Biochem. Biophys. Res. Commun.* **2010**, *393*, 377–383. [CrossRef] [PubMed]

34. Larsson, L.; Castilho, R.M.; Giannobile, W.V. Epigenetics and Its Role in Periodontal Diseases: A State-of-the-Art Review. *J. Periodontol.* **2015**, *86*, 556–568. [CrossRef] [PubMed]

35. Lavu, V.; Venkatesan, V.; Rao, S.R. The epigenetic paradigm in periodontitis pathogenesis. *J. Indian Soc. Periodontol.* **2015**, *19*, 142–149. [PubMed]

36. Cho, Y.-D.; Kim, P.-J.; Kim, H.-G.; Seol, Y.-J.; Lee, Y.-M.; Ku, Y.; Rhyu, I.-C.; Ryoo, H.-M. Transcriptomics and methylomics in chronic periodontitis with tobacco use: A pilot study. *Clin. Epigenet.* **2017**, *9*, 81. [CrossRef] [PubMed]

37. Xu, L.; Liu, Y.; Sun, Y.; Wang, B.; Xiong, Y.; Lin, W.; Wei, Q.; Wang, H.; He, W.; Wang, B.; et al. Tissue source determines the differentiation potentials of mesenchymal stem cells: A comparative study of human mesenchymal stem cells from bone marrow and adipose tissue. *Stem Cell Res. Ther.* **2017**, *8*, 275. [CrossRef] [PubMed]

38. Alongi, D.J.; Yamaza, T.; Song, Y.; Fouad, A.F.; Romberg, E.E.; Shi, S.; Tuan, R.S.; Huang, G.T.-J. Stem/progenitor cells from inflamed human dental pulp retain tissue regeneration potential. *Regen. Med.* **2010**, *5*, 617–631. [CrossRef] [PubMed]

39. Kern, S.; Eichler, H.; Stoeve, J.; Klüter, H.; Bieback, K. Comparative analysis of mesenchymal stem cells from bone marrow, umbilical cord blood, or adipose tissue. *Stem Cells* **2006**, *24*, 1294–1301. [CrossRef] [PubMed]

40. Strioga, M.; Viswanathan, S.; Darinskas, A.; Slaby, O.; Michalek, J. Same or not the same? Comparison of adipose tissue-derived versus bone marrow-derived mesenchymal stem and stromal cells. *Stem Cells Dev.* **2012**, *21*, 2724–2752. [CrossRef] [PubMed]

41. Moshaverinia, A.; Xu, X.; Chen, C.; Ansari, S.; Zadeh, H.H.; Snead, M.L.; Shi, S. Application of stem cells derived from the periodontal ligament or gingival tissue sources for tendon tissue regeneration. *Biomaterials* **2014**, *35*, 2642–2650. [CrossRef] [PubMed]
42. Moshaverinia, A.; Chen, C.; Xu, X.; Akiyama, K.; Ansari, S.; Zadeh, H.H.; Shi, S. Bone regeneration potential of stem cells derived from periodontal ligament or gingival tissue sources encapsulated in rgd-modified alginate scaffold. *Tissue Eng. Part A* **2013**, *20*, 611–621. [CrossRef] [PubMed]
43. Ansari, S.; Chen, C.; Xu, X.; Annabi, N.; Zadeh, H.H.; Wu, B.M.; Khademhosseini, A.; Shi, S.; Moshaverinia, A. Muscle tissue engineering using gingival mesenchymal stem cells encapsulated in alginate hydrogels containing multiple growth factors. *Ann. Biomed. Eng.* **2016**, *44*, 1908–1920. [CrossRef] [PubMed]

7

Honey-Based Templates in Wound Healing and Tissue Engineering

Benjamin A. Minden-Birkenmaier and Gary L. Bowlin * 🆔

Department of Biomedical Engineering, University of Memphis, 3806 Norriswood Ave., Memphis, TN 38152, USA; bmndnbrk@memphis.edu
* Correspondence: glbowlin@memphis.edu

Abstract: Over the past few decades, there has been a resurgence in the clinical use of honey as a topical wound treatment. A plethora of in vitro and in vivo evidence supports this resurgence, demonstrating that honey debrides wounds, kills bacteria, penetrates biofilm, lowers wound pH, reduces chronic inflammation, and promotes fibroblast infiltration, among other beneficial qualities. Given these results, it is clear that honey has a potential role in the field of tissue engineering and regeneration. Researchers have incorporated honey into tissue engineering templates, including electrospun meshes, cryogels, and hydrogels, with varying degrees of success. This review details the current state of the field, including challenges which have yet to be overcome, and makes recommendations for the direction of future research in order to develop effective tissue regeneration therapies.

Keywords: tissue engineering; tissue regeneration; electrospinning; cryogel; hydrogel; Manuka honey; chronic wound; Inflammation

1. Introduction

Honey has been used as a wound treatment by indigenous cultures around the globe for thousands of years. Archeological findings and early written works indicate that wounds were treated with honey by the ancient Egyptians, Greeks, and Romans, among others [1]. With the advent of antibiotics in the 1940s, honey fell out of favor as a wound treatment [2]. However, with the increasing prevalence of antibiotic-resistant bacteria, as well as new in vitro and in vivo data supporting honey's effectiveness in treating wounds and as a natural broad-band antibacterial agent, it has recently made a comeback in clinical medicine. Additionally, honey's ability to aid in situ cellularization and regeneration of implanted acellular tissue-engineered structures indicates its potential as a tissue engineering additive.

Honey is a natural substance produced by a variety of honeybee species around the world. First, the bees collect nectar from flowering foliage. This nectar is processed in an internal pouch called the crop, where a variety of enzymes break down sugars. The resulting solution is regurgitated by the bees into honeycomb within their hives, where liquid evaporation is enhanced by air currents created by the fanning of bee wings. The product is a highly concentrated viscous solution of floral sugars and proteins, enzymes, and amino acids derived from the bee crops [3]. These sugars are primarily fructose and glucose, with smaller amounts of maltose, sucrose, and isomaltose, and comprise approximately 80% of honey components, with water comprising <18% [4–6]. Glucose oxidase from the bee crop slowly breaks down glucose into gluconic acid, which lowers the pH of honey, and hydrogen peroxide, which helps kill bacteria [7]. In a wound site, the lower pH of honey (3.5–4) reduces protease activity, increases oxygen release from hemoglobin, and stimulates the activity of macrophages and fibroblasts, while the hydrogen peroxide content sterilizes the wound and stimulates vascular endothelial growth factor (VEGF) production [7]. Invertase, another enzyme from the bee crop, slowly divides sucrose into glucose and fructose, increasing the strength of the osmotic potential. In addition, flavenoids derived from the floral nectar sources neutralize free radicals created by the hydrogen peroxide [7]. Bees can also make honey from honeydews, a loose term which includes plant secretions and plant-sucking insect excretions [8]. These honeydew honeys have lower glucose and fructose content and higher levels of oligosaccharides [9]. Some research has also shown that honeydew honeys contain higher levels of phenolic contents, which have been shown to reduce MMP-9 expression in keratinocytes [10,11]. These findings indicate that honeydew honey may be a beneficial future focus of wound healing and tissue engineering research. However, as most honey used in tissue engineering research thus far originates with floral nectar, this review will focus on these nectar-based honeys.

Although some honey varieties have been shown to have beneficial effects in a wound site, most modern research has focused on a particular variety produced in New Zealand from the nectar of the *Leptospermum Scopartum* shrub, called Manuka honey. This honey contains the components of other honey varieties, but its unique component, methylglyoxal, acts as an additional antibacterial agent [12–15]. Several companies collect, pool, filter, and sterilize Manuka honey for clinical use, including ManukaGuard (located in New Zealand) and Medihoney (a subsidiary of Derma Sciences, Princeton, NJ, based in the United States). This collection and pooling of the honey helps limit batch-to-batch variability between hive locations and times of the year, while the filtration removes wax, dirt, and pollen particulates from the honey to reduce the potential to cause an allergic reaction. Although honey has been demonstrated to have antibacterial properties, these products are still sterilized via gamma irradiation or pasteurization to doubly ensure that no live bacteria or spores are present.

Honey performs several other functions as a wound covering. As a viscous fluid, its thick consistency forms a barrier between the wound and the external environment, protecting against bacteria and keeping the wound hydrated [16]. Its high concentration of sugars and other solutes creates a strong osmotic gradient that pulls fluid up through the subdermal tissue [17]. The water activity of honey, a measure of its osmotic potential, has been reported to range from 0.53 to 0.64 aw (activity of water, unitless) [18,19]. For reference, the water activity of distilled water is 1 aw, and substances with a lower water activity create a higher osmotic potential with water flowing from areas of high to low water activity. Water activity values below 0.91 aw inhibit bacterial growth [20]. The low water activity of honey causes fluid flow which flushes bacteria, debris, slough, and necrotic tissue out of the wound, and carries nutrients and oxygen from the deep tissue into the wound area. Additionally, the low pH of the honey increases tissue oxygenation, while the flavonoids and aromatic acids scavenge free radicals, preventing tissue damage and controlling inflammation [7,21]. The high sugar content of honey also provides an additional source of glucose for proliferating cellular components (i.e., fibroblasts and endothelial cells) in the area [16].

In addition to these other attributes, honey also has multiple antibacterial effects. These effects include the inhibition of bacterial growth as well as the direct killing of bacteria [22–25]. The osmotic potential of the honey crenates bacteria at the top of the wound, destroying them [15,26]. Although this osmotic potential was thought by some groups to be the main source of honey's antibacterial activity, studies have shown that honey maintains its antibacterial activity even when diluted by wound exudate [27,28]. An in vitro study by Cooper et al. found that dilutions of honey by factors of 7 to 14 maintain their bacterial inhibition, long past the dilution point where the osmotic potential of the solution would cease to be bactericidal [29]. It should be noted that this study lacked mechanistic controls of concentrated sugar solutions with equal water activity to honey, weakening their results. This lack of mechanistic controls is a general issue for some studies of honey's antibacterial effects, making it difficult to measure the contribution of each attribute to the overall antibacterial nature of honey. Nevertheless, honey has been shown to contain other components which contribute to its bactericidal effect. As mentioned above, honey contains hydrogen peroxide, with levels in the range of 12–72 µg/mL depending on the dilution of the honey and variety of the honey (it should be noted that some varieties of honey do not contain measurable levels of hydrogen peroxide, discussed further below) [22]. Hydrogen peroxide easily gives up one of its oxygen atoms to the surrounding environment, creating a free radical that causes oxidative damage to bacterial cell walls. Additionally, the presence of bee defensin-1 has been shown in some honey varieties, although the levels of this protein vary based on hive location [30,31]. Like other defensins, bee defensin-1 permeabilizes bacteria and inhibits their RNA, DNA, and protein synthesis [32]. As mentioned above, the glucose oxidase content of honey lowers its pH, which can also kill some bacteria [30]. Manuka honey, in particular, contains methylglyoxal, a compound which has been shown to damage bacteria flagella and thus limit their mobility and ability to adhere to surfaces [33]. However, Manuka honey has been shown to lack the defensin-1 content of other honey varieties, possibly due to decreased secretion by bees during the formation process [30,34]. The methylglyoxal content of Manuka honey has also been shown to inactivate defensin-1 when it is added to the honey, eliminating its contribution to the antibacterial effect [34]. Additionally, methylglyoxal crosslinks glucose oxidase, destroying its enzymatic activity and eliminating the hydrogen peroxide content in Manuka honey [35]. These studies utilized mechanistic controls of methyglyoxal alone or added to non-methylglyoxal-containing honeys to isolate the effect of the methylglyoxal on the other honey components. In at least one other study, however, the methylglyoxal component of honey has been neutralized to determine whether it is the sole contributor to the antibacterial effects of Manuka honey. Although this methylglyoxal-neutralized honey had decreased activity against *Staphylococcus aureus* and *Bacillus subtilis*, it did not have reduced activity against *Escherichia coli* or *Pseudomonas aeruginosa.* As such, other components besides methylgyoxal must contribute to the antibacterial activity of Manuka honey [22]. Manuka honey has been shown to be especially useful against antibiotic-resistant bacteria [12,36]. The many functions of Manuka honey thus not only clear wound debris, maintain hydration, control inflammation, and stimulate healing, but also sterilize the wound. Although a large number of groups researching honey as a wound treatment or in tissue-engineered templates have focused on Manuka honey, there are some notable studies which have examined other honey varieties and found them to have beneficial effects comparable to those of Manuka honey. Accordingly, while this paper includes a great deal of information about Manuka honey, its scope has been widened to include pertinent research into other varieties of honey.

As the focus of this review is on the use of honey in tissue engineering templates, it is not a comprehensive discussion of the components of honey, their contribution to its wound-healing mechanisms, or the entire body of research into honey's effects as a wound additive. Rather, these topics are summarized as they relate specifically to the incorporation of honey into tissue engineering templates. For more in-depth research into the components of honey, the reader should seek out "The components of honey and their effect on its properties: a review" by Thawley or "Major components of honey analysis by near-infrared transflectance spectroscopy" by Garcia-Alvarez et al. [37,38]. For a review of the anti-bacterial mechanisms of honey and quantitative studies of their contributions to the overall anti-bacterial nature of honey, readers should obtain "How honey kills bacteria" by Kwakman et al. or "Antibacterial components of honey" by Kwakman and Zaat [6,30]. If a more comprehensive review of the mechanisms of honey that contribute to wound healing is desired, the authors suggest "Honey: a potent agent for wound healing?" by Lusby et al. or "The evidence and the rationale for the use of honey as a wound dressing" by Molan [39,40].

2. In Vitro and In Vivo Evidence of the Beneficial Effects of Honey in Wounds

2.1. Antibacterial and Antibiofilm Effects of Honey

Studies have examined the anti-bacterial action of Manuka honey against a variety of pathogens. Sherlock et al. used agar plate well diffusion assays and a spectrophotometric minimum inhibitory concentration assay to demonstrate antibacterial effects. These effects were quantified for both Manuka honey and Ulmo honey, a strain from Chile. The results of these experiments, shown in Table 1, demonstrated that both Manuka and Ulmo honey significantly inhibited the growth of

E. coli, P. aeruginosa, and methicillin-resistant S. *aureus* (MRSA). Interestingly, the Ulmo honey was more effective against MRSA, although slightly less effective against *E. coli* and *P. aeruginosa* [41]. Jenkins et al. also reported that Manuka honey inhibits the growth of MRSA, and showed that the presence of honey causes a downregulation of universal stress protein A (UspA) in the MRSA, reducing its stress stamina response [12]. In addition to its effectiveness against MRSA, Cooper et al. showed that Manuka honey also inhibits the growth of at least seven different strains of vancomycin-resistant enterococci [15]. Manuka honey also has been shown to be effective against *Helicobacter pylori*, the cause of most stomach ulcers [42]. Research by Watanabe et al. in 2014 showed that Manuka honey inhibits influenza viral replication, enhancing the effects of antiviral drugs [43]. This work has been replicated with varicella and rubella viruses, indicating an exciting new avenue for the clinical use of Manuka honey [44,45].

Table 1. Honeys inhibit bacterial growth. Zones of inhibition (diameter, in mm) of different concentrations of Ulmo and Manuka honey against various strains of MRSA. Standard deviations are shown in parentheses. "-" indicates no inhibition at that concentration. "*" indicates a clinical isolate. Reproduced with permission from Sherlock et al. *Complementary and Alternative Medicine*; published by BMC, 2010.

Concentration	50% *v/v*		25% *v/v*		12.5% *v/v*		6.3% *v/v*	
Isolates	Ulmo	Manuka	Ulmo	Manuka	Ulmo	Manuka	Ulmo	Manuka
MRSA ATCC 43300	30 (1.7)	24 (1.5)	26 (0.6)	19 (2.1)	18 (0.6)	13 (1.0)	10 (0.6)	-
MRSA 0791 *	34 (1.5)	23 (1.2)	29 (1.7)	17 (1.7)	22 (2.1)	-	14 (2.5)	-
MRSA 28965 *	24 (1.0)	17 (1.7)	19 (1.5)	15 (2.0)	-	-	-	-
MRSA 01322 *	28 (5.8)	22 (1.0)	23 (4.2)	18 (0.6)	17 (2.9)	-	11 (2.0)	-
MRSA 0745 *	23 (2.7)	20 (1.7)	19 (2.1)	13 (1.7)	11 (2.7)	-	-	-
P. aeruginosa ATCC 27853	14 (2.3)	16 (7.8)	11 (1.0)	14 (6.9)	-	-	-	-
E. coli ATCC 35218	14 (1.5)	15 (2.5)	11 (1.7)	12 (2.9)	-	-	-	-

Investigators have also examined the ability of various honey types to inhibit biofilm formation or kill biofilm-embedded bacteria. Marckoll et al. tested the effects of Manuka honey and Norwegian Forest honey on biofilm-embedded MRSA, methicillin-resistant *S. epidermidis* (MRSE), extended-spectrum β-lactamase (ESBL) *Klebsiella pneumoniae*, and *P. aeruginosa*. This study found that the active components of the honey diffused through established biofilm matrices of all bacterial types and killed bacteria in a dose-dependent manner, with minimum concentrations between 6 and 12% of Manuka honey and 12 to 25% of Norwegian Forest honey killing biofilm-embedded bacteria depending on the bacterial type. While the presence of a biofilm did provide some protection to the MRSA, MRSE, and ESBL *Klebsiella*, no protection was observed in the *P. aeruginosa* biofilm [46]. Similarly, Bardy et al. tested the ability of Manuka honey and an Australian non-methylglyoxal-containing honey, Capilano honey, to inhibit biofilm formation of *S. aureus* strains isolated from clinical patients. Biofilm inhibition was found to be related to methyglyoxal content, with a minimum level of 0.53 mg/mL methyglyoxal in Manuka honey solution necessary for biofilm-cidal activity (a 33% *w/v* level of Manuka honey). By itself, the Capilano honey solutions did not inhibit biofilm formation, but when at least 1.05 mg/mL methylglyoxal was added they were able to become biofilm-cidal [47]. The difference in minimum methylglyoxal levels necessary to inhibit biofilm formation between these two honey types indicates that while methylglyoxal content is important, there is an additional unknown biofilm-cidal component present in Manuka honey that is not present in the Capilano variety.

Likewise, Alandejani et al. demonstrated the effectiveness of Manuka honey and Sidr honey, another methylglyoxal-containing variety, against biofilms of MRSA, methicillin-susceptible *S. aureus* (MSSA), and *P. aeruginosa*. Both honeys were tested at a 1:2 dilution level and found to inhibit the growth of most samples of each bacterial strain. However, no attempt was made to test further dilutions of these honeys or any non-methylglyoxal-containing honey variety [48]. In another attempt, Okhiria et al. tested concentrations of 0%, 20%, and 40% *w/v* Manuka honey on biofilms formed by six cultures of *P. aeruginosa* and found that biofilm shrinking only occurred at the 40% w/v level [49]. A more thorough study by Sojka et al. utilized a multispecies wound biofilm model containing *S. aureus*, *Streptococcus agalactiae*, *Enterococcus faecalis*, *P. aeruginosa* and *K. pneumoniae* to test non-diluted Manuka honey, Honeydew honey, and an artificial honey prepared from fructose, glucose, maltose, and sucrose (mechanistic control). While the artificial honey was somewhat effective against *P. aeruginosa*, decreasing the number of colony-forming units from around 10^6 c.f.u./mg to around 10^4 c.f.u./mg over a period of 48 h, it did not significantly kill or inhibit the other three bacterial strains. In contrast, both honey varieties significantly decreased *S. aureus* growth from around 10^6 c.f.u./mg to around 10^3 c.f.u./mg and decreased *S. agalctiae* and *P. aeruginosa* growth to around 0 c.f.u./mg over a 48-h period. This difference in bactericidal activity between the honey varieties and artificial honey indicates that while the osmotic pressure of the honeys plays a role in some anti-biofilm activity, it does not account for all of this activity in the honey. None of the natural or artificial honey types had an effect on *E. faecalis* growth, indicating that this bacterial strain was not susceptible to the antibacterial effects of the honey. The resilience of this strain to honey should be noted for future clinical application [50].

There is also in vivo evidence of the antibacterial effects of honey in wounds. A 2010 study by Moghazy et al. followed the treatment of 30 diabetic foot ulcers with commercial honey over a three-month period. A number of microorganisms were isolated from the ulcers at the beginning of the study, including *Staphylococcus aureus*, *E. coli*, *Proteus*, *Klebsiella*, and *Providencia*. All of these microorganisms were eradicated by the end of the three-month study. *Staphylococcus epidermidis*, a benign pathogen commonly found on human skin and thought to provide a reservoir of resistance genes to other infections, was isolated from 28 of the patients at the end of the study [51,52]. The presence of healthy *S. epidermidis* is a sign of healing in these wounds. While the results of this study are encouraging, it would have benefited from a non-honey treatment control group to establish an effective comparison to the current gold-standard treatments. In another study, Efem et al. used topical commercial honey treatment to treat 59 cases of non-healing ulcers. Swabs from 51 of the wounds before treatment indicated the presence of *P. pyocyanea*, *E. coli*, *S. aureus*, *Proteus mirabilis*, *Klebsiella*, *S. faecalis*, and *Streptococcus pyogenes*, while swabs performed after one week of honey treatment indicated the eradication of these microorganisms [53]. In addition to treatments with commercial store-bought honey, some studies have focused specifically on treatment with medical-grade Manuka honey. Gethin et al. compared Manuka honey treatment with a commercially available hydrogel dressing in 108 patients with sloughy infected venous leg ulcers. MRSA was identified in 16 of the wounds, 10 of which were treated with the honey while six were treated with the commercially available hydrogel. After four weeks of treatment, MRSA was eradicated in seven of the ten honey-treated wounds but only one of the six hydrogel-treated wounds [54]. These studies provide evidence supporting the use of Manuka honey as an anti-bacterial wound sterilizing agent.

As honey is cheaper than many antibiotics and has not yet been shown to induce resistant bacteria, it is likely to become a useful alternative to antibiotics in the field of wound care. However, it should be noted that biofilms of certain bacteria, such as *E. faecalis*, are resistant to the antibacterial effects of honey, which could complicate its use in clinical practice. Additionally, the literature concerning honey's efficacy against biofilms suggests that high concentrations (at least 33% w/v) of honey are necessary. This is complicated by cytotoxicity concerns discussed later in this review. Thus, care will have to be taken to tailor the amount and dilution of honey to specific situations in clinical practice. Larger, undiluted amounts of honey may be appropriate when fighting persistent infection, while smaller, more dilute amounts of honey are likely optimal when treating inflammation and inducing tissue infiltration and regeneration.

An additional benefit of using honey in therapeutic products is that these natural antibacterial properties give it an extremely long shelf life. In sealed containers, honey remains stable for hundreds or even thousands of years, and it is often used to increase the shelf life of other food products [55–59]. Many antibiotics have limited shelf lives even under refrigeration—for example, penicillin in solution has a shelf life of twelve months at 10 °C [60]. In contrast, honey's robust thermal stability allows it to go un-refrigerated and still maintain its properties indefinitely [61]. This is a major advantage, as it eliminates the need for a "cold chain" of constant refrigeration and, therefore, reduces costs substantially. The elimination of the cold chain is a particular benefit in rural areas or developing countries where there is less access to refrigeration and power interruptions can be frequent [62,63].

2.2. Immunomodulatory Effects of Honey

A number of studies have examined the effects of honey on the immune response, with results that paint an intriguing but incomplete picture. Tonks et al. tested the monocyte response to several honey types, including Manuka honey, and observed that all honey types tested caused an increase in the release of the inflammatory mediators tissue necrosis factor α (TNF-α), interleukin 6 (IL-6), and interleukin 1 (IL-1) over a 24-h culture period, as shown in Figure 1 [64]. Of the three honey varieties tested (all 1% v/v in culture medium), Manuka honey caused the lowest release of these three inflammatory cytokines, but these levels were still significantly higher than in the non-honey controls. Specifically, Manuka honey caused an increase of about 2000 pg/mL in TNF-α release, about 100 pg/mL in IL-1 release, and about 700 pg/mL in IL-6 release over 18 h of culture as measured relative to non-honey controls. This finding would seem to indicate that these honey varieties elicit an inflammatory reaction, in direct contrast to the studies discussed below.

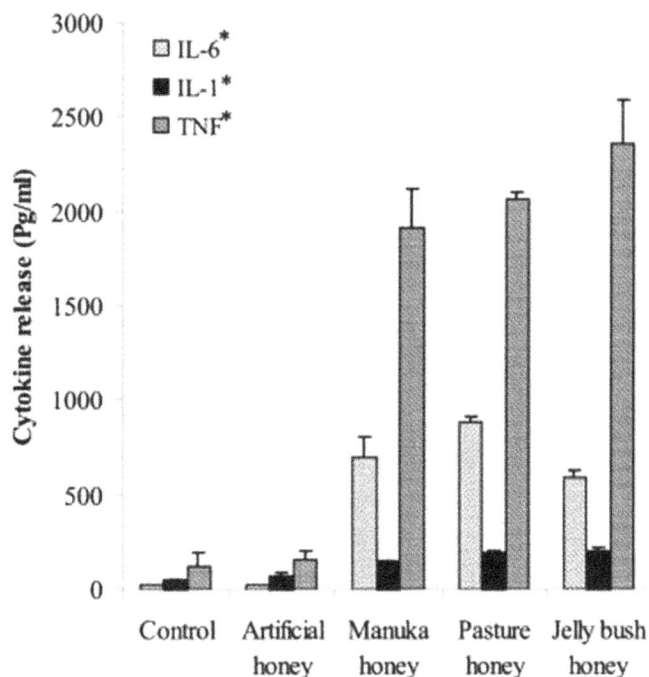

Figure 1. Honeys induce monocyte inflammation response. IL-6, IL-1, and TNF release from peripheral blood monocytes over 18 h in the presence of artificial honey (syrup control), Manuka honey, Pasture honey, and Jelly Bush honey. "*" indicates statistical significance ($p < 0.001$, analyzed by ANOVA with a Tukey pair-wise comparison). Reproduced with permission from Tonks et al., *Cytokine*; published by Elsevier, 2003.

In another study, Leong et al. examined the effect of 21 New Zealand honey types, including varieties of Manuka honey, on neutrophil superoxide production. Their results indicate that all honey types tested reduced superoxide production in a dose-dependent manner, and this decrease in superoxide production was independent of the methylglyoxal content of the honey samples. Cytotoxicity testing revealed that at the 50% inhibitory concentrations (IC_{50}, ranging from 3.1 mg/mL to 44.4 mg/mL depending on the honey variety) of honey on superoxide production, none of the honey varieties caused significant amounts of neutrophil death [13]. However, honey treatments in wounds typically involve direct application of honey to the wound at much higher concentrations than these IC_{50}s, throwing doubt into the relevance of this finding to wound treatment. This study also involved an in vivo murine test which measured the effect of topical application of these honey varieties on neutrophil recruitment to the site of arachidonic acid (inflammatory stimulus) application in a murine ear model. The results showed that several honey varieties, including Manuka honey, significantly decreased neutrophil recruitment to the site [13]. These results indicate an overall anti-inflammatory effect of Manuka honey on neutrophils, reducing their inflammatory superoxide production and attenuating their recruitment to a site of inflammation and, thus, correlate more closely than the results of the previous study with the clinical data showing that Manuka honey resolves inflammation [65,66].

Other in vivo evidence points to an anti-inflammatory effect of honey. In one example, rabbit wounds were treated with topical honey (type of honey not specified) and studied for 21 days. Histological examination of the wounds at 14 days revealed well-vascularized tissue with organized fibroblasts and collagen fibers with few inflammatory cells still present in the honey-treated group, while the non-honey group showed necrosis, uneven epithelialization, and a large neutrophil presence [66]. In a different study, Medhi et al. used a rat ulcerative colitis model to study the efficacy of rectally-applied Manuka honey to treat ulcerative colitis. Rats were administered intra-colonic 2,4,6-trinitrobenzene sulfonic acid (TBS) to induce colitis and then treated with Manuka honey at 5 g/kg body weight through a rubber tube inserted rectally. After 14 days, rats were sacrificed, and excised tissue was morphologically assessed. Histological sections of colon tissue were graded on a scale from 0 (no inflammation) to 3 (intensive inflammation). Treatment with Manuka honey decreased the sample scoring from approximately 1.8 (mean TBS control score) to approximately 0.2 (mean Manuka honey score), indicating almost no inflammation in the honey-treated samples [67]. This study indicates another promising use for Manuka honey in treating ulcerative colitis and other internal inflammatory diseases, and provides evidence of the general anti-inflammatory properties of honey which make it such an effective wound treatment. Clinical evidence also indicates that honey exhibits anti-inflammatory properties. For instance, in the 2010 diabetic foot ulcer study by Moghazy et al. that was discussed above, significantly decreased inflammation was observed in 27 of the 30 patients during and after the three-month honey treatment [51].

Together, this evidence seems to contradict the Tonks et al. study discussed above that showed that honey increases release of the inflammatory mediators TNF-α, IL-1, and IL-6 by monocytes. However, it should be noted that there are many other inflammatory and anti-inflammatory cytokines involved in the healing response that were not tested in the Tonks et al. study, and it only tested one concentration of honey rather than the gradient of honey concentrations present in a wound. Nevertheless, it is possible that these honeys cause a temporary increase in inflammatory cytokines in a wound site before later resolving that inflammation, or that by increasing inflammation the honey "shocks" the wound environment into quickly clearing infections to allow for inflammation resolution. Ultimately, future studies examining more honey concentrations and more of the relevant cytokines will be necessary to bridge the gap between these in vitro and in vivo findings. In addition, detailed time courses must be examined in order to understand the difference between how honey affects the different stages of inflammation in the wound site, especially the vast differences between the effects on the acute and chronic phases.

2.3. Wound Closure Effects of Honey

Studies have demonstrated that multiple varieties of honey promote wound closure. Ranzato et al. showed low concentrations (0.1% v/v) of a variety of honey types, including Manuka honey, increase closure rate in a keratinocyte scratch assay and promote fibroblast migration in a transwell insert chemotaxis assay. Specifically, 0.1% Manuka honey increased keratinocyte closure rate by 180%, and increased fibroblast migration by 150–240% (higher honey concentrations were not tested) [68]. It should be noted that this study did not make use of a sugar solution control, so it is unknown how much of this migration effect was caused by the sugar content of the honey as opposed to its other components. In the 21-day rabbit wound model study described above, faster and improved wound closure was observed in the honey-treated wounds. After 14 days, the non-honey wounds were covered by scabs and imperfect epithelialization, while skin repair in the honey-treated rabbits was perfect and detection of the injured area was difficult. Samples of the healed skin were excised and mechanically tested after 21 days, and the honey-treated rabbit skin had a significantly higher tensile yield strength (3.3 MPa) and ultimate strength (3.4 MPa) than that of the non-honey wounds (1.2 MPa and 2.3 MPa, respectively) [66]. Likewise, honey treatment in a rat dorsal wound model had similar effects. Topical application of honey to these wounds caused a 107% increase in salt-soluble collagen, a 117% increase in acid-soluble collagen, and a 109% increase in insoluble collagen after seven days relative to non-treated controls. Introduction of radiolabeled hydroxyproline one day before sacrifice allowed measurement of the collagen synthesis rate over a 24-h time period, and indicated a 124% increase in the acid-soluble collagen production rate and a 105% increase in the insoluble collagen production rate during this sixth day after wound creation relative to control, suggesting that the healing rate is increased at this time point by the honey treatment. The acid-soluble collagen extracted from honey-treated rats had a 122% increase in aldehyde content relative to that extracted from non-treated rats, indicating a higher degree of crosslinking in the wounds that were honey-treated. This was confirmed by an 11% drop in the solubility of the insoluble collagen of the honey-treated rats in the presence of urea. Interestingly, experimental groups of rats with honey administered orally and intraperitoneally showed higher degrees of collagen synthesis and crosslinking than the topical-administration group [69]. Although the administration of honey via the oral and intraperitoneal routes for wound healing has not been widely studied, these findings suggest it may, in fact, be more beneficial than the current topical administration model. The authors of this study suggest that the oral administration of the honey allows for greater nutrient uptake, which is an unsatisfactory explanation for these results as those nutrients would be processed and dispersed systemically and, thus, be unlikely to have a greater effect on the wound than topical administration. More exploration of the benefits of these routes of administration, including repeating this study, may be beneficial to confirm or disprove these potentially impactful findings.

Clinical evidence has also shown honey to improve wound closure. Numerous case studies have demonstrated beneficial effects of Manuka honey in the closure of various types of infected non-healing ulcers [70–74]. In the 2010 study by Moghazy et al. described earlier, ulcer size decreased in 28 of the 30 patients treated with honey, with complete healing in 13 of the patients after three months [51]. Likewise, in the study by Efem et al., it was described that the honey treatment caused more rapid wound debridement, promoted rapid epithelialization, and reduced edema, causing a faster healing rate and reduced morbidity. The author reports that within one week, sloughs, necrotic, and gangrenous tissues were separated from the ulcers enough to be lifted away by forceps without pain to the patients, while weeping ulcers were dehydrated and foul-smelling wounds were rendered odorless [53]. Unfortunately, while the progress of the wounds is described, no objective measurements of wound size or condition are included in this study, only general clinical observations of the wounds over time. In contrast, Jull et al. conducted an expansive review published in 2015 of 26 randomized or quasi-randomized trials evaluating honey as a treatment for a variety of wound types. As many of the trials examined in this review suffered from small sample sizes or reported insufficient data, few conclusions could be drawn. However, the authors did conclude that honey improves healing rates in partial thickness burn wounds relative to current gold-standard treatments, shortening healing time by about 4–5 days. They also found moderate evidence showing honey is more effective than standard antiseptic treatments in treating infected surgical wounds. However, there was insufficient evidence to make conclusions about the effects of honey in other wound types as of the publication of this review in 2015. More studies are necessary moving forward to statistically confirm the beneficial effects of honey in other varieties of wounds, such as chronically-inflamed wounds, pressure ulcers, Fournier's gangrene, and venous leg ulcers [75].

3. Cautionary Evidence of Cytotoxicity

While there are many studies that have demonstrated the potential benefits of honey in wound healing, less attention has been given to the counterproductive cytotoxic effects of high concentrations of honey. However, a few groups have studied these cytotoxic effects in various cell and animal models, and their data provides a cautionary window into the dangers of using high concentrations of honey in wounds or other therapeutic applications. An in vitro study by Sell et al. found that honey concentrations of 5% v/v or above were cytotoxic, killing almost 100% of the cells tested in fibroblast, pulmonary microvascular endothelial, and macrophage cultures after one day [65]. For reference, the Sherlock et al. study referenced earlier showed little inhibition of MRSA, *E. coli*, or *P. aeruginosa* growth at concentrations in the range of 6–12% v/v or below [41]. Marckoll et al. also found a minimum inhibitory concentration of Manuka and Norwegian Forest honey on a variety of bacterial biofilms to range between 6–12% for Manuka honey and 12–25% for Norwegian Forest honey. The cytotoxicity of honey has also been studied in vivo. A study in which 50% v/v Manuka honey was applied to chinchilla ears found that it caused severe inflammation and ototoxicity. Eight chinchillas had the honey solution applied to the round window membrane and the cochlea of one ear while the other ear received a sham treatment of normal saline solution. All eight chinchillas developed a head tilt and facial paralysis on the side of the experimental ear within 0–48 h of honey application, with a corresponding loss of balance and nystagmus. Extraction of the osseous bullae and cochleae showed that the honey-exposed bullae were soft and brittle and the cochleae were darker, compared to the control bullae and cochleae which were normal in color and consistency. Histological examination revealed a scarcity of cells and the creation of new vacuoles within the honey-exposed spiral ligaments, with damage to the organ of Corti and an excess of inflammatory cells found in the honey-exposed cochleae, whereas the saline-treated organs had normal appearance, architecture, and cellularity. Scanning electron microscope (SEM) images showed severe damage to the spiral ganglion and cochlear hair cells in the experimental ears, with no damage to the control ears, as shown in Figure 2 [76]. Possible causes of this damage are the low pH of the honey and its high osmolarity, although further testing with other acidic and hypertonic solutions will be necessary to confirm this theory. In a similar in vivo effort, Paramasivan et al. flushed ovine frontal sinuses with methylglyoxal concentrations ranging from 0.5 to 7.2 mg/mL, or with 16.5% w/v Manuka honey enriched with methylglyoxal in the same concentration range, twice daily for 14 days. Animals were sacrificed, and the tissue was analyzed by histology and tested for *S. aureus* biofilms which had been intentionally developed in the ovine sinuses before the study. The results indicated both the methylglyoxal alone and 16.5% Manuka honey enriched with methylglyoxal above 0.9 mg/mL eradicated the *S. aureus* biofilms, while honey/methylglyoxal treatment with less than or equal to 1.8 mg/mL methylglyoxal was non-irritating to the mucosa. However, methylglyoxal and honey/methylglyoxal treatment with methylglyoxal levels above 1.8 mg/mL caused cilia denudation and squamous metaplasia, indicating tissue damage [77]. These results point to methylglyoxal as a culprit of Manuka honey's cytotoxicity, although more testing in a variety of cell and animal models should be done to confirm this finding. If accurate, however, these results could indicate that other honey varieties without methylglyoxal may be optimal for applications in which a high honey concentration would be required.

The results of these papers provide compelling evidence of the cytotoxicity of Manuka honey at higher concentrations, and should prompt a re-assessment of the current use of Manuka honey in wound treatment. Since most clinical honey treatment involves directly applying undiluted honey to the wound, it is likely that honey's beneficial effects are at least somewhat counteracted by its cytotoxicity. Even accounting for dilution of the honey by wound exudate and excess liquid pulled from the deeper tissue by honey's high solute osmolarity, cells within the wound likely encounter honey concentrations at or above the 5% v/v cytotoxic concentration found by Sell et al. [65]. When designing tissue regeneration templates, this cytotoxicity must be accounted for to avoid killing infiltrating cells and impeding tissue ingrowth. Even in applications where tissue ingrowth is not necessary, such as a bone screw, care must be taken to avoid causing necrosis in the surrounding area. Thus, it is paramount that these tissue templates and devices that incorporate honey do so in a way that allows a low-level, controlled release to avoid cytotoxicity while prolonging the beneficial effects of the honey. However, as described above, higher levels of honey are necessary to impede or destroy bacterial biofilms. Therefore, in situations where a bacterial biofilm has been established, it may be necessary to first treat the wound with undiluted honey to eradicate this biofilm. Once the biofilm has been eliminated, a template can be applied that releases lower levels of honey over time to reduce inflammation and induce tissue regeneration and infiltration. It should be noted that certain beneficial effects discussed earlier, such as immunomodulation and promotion of wound closure, occur at honey concentrations below the cytotoxic level. However, the data does not currently exist to definitively state a therapeutic window for each of the beneficial effects of the honey discussed earlier. As such, the authors recommend that testing for cytotoxicity and desired therapeutic effects (fibroblast infiltration, immunomodulation, etc.) be performed for each honey-incorporating template using assays that take into account the expected microenvironment of the template upon implantation.

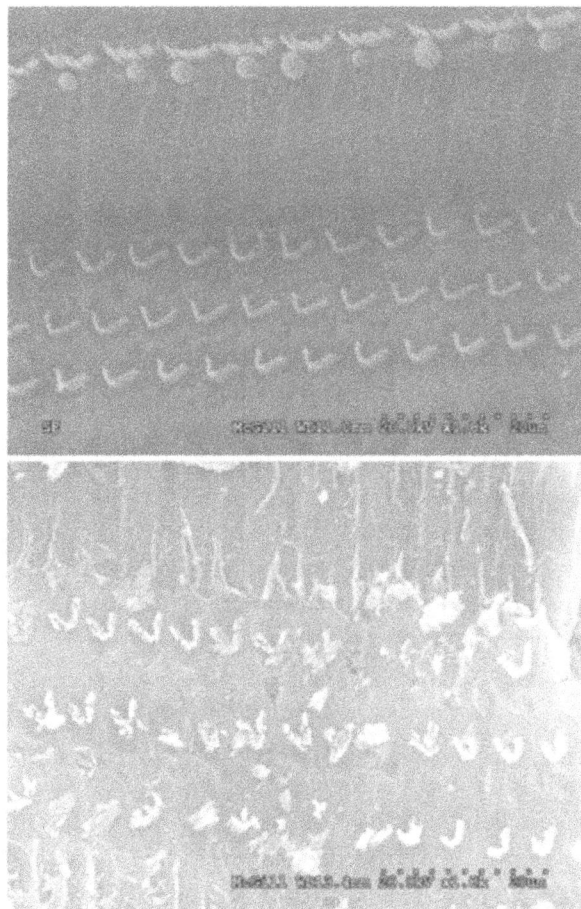

Figure 2. Undiluted honey damages cilia. (**Top**) SEM of saline-exposed chinchilla cochlea, in which normal inner and outer hair cells are observed. (**Bottom**) SEM of honey-exposed chinchilla cochlea in which the inner and outer hair cells have been damaged. Reproduced with permission from Aron et al., *Otolaryngology—Head & Neck Surgery*; published by BMC, 2012.

4. Honey in Tissue Engineering

Given the amount data supporting the use of honey as a wound treatment, the logical next step is to apply these findings to the field of tissue engineering and biomaterials. The implantation of a biomaterial within the body necessitates the creation of a wound, and the presence of these biomaterials provides a potential site for bacteria to deposit and fester after implantation. The antibacterial effects of honey, especially Manuka honey, could significantly reduce the rates of infection in biomaterials. Additionally, given the evidence showing that Manuka honey reduces inflammation and promotes fibroblast migration and collagen deposition, it is likely that it could promote tissue-material integration/regeneration and accelerate healing of the surrounding wound site [13,65,66,68,69]. An important consideration will be how to apply the honey to the biomaterial or incorporate it into the biomaterial to deliver appropriate concentrations of honey and achieve these optimal effects. It is likely that in most applications in which a bacterial biofilm is not present, a controlled-release profile will be desirable to avoid cytotoxic effects and prolong the presence of the honey in the region of interest/need. Thus, research has been focused into incorporating the honey throughout biomaterials to achieve this type of release. In the past decade, there have been numerous studies incorporating honey into a variety of biomaterial tissue templates for tissue regeneration.

4.1. Electrospun Templates

One of the first attempts to incorporate Manuka honey into an electrospun template was published in 2012 by Vadodaria et al. In this study, Manuka honey was combined with polyethylene oxide (PEO) into solutions which were then electrospun. SEM images showed that increasing Manuka honey content caused thicker, merged fibers, although these morphological changes could be somewhat compensated for by reducing the solution feed rate and increasing the applied voltage. Fourier-transform infrared spectroscopy (FTIR) revealed peaks indicating the presence of both PEO and Manuka honey in the fibers, while differential scanning calorimetry (DSC) indicated that increasing the Manuka honey content lowered the melting point of these templates [78]. This study did not include any experiments examining biocompatibility or cell behavior on the templates and did not include a honey release profile. Nevertheless, it established the basic parameters necessary for electrospinning Manuka honey into nanofibrous templates for use as delivery vehicles.

In another early study, Maleki et al. electrospun poly(vinyl alcohol) (PVA) templates containing various concentrations of Iran-Tabriz honey and dexamethasone, an anti-inflammatory drug. Both honey and dexamethasone decreased the fiber diameters of the templates in a dose-dependent fashion. Dexamethasone release studies showed a large burst release of dexamethasone within the first 10 min of soak due to the swelling of the PVA fibers, with no difference between the honey and non-honey dexamethasone templates [79]. Like the Vadodaria et al. paper, this effort did not conduct any cellular studies or bacterial inhibition assays, but it does indicate that honey can be incorporated into electrospun templates along with other additives. Unfortunately, although this paper did have release profiles of the dexamethasone from the templates, no honey release profile was included.

A more in-depth study of Manuka honey in electrospun templates was published by Minden-Birkenmaier et al. in 2015. In this study, solutions of poly(ε-caprolactone) (PCL), 1,1,1,3,3,3-hexafluoro-2-propanol (HFP), and various concentrations of Manuka honey were electrospun into fibrous templates which were then characterized with regards to fiber morphology, water vapor transmission rate (WVTR), permeability, mechanical properties, honey release, fibroblast response, and bacterial inhibition. Templates were also created replacing the Manuka honey with equivalent amounts of water to use as morphological controls. By sonicating the Manuka honey in the HFP before adding the PCL, and then electrospinning the resulting solution within 24 h, templates were created with equivalent fiber diameters, varying from 2 μm to 3.5 μm in diameter, up to a 20% v/v honey content. The water vapor transmission rate after a one-hour soaking period increased with increasing honey content, as did template permeability. Honey incorporation caused a decrease in the elastic moduli and peak stress of the templates, but there was no significant change in these properties over a 28-day soaking period. Glucose assays indicated that while up to 80% of the honey content of the templates was lost during a one-hour ethanol disinfection soak, the templates retained enough honey that they released significant amounts of honey over the following 24-h period of soaking, proportional to the amount incorporated into the scaffold. Fibroblast chemotaxis assays showed no effect of honey content in inducing chemotaxis towards the templates, indicating that the honey left in the templates to be released after the disinfection step is probably too low of an amount to induce chemotaxis. However, the 20% honey templates caused a significant increase in fibroblast proliferation and infiltration over PCL controls, indicating the potential of Manuka honey to improve template-tissue integration and regeneration (Figure 3). Bacterial inhibition studies of *E. coli* and *S. agalactiae* showed significant inhibition of both bacterial types by the 10% and 20% honey templates as expected due to the antibacterial properties discussed earlier, although this inhibition was significantly less than that of a sterile disc swabbed with pure Manuka honey. Together, these findings indicate the potential benefit of Manuka honey in improving cellular proliferation, cellular ingrowth, and bacterial inhibition associated with a tissue template [80]. However, the fact that 80% of the incorporated Manuka honey was removed during ethanol disinfection indicates that future studies may consider investigating core-shell electrospinning to protect the honey from leaching out during disinfection and provide for a more long-term, controlled release period. Other disinfection or sterilization methods, like gamma irradiation, should also be investigated as alternatives that may remove less of the incorporated honey.

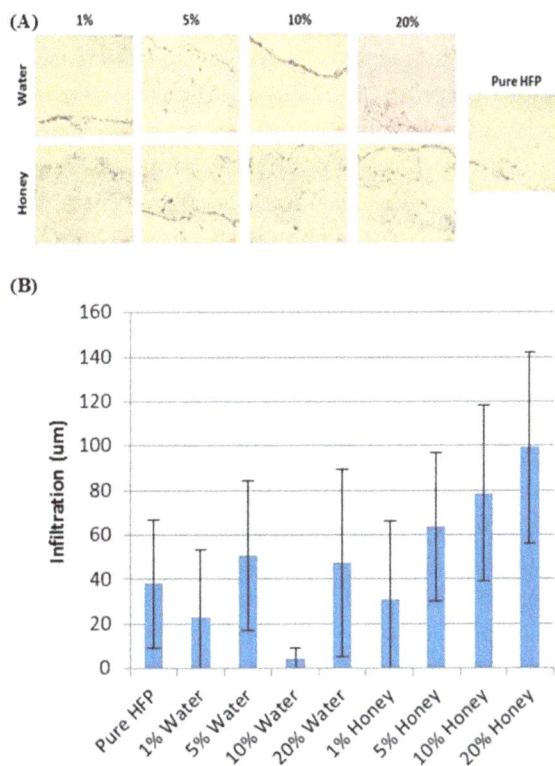

Figure 3. Honey induces fibroblast infiltration. (**A**) Representative images of H and E-stained honey templates and water controls after 28 days of fibroblast culture. (**B**) Cellular infiltration depth of the furthest 60 cells on each image. Reproduced with permission from Minden-Birkenmaier et al., *Journal of Engineered Fibers and Fabrics*; published by INDA, 2015.

The physical properties detailed in the Minden-Birkenmaier et al. study also indicate that the majority of the honey is likely sequestered to the fiber surface during the electrospinning process, where it is easily released once rehydrated, allowing for greater template permeability, but maintaining mechanical strength due to the PCL fiber core [80]. As expected, this morphology makes the templates hygroscopic, allowing them to soak up water, but changing their physical properties as they are exposed to ambient humidity or liquid water (as demonstrated by the water vapor transmission rate data discussed above). It will be important to take this hygroscopicity and the associated processing issues into account when producing and packaging honey-laden templates for clinical use. For instance, production and packaging in a low-humidity environment may be necessary to improve the shelf-life of future commercial honey-laden templates.

In a more recent effort, Balaji et al. combined Malaysian Tualang honey and papaya extract (PA) (also reported to have antimicrobial and anti-inflammatory properties [81,82]) into a N,N-dimethylformamide (DMF) solution along with polyurethane (PU), which was then electrospun into templates. Fiber diameter measurements showed that the honey and papaya extract reduced the template fiber diameter, from a mean diameter of 434 nm for PU controls down to 190 nm for the PA/honey templates; however, porosity only experienced a minor decrease. FTIR confirmed the presence of both the honey and the PA along with the PU in the fibers. Water absorption tests revealed that the presence of both honey and PA in the templates caused a three-fold increase in water uptake, indicating a potential benefit of these hydrophilic substances in absorbing wound exudate. Hemocompatibility studies demonstrated that the PA/honey templates had significantly greater adsorption of albumin, but significantly less adsorption of fibrinogen relative to PU controls, indicating a resistance to clotting. Activated partial thromboplastin (APTT) and prothrombin time (PT) assays likewise demonstrated that the PA/honey templates took longer to activate thromboplastin and prothrombin than the PU controls, 180 s (honey) versus 152 s (control) for APTT and 45 s (honey) versus 37 s (control) for PT. The PA/honey templates also had a decrease in the hemolytic percentage from the PU controls (2.7% for PU control and 0.9% for the honey template) indicating a reduction in red blood cell lysis. This hemocompatibility suggests a possible use of the template in vascular tissue engineering [83]. While the results of this study are impressive, particularly with regards to hemocompatibility, it would have been beneficial for separate templates containing different amounts of honey or PA to be tested, as the effects of the honey and the PA on hemocompatibility could have been isolated from each other. This separation of honey and PA would have allowed for a more robust study to indicate the true hemocompatibility potential of the honey and the PA alone.

Several studies have investigated combinations of silk fibroin and honey in electrospun templates. Kadakia et al. electrospun silk fibroin templates from HFP containing either poloxamer 407 (P407), a hydrophilic polymer used to improve cell adhesion, or Manuka honey. Fiber diameter measurements taken from SEM images revealed that the incorporation of P407 at either a 1:1 or 3:1 silk:P407 ratio (total polymer concentration of 10% w/v) significantly decreased fiber diameters (from 2.2 μm down to 1.8 μm), while 1% honey increased fiber diameters (from 4.4 μm up to 5.8 μm) and 5% honey decreased fiber diameters (down to 3.6 μm). Mechanical testing showed that increasing P407 and honey concentrations decreased the elastic moduli of the templates relative to silk control templates when dry. However, when the templates were hydrated, the honey templates had elastic moduli in the range of 5–9 MPa, above the 2–3 MPA range of the silk controls, indicating that the honey increased elasticity. While no difference was observed in the swelling of the silk/P407 templates and pure silk controls, the honey templates swelled to a significantly higher degree, with a swelling ratio of about 350% after four hours, while the silk fibroin control had a swelling ratio of approximately 240%. The water vapor transmission rate was observed to decrease from approximately 1750 g/m^2/day for the silk control template to approximately 1550 g/m^2/day with the incorporation of 1% honey and approximately 1400 g/m^2/day with the incorporation of 5% honey. Water contact angle measurements showed that the incorporation of p407 decreased the water contact angle from about 70° to about 45° for 3:1 silk:P407 and about 11° for 1:1 silk:P407. Surprisingly, the incorporation of honey increased the water contact angle from around 61° to approximately 67° for 1% honey and approximately 78° for 5% honey. Fibroblast experiments showed no increase in proliferation in the 1% honey templates, and impeded proliferation on the 5% honey templates, indicating a degree of cytotoxicity of the honey as discussed earlier in this review. However, fibroblasts infiltrated fully into all template types after 28 days, with no differences between groups, and no significant difference in hydroxyproline production was observed between the groups [84]. This study would have benefited from a glucose release assay to ascertain the amount of honey released during disinfection and subsequent culture. Given the cytotoxicity observed in the fibroblast culture experiments, it is speculated that templates with lower amounts of loaded honey or a lower, persistent release of honey over time would have performed better in the cellular studies.

In a similar effort, Yang et al. electrospun solutions of silk fibroin and poly(ethylene oxide) (PEO) with concentrations of 0%, 10%, 30%, 50%, and 70% w/v Manuka honey into nanofibrous templates. FTIR showed the presence of the Manuka honey in the fibers, and SEM images showed an increasing fiber diameter with honey concentration, from an average of 484 nm without honey to an average of 2229 nm with 70% w/v honey, as shown in Figure 4. Bacterial inhibition tests using *E. coli*, *S. aureus*, *P. aeruginosa*, and MRSA indicated that the templates retained the antimicrobial effects of the Manuka honey. Specifically, bacterial inhibition over 24 h of all four bacterial strains was approximately zero for the non-honey template, but increased to around 50% inhibition of *E. coli*, about 28% inhibition of *S. aureus*, about 57% inhibition of *P. aeruginosa*, and about 40% inhibition of MRSA for the 70% w/v honey template. Templates were also used to treat a mouse dorsal wound model over a 12-day period, and showed complete healing of the wounds treated with the 70% honey template, whereas wounds treated with a non-honey silk template or a commercial AquacelAg wound dressing had only around a 90% reduction in wound size over this timeframe [85]. The most novel part of this study was the use of deionized water and hydrophilic polymers in the electrospinning process, as opposed to the organic solvents used in the previously described studies. This water-based solution could potentially eliminate the sequestering of the honey to the outside of the fibers, reducing the mechanical strength of the template, but delaying the release of the honey over time. As such, this study would have benefited from a glucose release experiment showing the release profile of honey from the template to indicate if this delayed, controlled release is present.

4.2. Cryogels

Cryogels, fabricated by freezing a crosslinked polymer solution, have been investigated as templates for bone tissue engineering due to their porosity, elasticity, and ability to retain their three-dimensional architecture. As bone fractures or defects are often sites of biofilm formation and bacterial infection due to their open nature, the eradication of bacteria is of utmost importance. Thus, research has focused on incorporating honey into cryogels as an antimicrobial agent. In a 2017 study, Hixon et al. incorporated Manuka honey into cryogels formed from either gelatin or silk fibroin. While the silk cryogels had larger pores (average pore diameter 25–40 μm) than the gelatin cryogels (average pore diameter 17–20 μm), the incorporation of Manuka honey significantly decreased these pore diameters in the silk cryogels, but not in the gelatin cryogels. Honey decreased the swelling ratios of the gelatin cryogels, but not the silk ones. Ultimate compression testing indicated that honey significantly decreased the average peak stress in both cryogel types, and decreased the modulus of the gelatin cryogels, which could make the honey-incorporated cryogels less feasible in load-bearing bone tissue applications. Manuka honey incorporation had no significant effect on the proliferation of seeded MG-63 osteosarcoma cells, but increased cellular infiltration in the highest (10% v/v) honey concentration silk cryogel samples. Similar to the results observed in the electrospun honey templates discussed above, glucose release tests showed the bulk of the incorporated honey was released within the first hour of hydration, however, after this bulk release there was a consistent release of 0.03 mg/mL glucose per day throughout the 14-day soak period for both the gelatin and silk cryogels. The peracetic acid sterilization procedure also was shown to remove most of the incorporated honey from both polymer types. Thus, it may be beneficial to use other sterilization methods, such as gamma radiation, in the future to avoid leaching out the honey from these structures. Bacterial clearance tests showed that the incorporation of honey significantly increased bacterial clearance of both *E. coli* and *S. agalactiae*, and bacterial broth clearance and bacterial adhesion assays confirmed this trend. Honey incorporation did not alter mineralization of the cryogels by the MG-63 cells over a 28-day culture period [86]. While the mechanical testing data indicates that the presence of honey weakens these cryogels and makes them more brittle, their ability to inhibit bacterial growth and induce cellular infiltration suggests their potential usefulness in bone tissue engineering. This tradeoff of mechanical stability should be accounted for when designing future honey cryogel-based therapies. Efforts should focus on protecting the honey content from washing out during sterilization or as an initial bulk release, creating a more long-term, sustained release greater than one-to-two hours. Additionally, the utilization of other polymers should be explored as a means to maintain mechanical strength and elasticity even with the incorporation of honey.

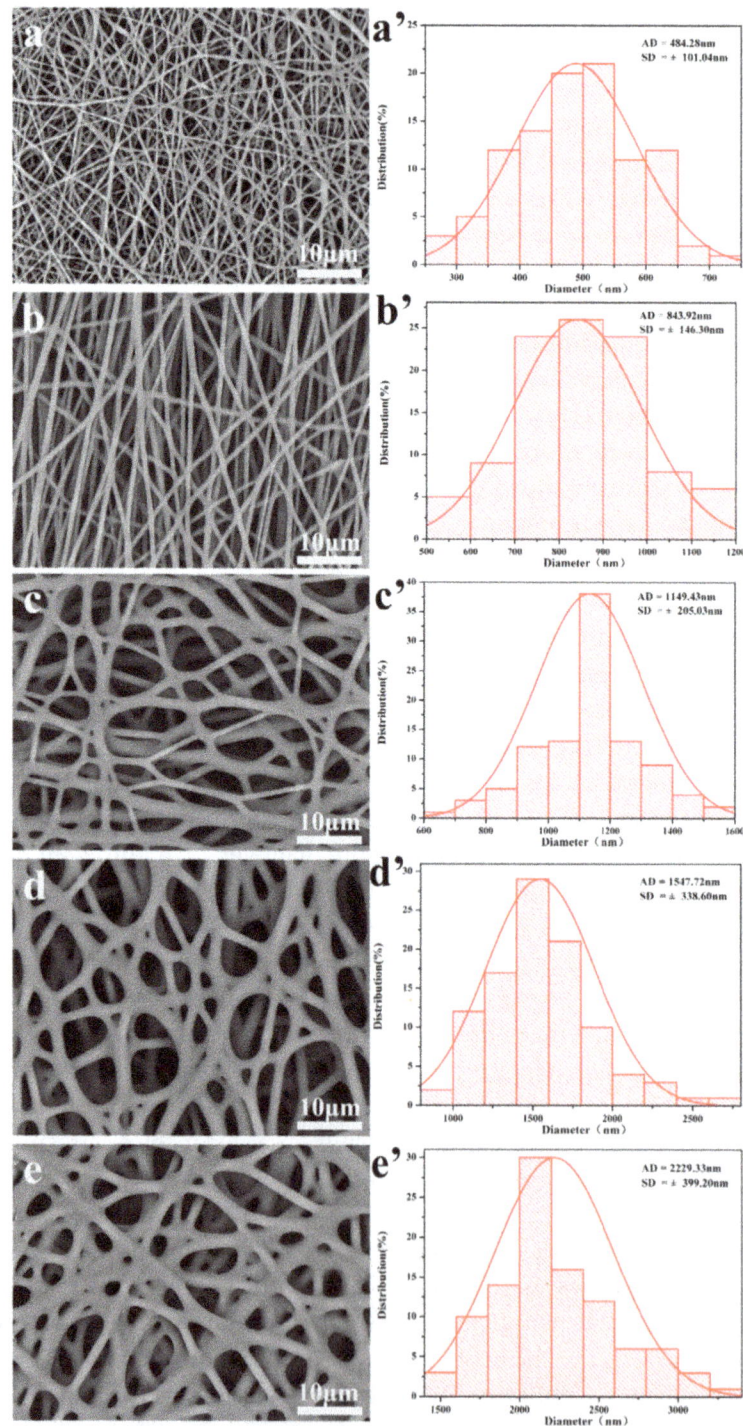

Figure 4. Honey increases fiber diameter. SEM images and histograms of fiber diameters of silk/PEO nanofibrous matrices spun with (**a,a'**) no Manukah honey; (**b,b'**) 10% *w/v* Manuka honey; (**c,c'**) 30% *w/v* Manuka honey; (**d,d'**) 50% *w/v* Manuka honey; and (**e,e'**) 70% *w/v* Manuka honey. Reproduced with permission from Yang et al., *Materials & Design*; published by Elsevier, 2017.

Although not addressed in this study, it has been shown that lowering the pH in the area around bone tissue can stimulate increased bone resorption and reduce mineral deposition by osteoclasts [87,88]. This effect is absent at or above a pH of 7.4, but is near-maximal at a pH of 7 [88]. Thus, there is a danger that the low pH of honey could impede bone regeneration rather than stimulate it. The study by Hixon et al. showed no effect of the honey on MG-63 osteosarcoma cell mineralization in vitro, but additional testing with non-cancerous osteoblasts and osteoclasts should be done. Given these well-documented effects of lowering the pH on bone resorption, it is speculated that honey-incorporated cryogels are may not be a useful bone-repair therapy.

In a subsequent study by Hixon et al., Manuka honey with various UMFs (unique manuka factor, a general quantification of bacterial inhibition) was incorporated into cryogels and electrospun templates, both fabricated from silk fibroin. The amount of honey in all constructs was kept constant at 5% v/v, while the UMF was varied by utilizing commercially available honeys rated with UMFs of 5+, 10+, 12+, 15+, and 20+. In general, UMF had no effect on the morphology of the cryogels or electrospun templates or on their ability to inhibit *E. coli* or *S. aureus*, and the electrospun templates had greater bacterial clearance (0.5–1 cm) than the cryogels (approximately 0.16 cm). The glucose release profiles from the cryogels and the electrospun templates were not statistically different, with the bulk of the glucose released within the first four days of soak maintaining a level of 0.4–0.6 mg/mL glucose in the surrounding solution [89]. Thus, it is unknown why the electrospun templates were more effective at clearing both types of bacteria. One explanation could be that different bactericidal components of the honey, such as the methylglyoxal, hydrogen peroxide, or the gluconic acid, are released at different rates or profiles than the glucose, and these rates may be different between the cryogels and the electrospun templates. Assays for these other components may be necessary to fully explain the dramatic difference in bacterial inhibition between these template types. Likewise, it is curious that the UMF of the honey used in these templates did not affect their bacterial clearance. Part of the problem may be that the exact UMF of each honey obtained was not listed, only that it was above the listed level (5+ means that it has a UMF of at least 5, not exactly 5). Thus, it is possible that the actual UMFs of the honeys did not vary as much as was thought based on their labels. It would be beneficial for future studies testing different UMFs to test the UMF in-house, rather than relying on the UMF rating of the commercial vendor. Thus, it is still unknown whether the UMF of the Manuka honey used affects bacterial clearance or other properties when incorporated into tissue templates.

4.3. Hydrogels

Hydrogels, highly absorbent networks of hydrophilic polymer chains, are often used as templates and drug delivery devices in tissue engineering due to their polymeric structure and the ability to control characteristics, such as pore size, water content, and degradation profile. Several groups have explored incorporating honey into hydrogels for use as wound coverings. In a 2012 study, Wang et al. incorporated Chinese Sunflower honey at 10% or 20% v/v into hydrogel sheets fabricated from chitosan and bovine gelatin. Swelling studies indicated that the presence of honey reduced the ability of the hydrogel to absorb fluid, with the 20% honey hydrogels swelling only around 250%, as compared to the 700% swelling of the non-honey control. Compression testing indicated that honey content also reduced the modulus of the hydrogel sheets from around 110 kPa to around 60–70 kPa for the 20% honey hydrogel and approximately 58 kPa for the 10% honey hydrogel. Antibacterial assays showed that the presence of honey significantly increased the inhibition of *S. aureus* and *E. coli* growth, with the 20% honey hydrogel causing almost 100% inhibition while the non-honey hydrogel caused approximately 20% inhibition of both bacterial types. In vivo oral toxicity tests were performed in mice, and dermal irritation and burn wound healing tests were performed in rabbits. As expected, the mice toxicology tests showed no toxic symptoms. After eight days of treatment in rabbit wounds, the honey hydrogel group averaged about 80% wound closure, while the ointment group had about 60% wound closure and the non-treated group had about 45% wound closure, as shown in Figure 5. Histological examination of the wounds after 12 days revealed that the untreated wounds were infected, contained a high amount of inflammatory cells, and had no hair follicles. The ointment-treated group had smaller ulcers than the non-treated group, but still contained acute inflammatory infiltrate that collected in small cysts underneath the regenerated epidermis. Both the ointment group and the honey hydrogel group showed epidermal healing, but the honey hydrogel group had less inflammatory infiltrate and also had proliferating hair follicles on the surface [90]. Although this study thoroughly characterized the in vitro and in vivo aspects of the honey hydrogel, the lack of a control non-honey hydrogel group in the animal studies call into question whether the honey improved the in vivo performance of the hydrogel. Additionally, the lack of a topical honey treatment group in these studies made it impossible to ascertain whether incorporation of the honey into a hydrogel improved healing over the current clinical method of treatment. Thus, the only conclusions that can be drawn from this section of the study are that this honey-containing hydrogel improves healing relative to an ointment treatment. A glucose release study would also improve this paper by showing whether the honey is released in a burst or released in a controlled fashion over days or weeks after implantation.

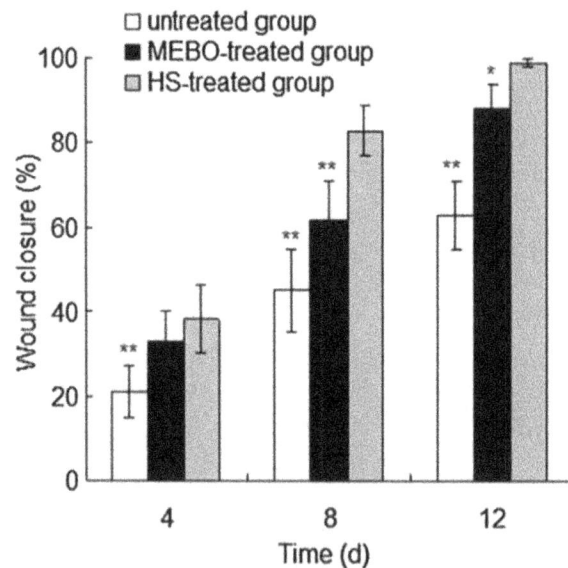

Figure 5. Honey hydrogel induces wound closure. Wound closure rates in rabbit wounds that were untreated, treated with a commercial wound ointment (MEBO), or treated with a honey-infused chitosan/gelatin hydrogel (HS) at 4, 8, or 12 days after beginning of treatment. "**" indicates statistical significance at $p < 0.005$, "*" indicates statistical significance at $p < 0.01$. Reproduced with permission from Wang et al., *Carbohydrate Polymers*, published by Elsevier, 2012.

More recently, Sasikala et al. incorporated Manuka honey into chitosan hydrogel films, also for use as wound dressings. Chitosan solutions containing 8% w/v Manuka honey were cast in Petri dishes and dried at 40 °C for 24 h. Honey increased the folding endurance of the samples with the honey samples surviving a mean of 289 folds, whereas the non-honey films survived a mean of 143 folds. This finding indicates greater flexibility of the honey hydrogels, which is likely a function of the hygroscopic effect of the honey. However, no effect of the honey was observed on the water vapor transmission rate of these films, which was curious given the results observed in electrospun scaffolds as detailed earlier in this review. As observed in the study discussed above, honey decreased the swelling ratios of the hydrogel films and increased the inhibition of *S. aureus* and *E. coli* growth. When these films were placed in a rat dorsal wound model, increased wound closure was observed in the honey samples relative to non-honey control films and cipladin ointment controls. Specifically, after 12 days of treatment, the honey hydrogel wound was 94% closed, the non-honey hydrogel wound was 78% closed, the ointment-treated wound was 86% closed, and the non-treated control wound was 64% closed [91]. The use of non-honey control hydrogel films in this animal study show the benefit of the honey to the wound healing process, which is an improvement over the Wang et al. study discussed above. However, many of the studies described in this paper seem to have been undertaken with a sample size of $n = 1$, as there are no standard deviations or standard errors reported. Although the methods section says that an ANOVA was performed on the wound closure data, it does not report which specific sample groups were significant from each other, casting doubt as to the scientific veracity of these findings. This study would benefit from being repeated more thoroughly so that its findings can be scientifically corroborated.

In 2008, Gethin et al. published a study in which 20 patients with chronic, non-healing leg ulcers were treated with Apinate, a commercially available Manuka honey hydrogel dressing made by Derma Sciences. This study focused on the effect of the Manuka honey hydrogel in lowering the wound pH, and the corresponding effect on wound size reduction. The Apinate dressing itself had a pH of 4.0, due to the acidity of the honey. After a two-week period, wounds treated with Apinate had a mean pH drop of 0.46, with a mean wound size reduction of 1 cm^2. A linear regression model was developed using the experimental data, showing a significant relationship between drop in pH and a reduction in wound size over the two-week period, with a one unit reduction in pH being associated with a decrease of 81% of wound size [3]. It is unknown how much of the healing effect was a function of the honey's pH, as opposed to its osmotic effects, bactericidal effects, or other properties detailed earlier in this review, or whether the pH is an effect of wound healing instead of a cause. However, a decrease in pH has been shown to increase oxygen saturation, reduce elastase activity, and kill certain bacteria, which all aid wound healing [92–94].

Giusto et al. have conducted research incorporating Manuka honey into pectin-based hydrogels. They report that honey-containing pectin hydrogels have superior bacterial clearance of *S. aureus* and *E. coli*, and demonstrate no cytotoxicity to fibroblasts [95,96]. Another study was conducted by Zhodi et al. in which Gelam honey, a honey produced in Malaysia, was incorporated into hydrogels made from polyvinyl pyrrolidone (PVP) and polyethylene glycol (PEG). Honey content significantly decreased the pH value of the hydrogels (from 5.3 to 4.3) and increased the swelling of the hydrogels by a factor of five relative to non-honey controls. A large-scale burn wound study was undertaken using 96 rats, six rats per experimental group. Wounds treated with the honey-containing hydrogels significantly decreased in size relative to non-honey hydrogel controls by days 21 and 28, with the honey-treated wounds averaging a 91% reduction in size compared to the 72% reduction in size of the control hydrogel wounds. Histological examination showed decreased inflammatory exudate by day seven and increased dermal repair and reepithelialization by day 21 in the honey-containing hydrogel wounds. These wounds also showed an increase in granulation tissue and capillary formation, as well as collagen synthesis. RNA extracted from the wound site showed that the honey-containing hydrogel treatment caused a significant decrease in IL-1α, IL-1β, and IL-6 expression relative to control hydrogels, a commercial Opsite film wound dressing, or non-treatment groups. Specifically, the honey hydrogel caused a drop from around 3.5% expression to 0.5% of IL-1α and IL-1β, and from about 3.5% expression to 0.1% expression of IL-6 mRNA after seven days, normalizing expression to a β actin control, as shown in Figure 6 [97]. This animal study is the most in-depth look at the in vivo effect of honey-containing hydrogels and demonstrates that the honey content reduces inflammatory cytokine output, reduces inflammatory exudate, increases the formation of granulation tissue, and increases the wound closure rate. Ideally, future studies of this type will also look at a greater number of relevant secreted factors, such as TNF-α, IL-8, MIP-1α, MIP-3α, VEGF, MMP-1, MMP-9, and Proteinase 3, among others. In this way, a more complete understanding of the wound environment could be ascertained.

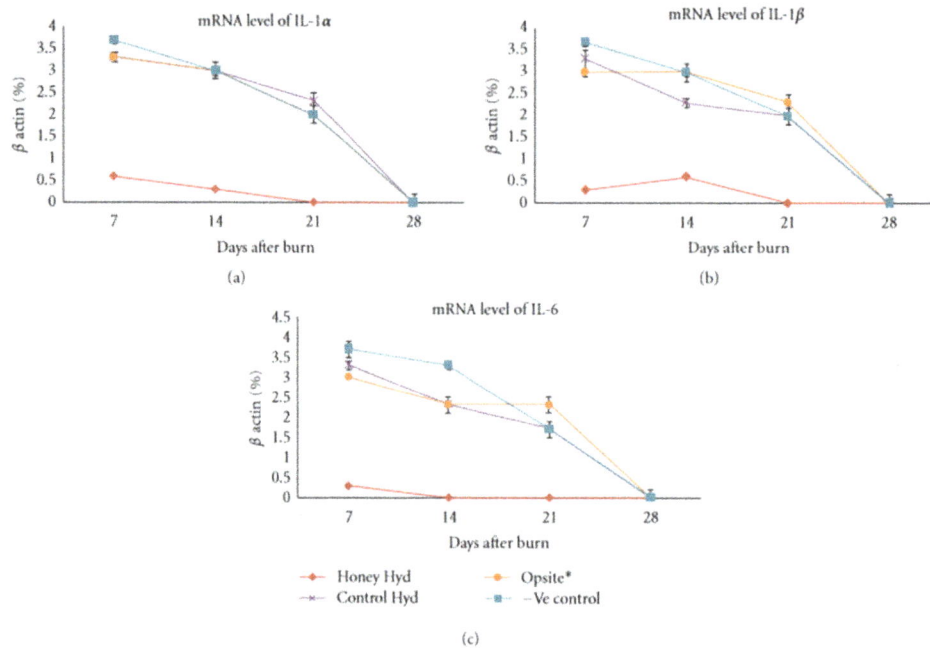

Figure 6. Honey reduces inflammatory cytokine expression. mRNA expression of (**a**) IL-1α, (**b**) IL-1β, and (**c**) IL-6 in a rat burn wound model treated with a control hydrogel, a commercial Opsite film dressing, the honey hydrogel, or non-treated (-Ve) control, normalized to β-actin.

5. Commercialization

Currently, several companies sell or are developing products which contain Manuka honey for wound care, as shown in Table 2. Derma Sciences, a tissue regeneration company based in Princeton, New Jersey, sells a line of Manuka honey products under the brand Medihoney®. In addition to pastes and gels combining Manuka honey with gelling agents to increase viscosity, this company also sells several variants of an alginate-based hydrogel sheet containing Manuka honey for use as wound coverings, including the Apinate dressing discussed earlier in the hydrogel section [98]. Several in vitro studies have confirmed the antibacterial effects of these products [99,100]. Furthermore, randomized controlled trials and case studies have demonstrated the antibacterial and wound healing effects of these products in the clinical setting [73,101–105]. Simon et al. detailed the treatment of various surgical wounds and drainage sites in pediatric oncology patients with a Medihoney® alginate wound covering, and reported that the honey reduced irritation and cleared infections. One acute lymphatic leukemia patient with a high level of immune suppression had a persistent deep surgical site infection that healed completely once treated with the Medihoney® [73]. Johnson et al. conducted a randomized controlled trial of topical Medihoney® application versus mupirocin in preventing catheter-associated infections and found that while the honey was comparable to the mupirocin in preventing infection, 2% of staphylococcal isolates were mupirocin-resistant. Thus, they concluded that honey represented a good alternative to the gold-standard antibiotic [101]. A wound-focused randomized controlled trial conducted by Robson et al. using topical Medihoney® observed about a 10% increase in the healing rate in the honey-treated wounds as opposed to conventionally-treated wounds, which was statistically significant [102]. A prospective observational study by Biglari et al. using Medihoney® focused specifically on chronic pressure ulcers and found that the honey eradicated bacterial growth in all 20 ulcers treated, with 90% of patients showing complete wound healing after four weeks [104]. Likewise, Smith et al. published a case series of topical Medihoney®-treated recalcitrant venous leg wounds that had failed to respond to four-layer compression, topical silver, nonadherent dressings, and antibiotic therapy. All 11 wounds treated had 100% closure by six weeks, with an average wound healing velocity of 0.25 cm^2/day [105]. In addition to proving the efficacy of the Medihoney® product line, specifically, these trials and case studies lend additional weight to the general benefits of Manuka honey on wounds.

Table 2. Summary of commercial wound-care products containing Manuka honey currently on the market.

Product Name	Company	Product Category
Medihoney® Paste	Derma Sciences	Paste, topical application
Manuka Honey (Paste)	Manuka Health	Paste, topical application
Activon Tube	Advancis Medical	Paste, topical application
Medihoney® Gel	Derma Sciences	Gel, topical application
ManukaApli	ManukaMed	Gel, topical application
Medihoney® Alginate Dressing	Derma Sciences	Composite Hydrogel
Manuka Health Wound Dressing	Manuka Health	Composite Hydrogel
Algivon, Algivon Plus	Advancis Medical	Composite Hydrogel
Medihoney® HCS Application	Derma Sciences	Honey-impregnated Dressing
Medihoney® Honeycolloid Application	Derma Sciences	Honey-impregnated Dressing
Manukahd	ManukaMed	Honey-impregnated Dressing
ManukaMed MedSaf	ManukaMed	Honey-impregnated Dressing
Manukahd Lite	ManukaMed	Honey-impregnated Dressing
Manukahd Lite rope	ManukaMed	Honey-impregnated Dressing
Actilite	Advancis Medical	Honey-impregnated Dressing
Algivon Tulle	Advancis Medical	Honey-impregnated Dressing

In addition to the Medihoney® line by Derma Sciences, there are a few other companies that sell Manuka honey-based therapeutic products. The New Zealand-based Manuka Health (Newmarket, Auckland) manufactures a glycerin-based honey-containing hydrogel sheet for wound covering [106]. Another New Zealand-based company, ManukaMed Ltd. (Solway, Masterton), sells variations of gauze-based, honey-impregnated fiber pad wound coverings [107]. The United Kingdom-based Advancis Medical (Kirkby-in-Ashfield, Nottinghamshire) makes several types of Manuka honey-impregnated wound coverings, including cellulose-based net dressings (Actilite and Activone Tulle) and alginate hydrogels (Algivon, Algivon Plus) [108]. Additionally, a Memphis, Tennessee-based company, SweetBio Inc. (Memphis, TN, USA), is developing a resorbable membrane for oral surgery, with manufactured prototypes currently undergoing testing [109]. Of these products, the only ones extensively studied in peer-reviewed literature are the Medihoney® coverings. Except the Sweetbio membrane, all of these products are designed as temporary wound coverings that must be physically replaced. While no company has published release profiles of the honey from their products, given the described methods of application it is likely that the honey is delivered at high concentrations as a bolus. As high concentrations of Manuka honey have been observed to be cytotoxic in vitro, there may be a potential to improve the healing outcomes of wounds treated with these products by diluting the honey or by attenuating the release to lower levels for prolonged amounts of time [13]. However, attenuating the honey concentration may diminish the antibacterial effects of the dressings. Thus, higher concentrations of honey should be used initially in infected wounds to ensure the killing of the invading bacteria. A two-step process may be optimal, in which heavily infected wounds are first treated by topical application of undiluted Manuka honey to eradicate the infection, and then the honey is removed and replaced by a controlled-release template to aid in tissue regeneration. Alternatively, delivery of lower levels of honey alongside an antibiotic could enable the eradication of bacteria and stimulation of healing without the undesirable cytotoxic effects of the honey.

6. The Future of Honey in Tissue Engineering

The greatest hurdles to be overcome in the development of honey-containing tissue engineering templates are the cytotoxicity of high concentrations of honey and the lack of prolonged, consistent release rates of the honey over time. In templates which rely on cellular infiltration and tissue ingrowth, cells may encounter higher concentrations of honey as they infiltrate into the honey-containing template than they would outside or adjacent to that template. Such templates will be surrounded by a honey gradient radiating away from them into the surrounding tissue, and migrating cells will encounter higher and higher honey levels as they move towards, and into, the template. Thus, it will be important to monitor honey levels not only in the template releasate, but within the template environment itself during in vitro honey release studies. While some of the studies discussed above did not see honey cytotoxicity as an impediment to cellular proliferation and infiltration, these studies used liquid ethanol disinfection or peracetic acid sterilization steps that washed away the majority of the honey from the templates before cell seeding [80]. In applications where templates are disinfected via ultraviolet or gamma radiation, or ethylene oxide sterilization, no honey will be removed before cell seeding or template implantation. Thus, cytotoxicity could impede cell infiltration and proliferation. The templates studied thus far tend to release their honey content in a bolus during the first day of soak or implantation. While this type of release may be acceptable for a wound covering that can be removed and replaced, templates that are surgically implanted and resorbed into the body must contain the entire honey amount necessary for the application. Thus, it is important that methods be developed to attenuate and delay the release of honey over a period of days to weeks. In electrospun fibers, this may be accomplished by the use of core-shell electrospinning, in which fibers are created with a core of one polymer type and a shell of another [110,111]. By encapsulating honey within the fiber cores, its release could be delayed over time, with either diffusion or polymer degradation controlling its release rate. Templates created via this method should be subjected to rigorous mechanical testing to ensure that incorporating a honey core does not cause the fibers to weaken or become too brittle for their intended use. As of yet, there are no published studies using core-shell electrospinning with honey. However, this technology is likely a next step for the field. As there is no equivalent to core-shell electrospinning in the field of hydrogel and cryogel fabrication, other methods must be used to attenuate the honey release. Possible techniques include increasing polymer molecular weight and concentration and increasing the crosslinking density to reduce liquid diffusion through the templates—however, this will likely decrease the swelling ratio of the templates which may be undesirable. These effects will likely have to be balanced to achieve the optimal honey release rate, water vapor transmission rate, and absorbance of these hydrogels and cryogels. Additionally, due to the reported decrease in compressive modulus and strength of these constructs with honey incorporation, care will have to be taken to make sure that their mechanical properties are not compromised for their intended application [86,90,112].

More study is also needed on the effects of honey on immune cells such as neutrophils and monocytes/macrophages. While Tonks et al. showed that Manuka honey causes an increase in the output of TNF-α, IL-1, and IL-6 by monocytes over a 24-h period, it would be informative to ascertain the effect on levels of other inflammatory, anti-inflammatory, and angiogenic signals, such as IL-8, VEGF, IL-4, IL-1ra, MIP-1α, MIP-3α, etc. Additionally, similar testing of neutrophil cytokine output would be helpful, as their inflammatory and anti-inflammatory effects as the first responding immune cells in wounds are being assessed with increased importance [113–115]. Further testing of the effects of honey on neutrophil superoxide output, chemotaxis, and NF-$\kappa\beta$ activation in the presence of different inflammatory and anti-inflammatory stimulators is needed to fully understand how honey affects the regulation of the wound environment by neutrophils. As noted earlier in this study, the current data suggests that high levels of honey (above 33% w/v for Manuka honey) are necessary to fight bacterial biofilms. However, the evidence also suggests that these high honey levels have the potential to cause significant cytotoxicity. Thus, in wounds with established biofilms, it is likely that a two-step treatment process will have to be implemented. First, a high concentration of honey can be applied topically to the wound to destroy the biofilm and eradicate the bacteria. After the infection has been eliminated, a tissue engineered template can be applied which releases lower levels of honey to reduce inflammation and aid in tissue regeneration, without causing cytotoxicity. In wounds without established biofilm-based infections, the first step may not be necessary, and the honey-eluting template may be directly applied to the area.

7. Conclusions

In vitro and in vivo evidence shows that honey, particularly Manuka honey, eliminates bacteria, resolves chronic inflammation, and promotes faster wound healing. Its potency against antibiotic-resistant bacteria, such as MRSA, makes it a particularly invaluable tool in an age where more strains of resistant bacteria are developing. As such, honey is a valuable addition to many tissue engineering templates in eliminating bacterial infection, aiding in inflammation resolution, and improving tissue integration with the template. Future research should focus on attenuating and prolonging the release of honey from the templates to avoid cytotoxicity and prolong the beneficial effects of the honey within the site.

Acknowledgments: The authors thank the University of Memphis and the Van Vleet Memorial Award.

Conflicts of Interest: Gary Bowlin has a financial interest in SweetBio Inc.

References

1. Forrest, R.D. Early history of wound treatment. *J. R. Soc. Med.* **1982**, *75*, 198. [PubMed]
2. Clardy, J.; Fischbach, M.A.; Currie, C.R. The natural history of antibiotics. *Curr. Biol.* **2009**, *19*, R437–R441. [CrossRef] [PubMed]
3. Gethin, G.T.; Cowman, S.; Conroy, R.M. The impact of Manuka honey dressings on the surface pH of chronic wounds. *Int. Wound J.* **2008**, *5*, 185–194. [CrossRef] [PubMed]
4. Ball, D.W. The chemical composition of honey. *J. Chem. Educ.* **2007**, *84*, 1643. [CrossRef]
5. White, J.; Doner, L.W. Honey composition and properties. *Beekeep. U. S. Agric. Handb.* **1980**, *335*, 82–91.
6. Kwakman, P.H.; Zaat, S.A. Antibacterial components of honey. *IUBMB Life* **2012**, *64*, 48–55. [CrossRef] [PubMed]
7. Molan, P.; Cooper, R.; Molan, P.; White, R. Why honey works. In *Honey in Modern Wound Management*; Wounds UK Ltd.: Aberdeen, UK, 2009; Volume 9, pp. 36–37.
8. Sanz, M.; Gonzalez, M.; De Lorenzo, C.; Sanz, J.; Martınez-Castro, I. A contribution to the differentiation between nectar honey and honeydew honey. *Food Chem.* **2005**, *91*, 313–317. [CrossRef]
9. Földházi, G. Analysis and quantitation of sugars in honey of different botanical origin using high performance liquid chromatography. *Acta Aliment.* **1994**, *23*, 299–311.
10. Majtan, J.; Bohova, J.; Garcia-Villalba, R.; Tomas-Barberan, F.A.; Madakova, Z.; Majtan, T.; Majtan, V.; Klaudiny, J. Fir honeydew honey flavonoids inhibit TNF-α-induced MMP-9 expression in human keratinocytes: A new action of honey in wound healing. *Arch. Dermatol. Res.* **2013**, *305*, 619–627. [CrossRef] [PubMed]
11. Meda, A.; Lamien, C.E.; Romito, M.; Millogo, J.; Nacoulma, O.G. Determination of the total phenolic, flavonoid and proline contents in Burkina Fasan honey, as well as their radical scavenging activity. *Food Chem.* **2005**, *91*, 571–577. [CrossRef]
12. Jenkins, R.; Burton, N.; Cooper, R. Effect of manuka honey on the expression of universal stress protein A in meticillin-resistant *Staphylococcus aureus*. *Int. J. Antimicrob. Agents* **2011**, *37*, 373–376. [CrossRef] [PubMed]
13. Leong, A.G.; Herst, P.M.; Harper, J.L. Indigenous New Zealand honeys exhibit multiple anti-inflammatory activities. *Innate Immun.* **2012**, *18*, 459–466. [CrossRef] [PubMed]
14. Adams, C.J.; Manley-Harris, M.; Molan, P.C. The origin of methylglyoxal in New Zealand manuka (*Leptospermum scoparium*) honey. *Carbohydr. Res.* **2009**, *344*, 1050–1053. [CrossRef] [PubMed]
15. Cooper, R.; Molan, P.; Harding, K. The sensitivity to honey of Gram-positive cocci of clinical significance isolated from wounds. *J. Appl. Microbiol.* **2002**, *93*, 857–863. [CrossRef] [PubMed]

16. Molan, P.C. Re-introducing honey in the management of wounds and ulcers-theory and practice. *Ostomy Wound Manag.* **2002**, *48*, 28–40.

17. Molan, P.C. Potential of honey in the treatment of wounds and burns. *Am. J. Clin. Dermatol.* **2001**, *2*, 13–19. [CrossRef] [PubMed]

18. Gleiter, R.; Horn, H.; Isengard, H.-D. Influence of type and state of crystallisation on the water activity of honey. *Food Chem.* **2006**, *96*, 441–445. [CrossRef]

19. Sundoro, A.; Nadia, K.; Nur, A.; Sudjatmiko, G.; Tedjo, A. Comparison of physical–chemical characteristic and antibacterial effect between Manuka honey and local honey. *J. Plast. Rekonstr.* **2012**, *1*. [CrossRef]

20. Rockland, L.B.; Beuchat, L.R. *Water Activity: Theory and Applications to Food*; M. Dekker: New York, NY, USA, 1987.

21. Molan, P.C. The evidence supporting the use of honey as a wound dressing. *Int. J. Lower Extrem. Wounds* **2006**, *5*, 40–54. [CrossRef] [PubMed]

22. Kwakman, P.H.; Te Velde, A.A.; De Boer, L.; Vandenbroucke-Grauls, C.M.; Zaat, S.A. Two major medicinal honeys have different mechanisms of bactericidal activity. *PLoS ONE* **2011**, *6*, e17709. [CrossRef] [PubMed]

23. Brudzynski, K.; Abubaker, K.; Castle, A. Re-examining the role of hydrogen peroxide in bacteriostatic and bactericidal activities of honey. *Front. Microbiol.* **2011**, *2*, 213. [CrossRef] [PubMed]

24. Roberts, A.E.; Maddocks, S.E.; Cooper, R.A. Manuka honey is bactericidal against Pseudomonas aeruginosa and results in differential expression of oprF and algD. *Microbiology* **2012**, *158*, 3005–3013. [CrossRef] [PubMed]

25. Wang, R.; Starkey, M.; Hazan, R.; Rahme, L. Honey's ability to counter bacterial infections arises from both bactericidal compounds and QS inhibition. *Front. Microbiol.* **2012**, *3*, 144. [CrossRef] [PubMed]

26. Molan, P.C. The antibacterial activity of honey: 2. Variation in the potency of the antibacterial activity. *Bee World* **1992**, *73*, 59–76. [CrossRef]

27. Efem, S. Clinical observations on the wound healing properties of honey. *Br. J. Surg.* **1988**, *75*, 679–681. [CrossRef] [PubMed]

28. Armon, P. The use of honey in the treatment of infected wounds. *Trop. Dr.* **1980**, *10*, 91. [CrossRef] [PubMed]

29. Cooper, R.; Molan, P.; Harding, K. Antibacterial activity of honey against strains of Staphylococcus aureus from infected wounds. *J. R. Soc. Med.* **1999**, *92*, 283–285. [CrossRef] [PubMed]

30. Kwakman, P.H.; te Velde, A.A.; de Boer, L.; Speijer, D.; Vandenbroucke-Grauls, C.M.; Zaat, S.A. How honey kills bacteria. *FASEB J.* **2010**, *24*, 2576–2582. [CrossRef] [PubMed]

31. Bachanová, K.; Klaudiny, J.; Kopernický, J.; Šimúth, J. Identification of honeybee peptide active against Paenibacillus larvae larvae through bacterial growth-inhibition assay on polyacrylamide gel. *Apidologie* **2002**, *33*, 259–269. [CrossRef]

32. Ganz, T. Defensins: Antimicrobial peptides of innate immunity. *Nat. Rev. Immunol.* **2003**, *3*, 710. [CrossRef] [PubMed]

33. Rabie, E.; Serem, J.C.; Oberholzer, H.M.; Gaspar, A.R.M.; Bester, M.J. How methylglyoxal kills bacteria: An ultrastructural study. *Ultrastruct. Pathol.* **2016**, *40*, 107–111. [CrossRef] [PubMed]

34. Majtan, J.; Klaudiny, J.; Bohova, J.; Kohutova, L.; Dzurova, M.; Sediva, M.; Bartosova, M.; Majtan, V. Methylglyoxal-induced modifications of significant honeybee proteinous components in manuka honey: Possible therapeutic implications. *Fitoterapia* **2012**, *83*, 671–677. [CrossRef] [PubMed]

35. Majtan, J.; Bohova, J.; Prochazka, E.; Klaudiny, J. Methylglyoxal may affect hydrogen peroxide accumulation in manuka honey through the inhibition of glucose oxidase. *J. Med. Food* **2014**, *17*, 290–293. [CrossRef] [PubMed]

36. Blaser, G.; Santos, K.; Bode, U.; Vetter, H.; Simon, A. Effect of medical honey on wounds colonised or infected with MRSA. *J. Wound Care* **2007**, *16*, 325–328. [CrossRef] [PubMed]

37. Thawley, A. The components of honey and their effects on its properties: A review. *Bee World* **1969**, *50*, 51–60. [CrossRef]

38. Garcia-Alvarez, M.; Huidobro, J.; Hermida, M.; Rodriguez-Otero, J. Major components of honey analysis by near-infrared transflectance spectroscopy. *J. Agric. Food Chem.* **2000**, *48*, 5154–5158. [CrossRef] [PubMed]

39. Molan, P.C. The evidence and the rationale for the use of honey as wound dressing. *Wound Pract. Res.* **2011**, *19*, 204–220.

40. Lusby, P.; Coombes, A.; Wilkinson, J. Honey: A potent agent for wound healing? *J. Wound Ostomy Cont. Nurs.* **2002**, *29*, 295–300. [CrossRef]

41. Sherlock, O.; Dolan, A.; Athman, R.; Power, A.; Gethin, G.; Cowman, S.; Humphreys, H. Comparison of the antimicrobial activity of Ulmo honey from Chile and Manuka honey against methicillin-resistant Staphylococcus aureus, Escherichia coli and Pseudomonas aeruginosa. *BMC Complement. Altern. Med.* **2010**, *10*, 47. [CrossRef] [PubMed]

42. Al Somal, N.; Coley, K.; Molan, P.; Hancock, B. Susceptibility of Helicobacter pylori to the antibacterial activity of manuka honey. *J. R. Soc. Med.* **1994**, *87*, 9–12. [PubMed]

43. Watanabe, K.; Rahmasari, R.; Matsunaga, A.; Haruyama, T.; Kobayashi, N. Anti-influenza viral effects of honey in vitro: Potent high activity of manuka honey. *Arch. Med. Res.* **2014**, *45*, 359–365. [CrossRef] [PubMed]

44. Zeina, B.; Othman, O.; Al-Assad, S. Effect of honey versus thyme on Rubella virus survival in vitro. *J. Altern. Complement. Med.* **1996**, *2*, 345–348. [CrossRef] [PubMed]

45. Shahzad, A.; Cohrs, R.J. In vitro antiviral activity of honey against varicella zoster virus (VZV): A translational medicine study for potential remedy for shingles. *Transl. Biomed.* **2012**, *3*. [CrossRef]

46. Merckoll, P.; Jonassen, T.Ø.; Vad, M.E.; Jeansson, S.L.; Melby, K.K. Bacteria, biofilm and honey: A study of the effects of honey on 'planktonic' and biofilm-embedded chronic wound bacteria. *Scand. J. Infect. Dis.* **2009**, *41*, 341–347. [CrossRef] [PubMed]

47. Jervis-Bardy, J.; Foreman, A.; Bray, S.; Tan, L.; Wormald, P.J. Methylglyoxal-infused honey mimics the anti-Staphylococcus aureus biofilm activity of manuka honey: Potential Implication in Chronic Rhinosinusitis. *Laryngoscope* **2011**, *121*, 1104–1107. [CrossRef] [PubMed]

48. Alandejani, T.; Marsan, J.; Ferris, W.; Slinger, R.; Chan, F. Effectiveness of honey on Staphylococcus aureus and Pseudomonas aeruginosa biofilms. *Otolaryngol. Head Neck Surg.* **2009**, *141*, 114–118. [CrossRef] [PubMed]

49. Okhiria, O.; Henriques, A.; Burton, N.; Peters, A.; Cooper, R. Honey modulates biofilms of Pseudomonas aeruginosa in a time and dose dependent manner. *J. ApiProd. ApiMed. Sci.* **2009**, *1*, 6–10.

50. Sojka, M.; Valachova, I.; Bucekova, M.; Majtan, J. Antibiofilm efficacy of honey and bee-derived defensin-1 on multispecies wound biofilm. *J. Med. Microbiol.* **2016**, *65*, 337–344. [CrossRef] [PubMed]

51. Moghazy, A.; Shams, M.; Adly, O.; Abbas, A.; El-Badawy, M.; Elsakka, D.; Hassan, S.; Abdelmohsen, W.; Ali, O.; Mohamed, B. The clinical and cost effectiveness of bee honey dressing in the treatment of diabetic foot ulcers. *Diabetes Res. Clin. Pract.* **2010**, *89*, 276–281. [CrossRef] [PubMed]

52. Otto, M. Staphylococcus epidermidis—The 'accidental' pathogen. *Nat. Rev. Microbiol.* **2009**, *7*, 555. [CrossRef] [PubMed]

53. Efem, S.E. Recent advances in the management of Fournier's gangrene: Preliminary observations. *Surgery* **1993**, *113*, 200–204. [PubMed]

54. Gethin, G.; Cowman, S. Bacteriological changes in sloughy venous leg ulcers treated with manuka honey or hydrogel: An RCT. *J. Wound Care* **2008**, *17*, 241–247. [CrossRef] [PubMed]

55. The Science Behind Honey's Eternal Shelf Life. Available online: https://www.smithsonianmag.com/science-nature/the-science-behind-honeys-eternal-shelf-life-1218690/ (accessed on 21 February 2018).

56. Frequently Asked Questions. Available online: https://www.honey.com/faq (accessed on 21 February 2018).

57. Antony, S.; Rieck, J.; Acton, J.; Han, I.; Halpin, E.; Dawson, P. Effect of dry honey on the shelf life of packaged turkey slices. *Poult. Sci.* **2006**, *85*, 1811–1820. [PubMed]

58. Ergun, M.; Ergun, N. Extending shelf life of fresh-cut persimmon by honey solution dips. *J. Food Process. Preserv.* **2010**, *34*, 2–14. [CrossRef]

59. Viuda-Martos, M.; Ruiz-Navajas, Y.; Fernández-López, J.; Pérez-Álvarez, J. Functional properties of honey, propolis, and royal jelly. *J. Food Sci.* **2008**, *73*, R117–R124. [CrossRef] [PubMed]

60. Randall, W.A.; Welcb, H.; Hunter, A.C. The stability of penicillin sodium held at various temperatures. *J. Pharm. Sci.* **1945**, *34*, 110–113. [CrossRef]

61. Dimins, F.; Kuka, P.; Kuka, M.; Cakste, I. The criteria of honey quality and its changes during storage and thermal treatment. *Proc. Latvia Univ. Agric.* **2006**, *16*, 73–78.

62. Bloom, B.R. Vaccines for the third world. *Nature* **1989**, *342*, 115. [CrossRef] [PubMed]

63. Wang, L.; Li, J.; Chen, H.; Li, F.; Armstrong, G.L.; Nelson, C.; Ze, W.; Shapiro, C.N. Hepatitis B vaccination of newborn infants in rural China: Evaluation of a village-based, out-of-cold-chain delivery strategy. *Bull. World Health Organ.* **2007**, *85*, 688–694. [CrossRef] [PubMed]

64. Tonks, A.J.; Cooper, R.; Jones, K.; Blair, S.; Parton, J.; Tonks, A. Honey stimulates inflammatory cytokine production from monocytes. *Cytokine* **2003**, *21*, 242–247. [CrossRef]

65. Sell, S.A.; Wolfe, P.S.; Spence, A.J.; Rodriguez, I.A.; McCool, J.M.; Petrella, R.L.; Garg, K.; Ericksen, J.J.; Bowlin, G.L. A preliminary study on the potential of manuka honey and platelet-rich plasma in wound healing. *Int. J. Biomater.* **2012**, *2012*. [CrossRef] [PubMed]

66. Oryan, A.; Zaker, S. Effects of topical application of honey on cutaneous wound healing in rabbits. *Transbound. Emerg. Dis.* **1998**, *45*, 181–188. [CrossRef]

67. Prakash, A.; Medhi, B.; Avti, P.; Saikia, U.; Pandhi, P.; Khanduja, K. Effect of different doses of Manuka honey in experimentally induced inflammatory bowel disease in rats. *Phytother. Res.* **2008**, *22*, 1511–1519. [CrossRef] [PubMed]

68. Ranzato, E.; Martinotti, S.; Burlando, B. Epithelial mesenchymal transition traits in honey-driven keratinocyte wound healing: Comparison among different honeys. *Wound Repair Regener.* **2012**, *20*, 778–785. [CrossRef] [PubMed]

69. Suguna, L.; Chandrakasan, G.; Joseph, K.T. Influence of honey on collagen metabolism during wound healing in rats. *J. Clin. Biochem. Nutr.* **1992**, *13*, 7–12. [CrossRef]

70. Gethin, G.; Cowman, S. Case series of use of Manuka honey in leg ulceration. *Int. Wound J.* **2005**, *2*, 10–15. [CrossRef] [PubMed]

71. Visavadia, B.G.; Honeysett, J.; Danford, M.H. Manuka honey dressing: An effective treatment for chronic wound infections. *Br. J. Oral Maxillofac. Surg.* **2008**, *46*, 55–56. [CrossRef] [PubMed]

72. Al-Waili, N.; Salom, K.; Al-Ghamdi, A.A. Honey for wound healing, ulcers, and burns; data supporting its use in clinical practice. *Sci. World J.* **2011**, *11*, 766–787. [CrossRef] [PubMed]

73. Simon, A.; Sofka, K.; Wiszniewsky, G.; Blaser, G.; Bode, U.; Fleischhack, G. Wound care with antibacterial honey (Medihoney) in pediatric hematology–oncology. *Support. Care Cancer* **2006**, *14*, 91–97. [CrossRef] [PubMed]

74. Okeniyi, J.A.; Olubanjo, O.O.; Ogunlesi, T.A.; Oyelami, O.A. Comparison of healing of incised abscess wounds with honey and EUSOL dressing. *J. Altern. Complement. Med.* **2005**, *11*, 511–513. [CrossRef] [PubMed]

75. Jull, A.B.; Cullum, N.; Dumville, J.C.; Westby, M.J.; Deshpande, S.; Walker, N. Honey as a topical treatment for wounds. *Cochrane Database Syst. Rev.* **2015**, *6*, CD005083. [CrossRef] [PubMed]

76. Aron, M.; Akinpelu, O.V.; Dorion, D.; Daniel, S. Otologic safety of manuka honey. *J. Otolaryngol.—Head Neck Surg.* **2012**, *41*, S21–S30. [PubMed]

77. Paramasivan, S.; Drilling, A.J.; Jardeleza, C.; Jervis-Bardy, J.; Vreugde, S.; Wormald, P.J. Methylglyoxal-augmented manuka honey as a topical anti–Staphylococcus aureus biofilm agent: Safety and efficacy in an in vivo model. In *International Forum of Allergy & Rhinology*; Wiley Online Library: Hoboken, NJ, USA, 2014; pp. 187–195.

78. Vadodaria, K.; Stylios, G.K. Ultrafine Web Formation from Bee's Sweet Treasure. In *Advanced Materials Research*; Trans Tech Publication: Stafa-Zurich, Switzerland, 2013; pp. 1784–1788.

79. Maleki, H.; Gharehaghaji, A.; Dijkstra, P. A novel honey-based nanofibrous scaffold for wound dressing application. *J. Appl. Polym. Sci.* **2013**, *127*, 4086–4092. [CrossRef]

80. Minden-Birkenmaier, B.A.; Neuhalfen, R.M.; Janowiak, B.E.; Sell, S.A. Preliminary Investigation and Characterization of Electrospun Polycaprolactone and Manuka Honey Scaffolds for Dermal Repair. *J. Eng. Fabr. Fibers (JEFF)* **2015**, *10*, 126–138.

81. Murthy, M.B.; Murthy, B.K.; Bhave, S. Comparison of safety and efficacy of papaya dressing with hydrogen peroxide solution on wound bed preparation in patients with wound gape. *Indian J. Pharmacol.* **2012**, *44*, 784. [CrossRef] [PubMed]

82. Sadek, K.M. Antioxidant and immunostimulant effect of Carica papaya Linn. aqueous extract in acrylamide intoxicated rats. *Acta Inform. Med.* **2012**, *20*, 180. [CrossRef] [PubMed]

83. Balaji, A.; Jaganathan, S.K.; Ismail, A.F.; Rajasekar, R. Fabrication and hemocompatibility assessment of novel polyurethane-based bio-nanofibrous dressing loaded with honey and Carica papaya extract for the management of burn injuries. *Int. J. Nanomed.* **2016**, *11*, 4339.

84. Kadakia, P.U.; Growney Kalaf, E.A.; Dunn, A.J.; Shornick, L.P.; Sell, S.A. Comparison of silk fibroin electrospun scaffolds with poloxamer and honey additives for burn wound applications. *J. Bioact. Compat. Polym.* **2016**, *33*. [CrossRef]

85. Yang, X.; Fan, L.; Ma, L.; Wang, Y.; Lin, S.; Yu, F.; Pan, X.; Luo, G.; Zhang, D.; Wang, H. Green electrospun Manuka honey/silk fibroin fibrous matrices as potential wound dressing. *Mater. Des.* **2017**, *119*, 76–84. [CrossRef]

86. Hixon, K.R.; Lu, T.; Carletta, M.N.; McBride-Gagyi, S.H.; Janowiak, B.E.; Sell, S.A. A preliminary in vitro evaluation of the bioactive potential of cryogel scaffolds incorporated with Manuka honey for the treatment of chronic bone infections. *J. Biomed. Mater. Res. Part B Appl. Biomater.* **2017**. [CrossRef] [PubMed]

87. Arnett, T.R. Extracellular pH regulates bone cell function. *J. Nutr.* **2008**, *138*, 415S–418S. [CrossRef] [PubMed]

88. Arnett, T.R. Acidosis, hypoxia and bone. *Arch. Biochem. Biophys.* **2010**, *503*, 103–109. [CrossRef] [PubMed]

89. Hixon, K.R.; Lu, T.; McBride-Gagyi, S.H.; Janowiak, B.E.; Sell, S.A. A Comparison of Tissue Engineering Scaffolds Incorporated with Manuka Honey of Varying UMF. *BioMed Res. Int.* **2017**, *2017*. [CrossRef] [PubMed]

90. Wang, T.; Zhu, X.-K.; Xue, X.-T.; Wu, D.-Y. Hydrogel sheets of chitosan, honey and gelatin as burn wound dressings. *Carbohydr. Polym.* **2012**, *88*, 75–83. [CrossRef]

91. Sasikala, L.; Durai, B. Development and evaluation of chitosan honey hydrogel sheets as wound dressing. *Int. J. Pharm. Biol. Sci.* **2015**, *6*, 26–37.

92. Leveen, H.H.; Falk, G.; Borek, B.; Diaz, C.; Lynfield, Y.; Wynkoop, B.J.; Mabunda, G.A.; Rubricius, J.L.; Christoudias, G.C. Chemical acidification of wounds. An adjuvant to healing and the unfavorable action of alkalinity and ammonia. *Ann. Surg.* **1973**, *178*, 745. [CrossRef] [PubMed]

93. Greener, B.; Hughes, A.; Bannister, N.; Douglass, J. Proteases and pH in chronic wounds. *J. Wound Care* **2005**, *14*, 59–61. [CrossRef] [PubMed]

94. Phillips, I.; Lobo, A.; Fernandes, R.; Gundara, N. Acetic acid in the treatment of superficial wounds infected by Pseudomonas aeruginosa. *Lancet* **1968**, *291*, 11–13. [CrossRef]

95. Giusto, G.; Beretta, G.; Vercelli, C.; Valle, E.; Iussich, S.; Borghi, R.; Odetti, P.; Monacelli, F.; Tramuta, C.; Grego, E. A simple method to produce pectin-honey hydrogels and its characterization as new biomaterial for surgical use. *J. Biomed. Mater. Res. Part B* **2018**. under review.

96. Giusto, G.; Beretta, G.; Vercelli, C.; Valle, E.; Iussich, S.; Borghi, R.; Odetti, P.; Monacelli, F.; Tramuta, C.; Grego, E. Pectin-honey hydrogel: Characterization, antimicrobial activity and biocompatibility. *Biomed. Mater. Eng.* **2018**, *29*, 347–356. [CrossRef] [PubMed]

97. Mohd Zohdi, R.; Abu Bakar Zakaria, Z.; Yusof, N.; Mohamed Mustapha, N.; Abdullah, M.N.H. Gelam (*Melaleuca* spp.) honey-based hydrogel as burn wound dressing. *Evid.-Based Complement. Altern. Med.* **2011**, *2012*. [CrossRef]

98. Medihoney. Available online: http://www.dermasciences.com/medihoney (accessed on 4 December 2017).

99. Müller, P.; Alber, D.G.; Turnbull, L.; Schlothauer, R.C.; Carter, D.A.; Whitchurch, C.B.; Harry, E.J. Synergism between Medihoney and rifampicin against methicillin-resistant Staphylococcus aureus (MRSA). *PLoS ONE* **2013**, *8*, e57679. [CrossRef] [PubMed]

100. Cooper, R.; Jenkins, L.; Hooper, S. Inhibition of biofilms of Pseudomonas aeruginosa by Medihoney in vitro. *J. Wound Care* **2014**, *23*, 93–96. [CrossRef] [PubMed]

101. Johnson, D.W.; van Eps, C.; Mudge, D.W.; Wiggins, K.J.; Armstrong, K.; Hawley, C.M.; Campbell, S.B.; Isbel, N.M.; Nimmo, G.R.; Gibbs, H. Randomized, controlled trial of topical exit-site application of honey (Medihoney) versus mupirocin for the prevention of catheter-associated infections in hemodialysis patients. *J. Am. Soc. Nephrol.* **2005**, *16*, 1456–1462. [CrossRef] [PubMed]

102. Robson, V.; Dodd, S.; Thomas, S. Standardized antibacterial honey (Medihoney™) with standard therapy in wound care: Randomized clinical trial. *J. Adv. Nurs.* **2009**, *65*, 565–575. [CrossRef] [PubMed]

103. Johnson, D.W.; Clark, C.; Isbel, N.M.; Hawley, C.M.; Beller, E.; Cass, A.; De Zoysa, J.; McTaggart, S.; Playford, G.; Rosser, B. The honeypot study protocol: A randomized controlled trial of exit-site application of medihoney antibacterial wound gel for the prevention of catheter-associated infections in peritoneal dialysis patients. *Perit. Dial. Int.* **2009**, *29*, 303–309. [PubMed]

104. Biglari, B.; Vd Linden, P.; Simon, A.; Aytac, S.; Gerner, H.; Moghaddam, A. Use of Medihoney as a non-surgical therapy for chronic pressure ulcers in patients with spinal cord injury. *Spinal Cord* **2012**, *50*, 165–169. [CrossRef] [PubMed]

105. Smith, T.; Legel, K.; Hanft, J.R. Topical Leptospermum honey (Medihoney) in recalcitrant venous leg wounds: A preliminary case series. *Adv. Skin Wound Care* **2009**, *22*, 68–71. [CrossRef] [PubMed]

106. Wound Dressing with Manuka Honey. Available online: http://honeywoundcare.com/product-range/product-61/Wound-Dressing-with-Manuka-Honey (accessed on 4 December 2017).

107. ManukaMed. Available online: https://shop.manukamed.com/products/manukahd (accessed on 4 December 2017).

108. Activon—Manuka Honey Dressings. Available online: http://www.advancis.co.uk/products/activon-manuka-honey (accessed on 4 December 2017).

109. What We Do. Available online: http://sweetbio.com/what-we-do/ (accessed on 4 December 2017).

110. Sun, Z.; Zussman, E.; Yarin, A.L.; Wendorff, J.H.; Greiner, A. Compound core–shell polymer nanofibers by co-electrospinning. *Adv. Mater.* **2003**, *15*, 1929–1932. [CrossRef]

111. Zhang, Y.; Huang, Z.-M.; Xu, X.; Lim, C.T.; Ramakrishna, S. Preparation of core—Shell structured PCL-r-gelatin bi-component nanofibers by coaxial electrospinning. *Chem. Mater.* **2004**, *16*, 3406–3409. [CrossRef]

112. Wang, P.; He, J.-H. Electrospun polyvinyl alcohol-honey nanofibers. *Therm. Sci.* **2013**, *17*, 1549–1550. [CrossRef]

113. Tamura, D.Y.; Moore, E.E.; Partrick, D.A.; Johnson, J.L.; Offner, P.J.; Silliman, C.C. Acute hypoxemia in humans enhances the neutrophil inflammatory response. *Shock* **2002**, *17*, 269–273. [CrossRef] [PubMed]

114. McCourt, M.; Wang, J.H.; Sookhai, S.; Redmond, H.P. Proinflammatory mediators stimulate neutrophil-directed angiogenesis. *Arch. Surg.* **1999**, *134*, 1325–1331. [CrossRef] [PubMed]

115. Chavakis, T.; Cines, D.B.; Rhee, J.-S.; Liang, O.D.; Schubert, U.; Hammes, H.-P.; Higazi, A.A.-R.; Nawroth, P.P.; Preissner, K.T.; Bdeir, K. Regulation of neovascularization by human neutrophil peptides (α-defensins): A link between inflammation and angiogenesis. *FASEB J.* **2004**, *18*, 1306–1308. [CrossRef] [PubMed]

8

Stem Cells and Engineered Scaffolds for Regenerative Wound Healing

Biraja C. Dash [1,*] [iD], Zhenzhen Xu [1], Lawrence Lin [2], Andrew Koo [1], Sifon Ndon [1], Francois Berthiaume [3], Alan Dardik [4] and Henry Hsia [1,*]

[1] Department of Surgery (Plastic), Yale School of Medicine, New Haven, CT 06510, USA; zhenzhen.xu@yale.edu (Z.X.); andrew.koo@yale.edu (A.K.); sifon.ndon@yale.edu (S.N.)

[2] Department of Public Health Studies, Johns Hopkins University, Baltimore, MD 21218, USA; llin31@jhu.edu

[3] Department of Biomedical Engineering, Rutgers University, The State University New Jersey, Piscataway, NJ 08901, USA; fberthia@soe.rutgers.edu

[4] Department of Surgery (Vascular), Yale School of Medicine, New Haven, CT 06510, USA; alan.dardik@yale.edu

* Correspondence: biraja.dash@yale.edu (B.C.D.); henry.hsia@yale.edu (H.H.)

Abstract: The normal wound healing process involves a well-organized cascade of biological pathways and any failure in this process leads to wounds becoming chronic. Non-healing wounds are a burden on healthcare systems and set to increase with aging population and growing incidences of obesity and diabetes. Stem cell-based therapies have the potential to heal chronic wounds but have so far seen little success in the clinic. Current research has been focused on using polymeric biomaterial systems that can act as a niche for these stem cells to improve their survival and paracrine activity that would eventually promote wound healing. Furthermore, different modification strategies have been developed to improve stem cell survival and differentiation, ultimately promoting regenerative wound healing. This review focuses on advanced polymeric scaffolds that have been used to deliver stem cells and have been tested for their efficiency in preclinical animal models of wounds.

Keywords: chronic wound; scaffold; natural polymer; synthetic polymer; stem cell; surface modification

1. Introduction

Wounds result from a disruption of the normal architecture of the skin [1]. Wound healing involves the coordination of many distinct but spatiotemporally overlapping physiological processes aimed at restoring the structural and functional integrity of the skin as a barrier to external stressors. This includes hemostasis, inflammation, cellular proliferation, angiogenesis, extracellular matrix (ECM) deposition, scar formation, and remodeling. These processes are highly regulated by the secretion of various cytokines, chemokines, and growth factors [2–7]. Disruptions in signaling during any of these stages can lead to chronic wound formation.

Blood vessels are damaged during wound formation, initiating the clotting cascade. The resulting fibrin matrix serves as a site for cellular attachment, as well as a reservoir for cytokines and growth factors needed during latter phases of healing [2,4]. Platelets release various pro-inflammatory cytokines including platelet derived growth factor (PDGF), transforming growth factor beta (TGFβ), and fibroblast growth factor (FGF) [6]. Polymorphonuclear cells, followed by macrophages, migrate from the intravascular space to the site of injury. Together, these inflammatory cells phagocytose bacteria and cellular debris while macrophages also release matrix metalloproteinases (MMPs) that degrade ECM. They also secrete cytokines including interleukin 1 (IL-1), interleukin 6 (IL-6), vascular endothelial growth factor (VEGF), and tumor necrosis factor alpha (TNFα), which promote further cellular recruitment and proliferation of keratinocytes and fibroblasts [2,4,6].

Fibroblasts migrate to the wound matrix and release collagen, fibronectin, and proteoglycans, replacing the initial fibrin matrix with new ECM. In response to the hypoxic wound environment, they too release cytokines, including fibroblast growth factor (FGF), hepatocyte growth factor (HGF), TGFβ, EGF, and VEGF. In conjunction with signaling factors released during earlier stages, these promote the formation of new blood vessels, a critical component of acute wound healing [7]. Re-epithelialization then occurs as epithelial cells migrate from the periphery to the center of the wound. In full thickness wounds, stem cells from surrounding hair follicles and the interfollicular epidermis also migrate to the wound, contributing to the process of re-epithelialization [8]. Later phases of wound healing are characterized by further ECM deposition, wound contraction and scar formation, and tissue remodeling.

Progression from an acute to chronic wound results from various local or systemic factors that alter the wound microenvironment. Local factors (e.g., tissue ischemia, infection, the presence of necrotic tissue, and ionizing radiation), as well as systemic factors (e.g., diabetes, smoking, and malnutrition) impair wound healing through prolonged inflammation, reduced angiogenesis, and a decrease in growth factor signaling. Additionally, chronic wounds have been found to have increased expression of MMPs causing excessive wound matrix degradation and prolonged healing, as is the case in diabetic patients [9].

Treating wounds and their associated complications places a large financial burden on the healthcare system, which exceeded $25 billion annually in the US by 2009 [10]. Traditional wound care involves infection control, debridement, selecting appropriate dressings to maintain a favorable wound healing environment, and addressing the underlying cause, such as ischemia or diabetes. However, the efficacy of current treatment modalities is limited for complex wounds. Recent advancements in the field of regenerative medicine, such as growth factor delivery and cell therapy, provides additional therapeutic options to potentially facilitate faster wound healing and restoration of normal skin architecture [7,11,12].

Among cell-based therapy, the use of stem cells for their regenerative wound healing potential is gaining widespread recognition. Many types of adult stem cells have been used in clinical trials for wound healing, but none has been approved yet [12–15]. Major limitations to clinical translation of stem cell-based therapies include stem cell immunogenicity and their reduced survival and paracrine activity in vivo. Polymeric biomaterial systems, especially biopolymer-based and/or surface modified scaffolds have been reported to be immunomodulatory by modulating innate immune response from inflammatory M1 phenotype towards a regenerative M2 phenotype. Furthermore, these scaffolds have shown promotion of wound healing by remodeling the nondermal tissue and recruiting endogenous stem cells to the chronic wound site [7,11,16,17]. These polymeric scaffolds have been developed to reduce the immunogenic effect of these stem cells and protect and improve their regenerative capacity within the chronic wound environment.

This review focuses primarily on various polymeric biomaterial-based delivery vehicles that have been developed to date for stem cell delivery to wounds. Our goal is to discuss stem cell delivery systems that have been specifically tested in animal models of wounds. Here, we first describe various stem cells that have been utilized for wound healing applications and discuss their delivery by natural and synthetic polymer-based scaffolds to acute and chronic wounds. Furthermore, we have a section on surface modification describing different modification strategies of these scaffolds to promote stem cell survival and wound healing.

2. Stem Cells for Regenerative Wound Healing

Stem cells have self-renewal and multipotency abilities that may help in creating advanced tissue engineered skin substitutes. Among the main sources of cells that may be used to engineer such substitutes are adult stem cells, embryonic stem cells (ESCs), and induced pluripotent stem cells (iPSCs). These cells have the potential to secret paracrine factors, making them an attractive option for the treatment of acute and chronic wounds [18]. So far, adult stem cells, especially mesenchymal stem cells

(MSCs), have been widely used for regenerative healing [12]. MSCs can self-renew and have the ability to differentiate into osteoblastic, adipocytic, and chondrocytic lineages. MSCs can be isolated from various tissues, including bone marrow, adipose tissue, umbilical cord blood, nerve tissue, and skin dermis. They modulate wound healing by secreting paracrine factors such as VEGF, stromal cell derived factor (SDF-1), epidermal growth factor (EGF), keratinocyte growth factor (KGF), insulin like growth factor (IGF); enzymes such as matrix metalloproteinase-9 (MMP-9); and immunomodulatory cytokines including interferon-λ, TNF-α, IL-1α and IL-1β [18,19]. Some other adult stem cells that have been used for wound repair are hair follicle stem cells (HFSCs), epidermal stem cells, and unrestricted somatic stem cells (USSC) [20–22].

ESCs, derived from the inner cell mass of the blastocyst, have the ability to differentiate into any cell type, including skin cells [23–25], and have immense regenerative potential. Aside from ethical concerns and the substantial legal restrictions of the use of ESCs in any capacity, another major limitation of using ESC-derived cells for regenerative wound healing is their allogeneic and immunogenic nature [26].

iPSCs are the newest class of pluripotent stem cells first developed in 2006 by Takahashi and Yamanaka [27]. Like ESCs, iPSCs can differentiate into all types of cells, but they can be derived from adult somatic cells of the body through the induced expression of four factors: Oct-3/4, Sox2, c-Myc, and KLF4 [27]. This revolutionary technology allows for generation of autologous pluripotent stem cell populations, thereby avoiding immunogenicity and the ethical issues associated with human ESCs [28]. iPSCs have already been used to derive skin cells—including folliculogenic human epithelial stem cells, fibroblasts, and keratinocytes—to engineer skin substitutes [29–32]. The use and application of stem cells in regenerative wound healing have been extensively reviewed in several recent clinical and scientific publications [8,12,14,15].

3. Polymer-Based Biomaterials for Stem Cell Delivery for Regenerative Wound Healing

Over the last few decades, stem cell delivery using polymer based biomaterials has been the subject of intense investigation [33,34]. These biomaterials can be fabricated using natural polymers such as hyaluronan, chitosan and alginate, collagen, elastin, fibrin, and silk [35,36] and genetically engineered peptides [37,38] or synthetic polymers including poly(lactic-*co*-glycolic) acid, polyanhydrides, polyethylene glycol, and others. These different polymers have specific physio-chemical characteristics and functional groups that allow for precise control of the creation of biomaterials with desired properties for a wide range of cell therapy applications. However, it is to be noted that the most important features critical to the success of any polymer are bioactivity, biodegradability, and biocompatibility [35].

3.1. Natural Polymer-Based Biomaterials

Natural polymers such as collagen, fibrinogen, tropoelastin, hyaluronic acid, glycosaminoglycans, etc., are found in the extracellular matrix. Scaffolds fabricated using these polymers are biocompatible, bioactive, assist in cell attachment to cell surface receptors, and provide a niche to control cell function [34]. Some other natural polymers that have been used in designing biomaterials include plant, insect, or animal derived components, such as cellulose, chitosan, silk fibroin, etc., and have cell attachment sites and other properties to provide favorable microenvironments for stem cell delivery. Refer to Table 1 for further information on various biopolymers and their use in wound healing applications. Some of the disadvantages of these polymers include difficulty of sterilization and purification, high lot-to-lot variability, and a high potential for pathogen contamination during the isolation process. Additionally, there is limited ability to control physio-chemical properties and degradation rates of scaffolds made using these materials [39–41].

Table 1. Biological polymers for wound healing application.

Biological Polymer	Structure	Selective Wound Healing Application	Important Findings	Reference
Collagen	A fibrous triple helical protein. Collagen type I, a major subtype consists of two alpha 1 units and one alpha 2	3D scaffold Hydrogel Composite material	MSCs within the scaffolds greatly ameliorated the quality of regenerated skin, reduced collagen deposition. Enhanced reepithelization, increased neo-angiogenesis, and promoted a greater return of hair follicles and sebaceous glands. The mechanisms involved in these beneficial effects were likely related to the ability of MSCs to release paracrine factors modulating the wound healing response. Self-assembled ASC spheroids on chitosan-hyaluronan membranes expressed more cytokine genes (fibroblast growth factor 1, vascular endothelial growth factor, and chemokine [C-C motif] ligand 2) as well as migration-associated genes (chemokine [C-X-C motif] receptor type 4 and matrix metalloprotease 1. Spheroids combined with the use of biomaterials can enhance skin wound healing and more capillary formation. This study shows the potential use of biomaterial-derived 3D MSC spheroids in wound treatment. Increased the recruitment of provascular circulating bone marrow–derived mesenchymal progenitor cells in vivo. Significant increase in BM-MPC migration, proliferation, and tubulization when exposed to hydrogel-seeded ASC conditioned medium. BM-MPC expression of genes related to cell stemness and angiogenesis was also significantly increased following exposure to hydrogel-seeded ASC conditioned medium. ASC-seeded hydrogels improve both progenitor cell recruitment and functionality to effect greater neovascularization.	[42–44]
Gelatin	A hydrolytic byproduct of collagen	Microgel Composite material; nanofibers; electrospun scaffold	Biodegradable gelatin microgels (GMs), as 3D micro-scaffolds, could provide suitable microenvironment for stem cell proliferation. GMs could greatly improve growth factor secretion from hASCs and might be an enhanced strategy to promote hASCs-assisted wound healing. GMs could protect hASCs in the micro-niches and exhibit excellent injectability through syringe without obvious cell damages. hASCs delivered via GMs assisted injection could retain hASCs in situ and improve cell survival in wounds. Besides, hASCs retention improved by GMs could enhance secretion of positive tissue growth factors for wound healing, indicating an advanced injection strategy superior to free cell suspension injection GMs assisted cell delivery could accelerate wound healing by constant modulation of local growth factor expression to enhance regeneration and vascularization Treated wounds closed much faster, with increased re-epithelialization, collagen formation, and angiogenesis in vivo. USCs could secrete VEGF and TGF-β1. USC-conditioned medium enhanced the migration, proliferation, and tube formation of endothelial cells in vitro The composite nanofibrous scaffold was found to be biocompatible. Electrospun scaffold containing sericin promoted epithelial differentiation of hMSC.	[45–47]

Table 1. *Cont.*

Biological Polymer	Structure	Selective Wound Healing Application	Important Findings	Reference
Elastin	An elastic protein made up of water soluble tropoelastin	Electrospun scaffold	Electrospun tropoelastin membranes form stable structures that retain their integrity and strength in tissue culture medium. ADSCs rapidly proliferate on the scaffold and secrete an ECM that eventually covers the entire scaffold in vitro. The populated scaffold is well tolerated in a murine excisional wound model. Wounds treated with ADSC-populated tropoelastin scaffolds showed greater rate of closure and restoration of normal epithelium.	[48]
Fibrin	A fibrous non-globular protein produced by the cleavage of fibrinogen.	Gel Composite material	Enhanced wound healing for the scaffold containing ADSC and keratinocyte. Total epithelialization and higher collagen deposition and higher vascularization. Co-administration of FbnE enriched scaffold (SM) with CB-EPC accelerated wound closure and vascularization compared to FbnE scaffold alone. No differences in number of pericytes and myofibroblasts seen. No comparison to Integra treated mice as all died before study was complete.	[49,50]
Laminin	A heterotrimeric glycoprotein and binds to ECM proteins and cell membranes	Surface modification	The biomimetic collagen scaffold increases VEGF secretion from MSC in vitro. Activated MSC in collagen scaffolds increase wound healing in vivo. Activated MSCs increase wound healing in a splinted back wound model. Laminin improves wound healing efficiency. MSC delivered topically increase wound healing.	[51]
Silk protein	Extracted from cocoon of silk worms. It contains fibrous protein fibroin and water soluble sericin protein.	Scaffold Composite material Film	Scaffolds significantly improved tissue regeneration, reducing the wound area. Decellularized patches are almost as effective as cellularized patches in the treatment of diabetic wounds Scaffolds improve healing through the release of angiogenic and collagen deposition stimulating molecules. Silk Fibroin can be used as a scaffold and is biocompatible, biodegradable and has excellent tensile strength The scaffold can support adipose derived stem cells for skin tissue engineering. Added Pectin and Glycerol can benefit scaffold by promoting protein conformation transition and biomaterial flexibility, respectively. BDNF-induced proliferation and migration of MSCs. BDNF stimulation affects the ability of MSCs to secrete IL-8, NGF, and MMP-9 and that this process depends largely on the Akt signaling pathway. The upregulation of NGF, IL-8, and MMP-9 in the BDNF-CM group contributed to angiogenesis. The BDNF-CM-modified materials also significantly accelerated wound healing. BDNF promotes angiogenesis and enhances the milieu-dependent endothelial differentiation of MSCs in ischemic ulcers.	[46,52–54]

Table 1. *Cont.*

Biological Polymer	Structure	Selective Wound Healing Application	Important Findings	Reference
Hyaluronic acid/Hyaluronan	An anionic nonsulfated glycosaminoglyacan that helps in cell migration and proliferation	Surface modification Composite material	Scaffolds seeded with VEGF165-modified rHFSCs, resulted in promotion of angiogenesis during wound healing and facilitation of vascularization in skin substitutes. Increased VEGF165 level in the repair microenvironment improved vascularization ability. It is believed that HFSCs secrete a variety of cytokines to promote wound healing. Subcutaneous implantation showed that vascularization capacity of non-seeded SJS and ADM were greater than that of Co-CS-HA in subcutaneous wounds. ADSCs cultured on SIS secreted more VEGF compared to those seeded on ADM and Co-CS-HA. ADSC-seeded scaffolds enhanced angiogenesis and wound healing rate compared to non-seeded scaffold in vivo mouse models. ADSC-SIS and ADSC-ADM had greater microvessel densities than ADSC-*co*-CS-HA in vivo.	[20,55]
Alginate	An anionic polysaccharide consisting of homopolymeric blocks of (1-4)-linked β-D-mannuronate (M) and C-5 epimer α-L-guluronate (G) residues.	Hydrogel Composite material	Humans MSCs remained viable for the duration of 6 weeks within the gels. Human VEGF and bFGF was found in quantifiable concentrations in cell culture supernatants of gels loaded with MSCs and incubated for a period of 6 weeks. Conditioned medium from mesenchymal stromal cells stimulates migration of dermal fibroblasts in scratch assays. Conditioned medium of mesenchymal stromal cells induces alterations in the expression of genes involved in wound healing. Encapsulated mesenchymal stromal cells retain stem cell characteristics and remain viable during long-term encapsulation. Encapsulated mesenchymal stromal cell-derived conditioned medium stimulates migration of dermal fibroblasts and induces alterations in the expression of genes involved in wound healing. Cells attached proliferated on the porous membrane. Accelerated wound healing.	[56,57]
Chitosan	A linear polysaccharide consisting of β-(1→4)-linked D-glucosamine and N-acetyl-D-glucosamine.	Membrane Scaffold Composite material	Conditioned medium from mesenchymal stromal cells stimulates migration of dermal fibroblasts in scratch assays. Conditioned medium of mesenchymal stromal cells induces alterations in the expression of genes involved in wound healing. Encapsulated mesenchymal stromal cells retain stem cell characteristics and remain viable during long-term encapsulation. Encapsulated mesenchymal stromal cell-derived conditioned medium stimulates migration of dermal fibroblasts and induces alterations in the expression of genes involved in wound healing. The scaffold was biocompatible UC-MSC differentiated to epidermis and positive for the epidermal markers cytokeratin 19 and involucrin at 14 days. The constructed epidermis substitutes helped rapid wound healing. Better cell adhesion, growth, and proliferation inside the modified scaffolds. Showed a good resilience and compliance with movement as a skin graft. All scaffolds, especially those with stem cells, exhibited pronounced effects on wound closure. The reconstructed skin in grafted groups demonstrated an intact epithelium with the formation of new hair follicles and sebaceous glands, which were reminiscent of the structures of normal skin.	[58-60]

Table 1. *Cont.*

Biological Polymer	Structure	Selective Wound Healing Application	Important Findings	Reference
Pullulan	A polysaccharide consisting of maltotriose units, connected by an α-1,6 glycosidic bond.	Hydrogel Composite material	Described above. Hydrogel seeding of ASCs resulted in the enhanced expression of multiple stemness and angiogenesis-related genes (Oct4, Vegf, Mcp-1, and Sdf-1) in vitro. ASCs seeded within hydrogel scaffolds showed minimal proliferation and maintained baseline levels of metabolic activity. Hydrogel delivery improved ASC survival in vivo. Resulted in accelerated wound closure and increased vascularity in splinted murine wounds.	[44,61]
Xanthan	A hetero-polysaccharide with main containing glucose units and side chain of trisaccharides	Membrane	Described above.	[58]
Gellan gum	A water soluble anionic polysaccharide with repeated tetrasaccharide units containing two D-glucose and one of each L-rhamnose and D-glucuronic acid	Hydrogel	The hydrogels absorbed early inflammatory cell infiltrate and led to formulation of granulation tissue in vivo. Improved wound closure, re-epithelialization, and matrix remodeling. Promoted superior neo-vascularization.	[62]

Collagen, one of the major ECM-based proteins, has been used widely to fabricate scaffolds for stem cell delivery. Bone marrow derived MSCs (BM-MSCs), adipose derived MSCs (ADSCs), and other adult stem cells have been delivered using collagen scaffolds. O'Loughlin et al. fabricated a collagen scaffold to deliver BM-MSCs to a diabetic rabbit ear ulcer wound model. Topical application of the MSC-collagen treatment in a dose-dependent fashion significantly increased wound closure rates compared to that of untreated groups. Higher cell doses yielded better wound closure and faster rates of wound healing. The wound closure was associated with increased angiogenesis in the MSC-collagen groups [63]. A similar study by Kim et al. showed delivery of BM-MSCs in collagen gel scaffolds to full thickness skin wounds of rats. The wound size was significantly reduced in rats treated with MSC-collage scaffold compared to collagen alone on day 7. There was no difference by day 14, suggesting that MSCs appear to accelerate early wound healing. MSC-collagen treated mice also had significantly more expression of VEGF and more intense early expression of MMP-9 compared to collagen treatment without MSCs, indicating a possible mechanism of MSCs' effects on wound healing [64]. Similarly, other stem cells such as ADSCs and Leucine Rich Repeat Containing G Protein-Coupled Receptor 6 (LGR6+) epithelial stem cells have been delivered to wounds using collagen scaffolds [65,66].

Fibrin, another ECM-based protein, has been shown to form gels by reacting with thrombin. Although not investigated as extensively as collagen, fibrin has also been studied as a potential scaffold material for the delivery of stem cells. Ozpur et al., described a method of creating artificial skin tissue by using fibrin gels seeded with keratinocytes and ADSCs. This skin substitute promoted healing by complete re-epithelialization of the wound with no contraction and increased vessel density and collage deposition [49]. Another group, Falanga et al., also constructed gels to carry MSCs. In in vivo experiments, diabetic mice with full thickness wounds demonstrated accelerated healing with topical application of MSC and fibrin compared to fibrin alone. In human patients, acute wounds treated with fibrin gel demonstrated healing and resurfacing. Treatment of human chronic wounds with fibrin and MSC resulted in increased wound closure in wounds that were previously refractory to other therapies. A significant correlation was found between the number of MSCs applied and the subsequent decrease in size of the chronic wounds [67].

Another important ECM bio-macromolecule is elastin, which has been used for various tissue engineering applications, including in the form of tropoelastin [68,69]. In the work of Machula et al., human ADSCs were seeded on recombinant tropoelastin-based electrospun scaffolds. The ADSC-seeded scaffold formed a stable structure with high biocompatibility. The scaffold demonstrated rapid cell proliferation with deposition of new ECM. Experimentation with the seeded scaffold in vivo with a murine excisional wound model revealed greater would closure as well as greater epithelial thickness [48].

Silk protein is a natural fibrous protein that has been used for various tissue engineering applications. In a study by Navone et al., AD-MSCs were obtained from human lipoaspirates and tested for the ability to differentiate both before and after seeding on silk fibroin (SF) scaffolds. Sheets of the silk fibroin scaffold were prepared through electrospinning and were loaded with MSCs to generate cellularized scaffolds (AD-MSCs-SF). The SF scaffolds allowed for growth of AD-MSCs as well as retention of differentiation capacity. Evaluation of both the cellularized and decellularized scaffolds in diabetic mice revealed improved tissue regeneration and enhanced expression of angiogenesis-related genes, corroborated by in vitro analysis. The ability of the scaffolds to affect migration of human umbilical vein endothelial cells (HUVEC), keratinocytes, and dermal fibroblasts was also observed, with enhanced HUVEC migration and release of angiogenic factors. Interestingly, the decellularized scaffold demonstrated comparable efficacy to the cellularized scaffold, showing potential for future application in wound care [52].

Hyaluronic acid (HA) or hyaluronan is another major component of ECM. It contains sites for cell adhesion and has been widely used for surface modification of scaffolds and/or hydrogels to enhance cell survival and proliferation. Recently Gerecht et al. used acrylated HA (AHA) to form a smart hydrogel containing cell adhesive peptide and MMP-sensitive peptide cross-linker. The AHA hydrogel was used to deliver early vascular cells derived from human iPSC (hiPSC). The vascularized scaffold showed enhanced angiogenesis and wound closure in diabetic immunodeficient mice [70].

Chitosan, xanthan, and carboxymethylcellulose are polysaccharides that have been used to culture stem cells. More information regarding other polysaccharide-based polymers can be found in Table 1. Bellini et al. formulated a 3D membrane structure comprised of chitosan and xanthan, which showed adherence of MSCs and demonstrated improved wound healing in rat models of wound healing [58]. Similarly, Rodrigues et al. extracted rat ADSCs from adipose tissue and combined them with a sodium carboxymethylcellulose (CMC) scaffold. Evaluation of CMC toxicity in vitro revealed no significant increase in lactate dehydrogenase activity that would suggest increased cell death. The results of in vivo testing of CMC associated with ADSCs in a rat wound model revealed increased cell proliferation rates of the granulation tissue as well as greater epithelium thickness when compared to untreated rats. However, there was no increase in collagen fibers, nor were there changes in the speed of wound closure [71].

3.2. Synthetic Polymer-Based Biomaterials

Synthetic polymers are an alternative to natural materials and have well-defined and tunable chemical and mechanical properties. They present some advantages over natural polymer based biomaterials including their ability to be fabricated to have specific degradation rates. Thus a wide array of synthetic polymers has been explored for almost all tissue engineering applications [36]. Refer to Table 2 for a list of various synthetic polymers and their application in wound healing.

Table 2. Synthetic polymers for wound healing application.

Synthetic Polymer	Structure	Selective Wound Healing Application	Important Findings	Reference
Poly(L-lactic acid) (PLA or PLLA)	A biodegradable and thermoplastic polymer synthesized using monomers of lactic acid or lactide.	Scaffold	Scaffolds had a high porosity and a 50–75% increase in swelling, along with complete protein release in the presence of phosphate-buffered saline. Accelerated wound re-epithelialization in mouse model Maintained optimal hydration of the exposed tissues and decreased wound healing time in vivo.	[72]
Poly(ethylene glycol) (PEG)	A hydrophilic polymer synthesized by anionic ring-opening polymerization of ethylene oxide.	Scaffold Hydrogel Composite material	Polymer network/porous scaffold helps cells from oxidative stress. The implant showed fibroblast proliferation, collagen deposition, and anti-oxidant enzyme activity. Enhanced wound healing by enhanced engraftment and increased vascularization. Decreased proinflammatory cytokines and increased anti-inflammatory cytokines. Wounds treated with MDSC and PEG-PLGA-PEG showed enhanced wound closure rate, epithelium migration, and collagen deposition. There was increased engraftment of MDSCs into the wound bed compared to controls (MDSC treatment without hydrogel and MDSC with control dressing). In wounds, 25% MDSCs differentiated into fibroblasts, 10% into myofibroblasts, 10% into endothelial cells, and none into macrophages. Rat excision wounds treated with PEGylated fibrin-collagen bilayer hydrogels show decreased wound contraction over time. They show faster wound closure in comparison with control Within bilayer hydrogels, dsASCs proliferate, differentiate, maintain a spindle-shaped morphology in collagen, and develop tubular microvascular networks in PEGylated fibrin.	[73–75]
Poly(lactic-co-glycolic acid) (PLGA)	A biodegradable copolymer of glycolic and lactic acid.	Scaffold	Scaffolds seeded with VEGF-transfected stem cells led to increased blood vessel migration into the constructs compared to control cells or cells transfected with VEGF using a commercial reagent. There was increased endothelial cell density compared to the controls.	[74,76]
Polyurethane	A polymer synthesized by reacting poly-isocyanate and polyol. Contains urethane to join organic units.	Composite material	Described above.	[73]
Poly(N-isopropylacrylamide) (PNIPAM)	A thermoresponsive polymer synthesized using free radical polymerization of N-isopropylacrylamide.	Thermosensitive hydrogel	Hydrogel and BMSC combination therapy promoted wound contraction. The hydrogel inhibited chronic inflammation. Hydrogel and BMSCs combination therapy promoted the formation of granulation tissue. Hydrogel and BMSCs combination therapy promoted keratinocyte proliferation and differentiation. Hydrogel and BMSCs combination therapy improved the quality of wound healing.	[77]
Polycaprolactone (PCL)	A biodegradable polyester prepared by ring opening polymerization of ε-caprolactone.	Composite material	ADSCs differentiated into epidermal-like structures Observed higher microvessel density in rat skin tissue injury models. Improved healing was observed in vivo. ADSCs-PLCL/P123 scaffolds with the thickness of 150–250 μm match well with the epidermis layer (200–250 μm).	[47,78]
PAA-poly(amidoamine)	A dendrimer with repetitively branched subunits of amide and amine	Hydrogel	Described above.	[77]

Chen et al., reported a hydrogel of poly N-isopropylacrylamide (PNIPAM) and poly(amidoamine) (PAA). The PNIPAM-PAA hydrogel was used to deliver BM-MSCs to ulcers in a diabetic mice. The hydrogel alone and with BM-MSCs showed reduction of M1 inflammatory macrophages. BM-MSC-laden hydrogels induced rapid healing of diabetic ulcer by increasing angiogenesis, granulation tissue formation, ECM secretion, wound contraction, and re-epithelialization [77].

Poly(vinyl alcohol) (PVA) is a synthetic polymer that has been widely used for drug delivery and tissue engineering application [79]. In one study, a crosslinked PVA membrane was used for the delivery of Wharton's jelly MSCs (WJ-MSCs) [80]. The PVA membrane supported cell adhesion and proliferation. WJ-MSCs delivered using PVA membranes showed improved healing in a canine wound model.

Poly-L-lactic acid (PLLA) and poly lactic-co-glycolic acid (PLGA) are biodegradable synthetic polymers that have been used to deliver stem cells to wounds [72,76]. Kim et al. studied the effect of umbilical cord derived endothelial progenitor cells (CB-EPCs) in a PLLA scaffold on wound healing in vivo. These biodegradable PLLA scaffolds were modified with Arg–Gly–Asp (RGD) peptide to promote cell adhesion. The RGD-g-PLLA scaffold was shown to successfully support the growth and endothelial functions of these CB-EPCs in vitro. In murine dermal wound models, the CB-EPC-seeded scaffold promoted vascular regeneration and was superior to conventional intradermal CB-EPC injection treatments in terms of localization and survival/retention of EPCs, as well as neovascularization at the wound bed. Similarly, another work by Yang et al. studied the effect of scaffolds seeded with VEGF-enhanced stem cells on wound healing in mice. Scaffolds made using PLGA and PLLA were used to deliver VEGF enhanced human MSCs. Subcutaneous implantation of the scaffold plus enhanced human MSCs demonstrated increased blood vessel migration into the constructs compared to implantation of stem cells transfected by a control plasmid and those transfected with VEGF using a commercial reagent [76].

Some other biodegradable synthetic polymers that have been used for regenerative applications are polycaprolactone (PCL) and polyethyleneglycol (PEG). Bahrami et al., fabricated an electrospun nanofibrous scaffold using PCL to deliver USSCs. The scaffold, which was modified with laminin, showed good cell viability, and in vivo study showed increased wound closure and a thin epidermis with recovered skin appendages in the dermal layer [22]. Similarly, Gu et al., fabricated a scaffold using PCL and poloxamer (PLCL/P123) to deliver rat ADSCs. The ADSC-seeded scaffolds were evaluated in vivo through a rat skin tissue injury model, producing the highest percentages of wound closure when compared to the PLCL/P123 scaffold alone or with a control petrolatum gauze, indicating enhanced wound healing. Rats treated with the ADSC-seeded scaffolds also exhibited greater microvessel density. The regenerated skin demonstrated a thick and integrated epidermal structure [78].

Geesala et al. produced a semi-inter penetrating network porous 3D scaffold using PEG-polyurethane (PEG-PU) and PEG-dimethylether (PEGDME). The scaffold was also shown to be thermostable, biodegradable, cytocompatible, and protective against oxidative stress. The scaffold loaded with BM-MSCs promoted healing with an increase in fibroblast proliferation, collagen deposition, and anti-oxidant enzyme activities. Furthermore, it reduced proinflammatory cytokines IL-1b, 6, 8, and TNF-α and upregulated anti-inflammatory cytokines IL-10 and 13. Increased engraftment was seen by expression of Sca-1+Lin-CD90+CD133+. Improved neovascularization was also evident, with increased expression of PECAM, VEGFR3, neuropilin 2, and Tie2 [73]. In the work of Dong et al., an injectable hydrogel was prepared using a thermoresponsive hyperbranched copolymer of poly(ethylene glycol) methyl ether methacrylate-co-2-(2-methoxyethoxy) ethyl methacrylate-co-poly(ethylene glycol) diacrylate (PEGMEMA-MEO$_2$MA-PEGDA). The hydrogel supported cell viability and proliferation of rat ADSCs. In vivo testing in a rat dorsal full-thickness wound model revealed that the hydrogel system improved angiogenesis [81]. A modification of the PEGMEMA-MEO$_2$MA–PEGDA hydrogel with HA showed a marked increase in secretion of growth factors over a 7-day period [82]. Lee et al. investigated wound healing in diabetic mice using muscle-derived stem cells (MDSCs) and a thermosensitive hydrogel made of the triblock co-polymer PEG-PLGA-PEG. In vivo, wounds of diabetic mice treated with MDSCs in the PEG-PLGA-PEG hydrogel showed increased epithelium migration, collagen deposition, engraftment of MDSCs, and enhanced wound closure rates within the first ten days after treatment [74].

Though these synthetic polymers allow for engineering of biomaterials with tunable properties (e.g., chemical and mechanical properties and degradation rates), disadvantages for choosing such materials include poor bioactivity due to a lack of cell attachment sites and acidic byproducts that can trigger an immune response [39–41,83]. It is thus critical to modify synthetic materials with biological or chemical entities to achieve an appropriate cellular response.

4. Modifications to Biomaterials

The interaction between cells and the biomaterial surface dictates the survival and function of stem cells within scaffolds. Surface modification of a biomaterial by incorporating suitable biological and chemical cues can ultimately control stem cell activity by manipulating the signal transduction pathways in stem cells after its attachment on the surface [39,40]. Surface modification of scaffolds is therefore an important aspect in tissue engineering in order to control cellular behavior [83]. This section will discuss different methods that have been utilized to modify scaffolds for stem cell delivery for wound healing (Table 3).

Modifications to collagen scaffolds have been made to replicate the ECM environment of wounds by addition of biomacromolecules such as glycosaminoglycans (GAGs) or laminin. Liu et al. (2008) created collagen–GAG scaffolds to study wound healing in porcine models. In in vivo partial thickness burn wounds, treatment with MSC-seeded scaffolds showed improved healing and keratinization, less wound contraction, and more vascularization compared to scaffold alone or no treatment. Labeled MSCs were also detected in the epidermal and dermal components of the wound bed, indicating that they had migrated from the scaffold and were integrated into the neoepidermis and neodermis [84]. In another study, Assi et al. used a collagen scaffold to deliver BM-MSCs to treat diabetic ulcers [51]. Laminin was added for functionalization of the scaffold. The functionalized collagen scaffold promoted survival of MSCs and increased secretion of VEGF. Laminin further enhanced the healing effect of the scaffold.

Catanzano et al., used an alginate (ALG)-HA hydrogel for dermal regeneration [85]. HA of 10–20% of the ALG weight was incorporated in a physically crosslinked ALG hydrogel. The integration of HA showed therapeutic efficacy of the hydrogel by significantly promoting gap closure in an in vitro study and by maintaining cell survival of ADSCs and keratinocytes. In an in vivo rat excisional wound model, the use of the alginate-HA hydrogel promoted wound closure compared to ALG alone. In another work, Schmitt et al. synthesized calcium alginate gels by both internal and external crosslinking, and incorporated PEG 300,000 and HA for cell adhesion [56]. They observed cell survival within the gel for 6 weeks and secretion of paracrine factors such as VEGF and bFGF. They theorized that the gel can be used for wound healing applications. In another study by Cerqueira et al., human ADSCs and microvascular endothelial cells (MECs) were obtained from human adipose tissue [62]. The cells were cultured and combined with gellan gum–hyaluronic acid (GG–HA) hydrogels. The hydrogels demonstrated sponge-like properties with composition and physical properties similar to the ECM of the skin. The GG-HA hydrogels promoted neovascularization that was further augmented by the inclusion of MECs. The GG-HA hydrogels absorbed early inflammatory cell infiltrates and led to the formulation of granulation tissue, with greater epidermal thickness observed in mice implanted with hydrogels than those without. Hydrogel degradation was observed, along with improvements in wound closure, re-epithelialization, and matrix remodeling.

Silk protein sericin has widely been used for various biomedical application [86]. Recently, sericin has been used to modify collagen scaffolds for the culture of stem cells. In one of the studies, the researchers showed that the addition of sericin yielded an increased rate of proliferation of ADSCs seeded within collagen gel, compared to a pure collagen control. Sericin also stimulated the overexpression PPARγ2 [87]. In another work by Kim et al., MSCs were seeded on a collagen scaffold modified by silkworm gland hydrolysate (SSGH). The scaffold demonstrated cytocompatability when evaluated in vitro with MSCs. The SSGH/collagen scaffold also demonstrated antioxidant properties, mitigating cell damage induced by oxidative stress. Experimentation with the SSGH/collagen scaffold in a mouse full-thickness excisional wound model revealed accelerated wound re-epithelialization and decreased wound-healing time while maintaining optimal tissue hydration. The observations suggest that the SSGH/collagen scaffold may stimulate growth factor production in healthy cells found in the wound site [88]. In another work, Bhowmick et al. also generated electrospun nanofibrous scaffolds with gelatin, HA, chondroitin sulfate, and sericin. Sericin modified scaffold promoted cell viability and proliferation of human fibroblast cells, keratinocytes, and MSCs. Further co-culture of keratinocytes and hMSCs showed differentiation of hMSCs towards the epithelial lineage. The scaffold thus is a promising treatment for dermal regeneration [46].

Table 3. Surface modifications of scaffolds for wound healing application.

Biopolymers for Surface Modification	Effect on Stem Cells and Wound Healing
Glycosaminoglycan	Promoted MSC survival. Improved healing, keratinization and vascularization [84].
Laminin	Enhanced MSC survival and VEGF secretion. Promoted healing [51].
HA	Maintained cell survival of ADSC and keratinocytes and improved wound closure [85]. Enhanced cell adhesion and survival and secretion of paracrine factors such as VEGF and bFGF [56]. Enhanced neovascularization, wound closure, re-epithelialization, matrix remodeling and reduced inflammation [62].
Sericin/Silk derivative	Increased cell proliferation of ADSC and maintained adipogenicity of the cells by stimulating the expression of PPARγ2 [87]. Reduced oxidative stress in the cells, enhanced re-epithelialization and wound closure [88]. Promoted cell viability and proliferation of MSCs and keratinocytes and fibroblasts and helped differentiation of MSCs to epithelial lineage [46].
Fibrin	Fibrin fragment E promoted cell adhesion and differentiation of cord blood epidermal progenitor cells to endothelial cells and enhanced vascularization and wound closure [50]. PEGylated fibrin promoted cell proliferation of ADSCs and tubular microvascular formation in the scaffold and enhance wound closure, re-epithelialization [75].

In another work, Fibrin fragment E (FbnE) was used for the modification of an alginate scaffold and used for culturing cord blood epidermal progenitor cells (CB-EPCs). FbnE was shown to promote adhesion and endothelial differentiation of CB-EPCs in vitro. In vivo, co-administration of the FbnE-enriched scaffold with CB-EPCs accelerated wound closure and vascularization compared to administration of the FbnE scaffold alone [50].

In a work by Natesan et al., PEGylated fibrin was used along with collagen to fabricate a bilayer hydrogel to deliver ADSCs. The authors found that the ADSCs successfully proliferated and differentiated in the bilayer hydrogel in vitro. They formed dense tubular microvascular networks in the PEGylated fibrin-based hydrogel. The bilayer hydrogel decreased wound contraction and increased wound closure time compared to the saline-treated control. In addition, rats treated with the ADSCs–bilayer hydrogel showed a significant increase in granulation tissue formation, re-epithelialization of wound margins, and better progression of the epithelial margin toward the center of the wound [75]. Similarly, Xu et al. utilized a gelatin modified PEG hydrogel. MSCs containing gelatin/PEG hydrogel were applied to a full thickness wound model in rats. By day 7, the seeded gels showed significantly accelerated wound closure and reepithelialization, and improved angiogenesis and granulation tissue formation [89].

5. Conclusions and Future Directions

Advancements in the field of stem cell biology and biomaterials have created exciting opportunities for wound healing therapy. Polymeric biomaterials can now be rationally designed to provide a niche for the stem cells to survive and function. The purpose of this review was to detail some of the most commonly used biomaterial scaffolds for transplantation of stem cells for wound healing applications. It sought to detail the types of materials available and highlight their unique properties with respect to stem cell delivery with an emphasis on how they perform in vivo in preclinical models of wounds. We believe this will help readers choose a material that would best suit their specific wound healing needs. This review also touched on how modifications to biomaterials can play a large role in further promoting stem cell function and regenerative wound healing. Other considerations when designing such scaffolds including the method of fabrication and composite scaffolds (e.g., blending of two or more different polymers to fabricate composite scaffolds to further achieve additional desired properties and benefits) are beyond the scope of this review. Ultimately there are a number of significant challenges researchers must overcome before a stem cell and biomaterial-based option for wound healing therapy can be used in the clinical setting. In addition to overcoming the difficulty of obtaining a renewable source of stem cells in large quantities, major advances in both the understanding of the local cues necessary for stem cell survival and function and development of biomaterials necessary to promote these functions is key. These polymer-based biomaterials should be tested in a variety of in vivo and in vitro studies prior to their application with the stem cells on a clinical level. The future lies in utilizing high throughput arrays to test the functionality of new biomaterials for their usage in stem cell delivery in short periods of time while wasting few materials, ultimately allowing for a more rapidly developed end product. Further progress in this field involves utilizing a hybrid approach to produce personalized tissue engineered constructs by using patient specific cells, biomimetic matrices (e.g., collagen, gelatin, chitosan, Fibrin etc.) and bioactive stimuli (e.g., Fibrin, laminin, silk protein, GAG etc.) to promote a regenerative healing (Figure 1). Furthermore, multiple cell types may also be employed with spatial control to generate skin tissues ex vivo, along with the requisite vascular supply. These advancements will lead to more clinical trials and finally allow the translation of stem cell therapy-based regenerative wound healing from bench-side research into bedside treatments for patients.

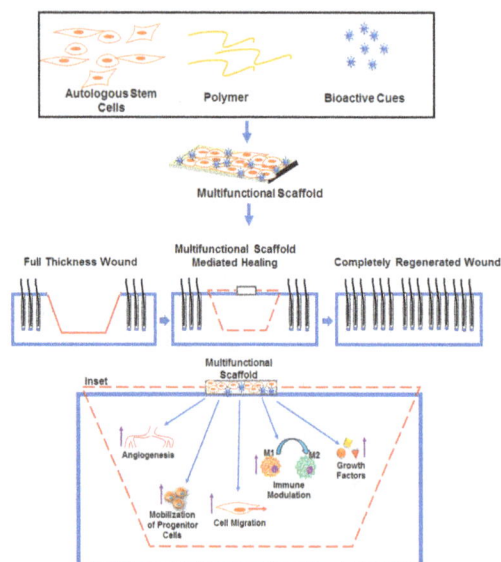

Figure 1. Schematic showing development of a personalized regenerative treatment modality for chronic wound patients by fabricating a multifunctional scaffold system using autologous stem cells, different bioactive cues, and polymer of interest. The scaffold when implanted into a full thickness wound would promote regenerative healing (scar-free) by providing a vascular bed, modulating inflammatory M1 macrophages to a pro-healing M2 macrophages, inducing movement of progenitor cells and increased migration of fibroblast and keratinocytes.

Acknowledgments: The authors thank the support from Yale Department of Surgery; NIH/NIBIB; R21EB021570; USAMRMC/DoD SC10029; Ohse grant.

Conflicts of Interest: The authors declare no conflict of interest.

References

1. Atiyeh, B.S.; Ioannovich, J.; Al-Amm, C.A.; El-Musa, K.A. Management of acute and chronic open wounds: The importance of moist environment in optimal wound healing. *Curr. Pharm. Biotechnol.* **2002**, *3*, 179–195. [CrossRef] [PubMed]
2. Eming, S.A.; Martin, P.; Tomic-Canic, M. Wound repair and regeneration: Mechanisms, signaling, and translation. *Sci. Transl. Med.* **2014**, *6*, 265sr266. [CrossRef] [PubMed]
3. Sorg, H.; Tilkorn, D.J.; Hager, S.; Hauser, J.; Mirastschijski, U. Skin wound healing: An update on the current knowledge and concepts. *Eur. Surg. Res.* **2017**, *58*, 81–94. [CrossRef] [PubMed]
4. Werner, S.; Grose, R. Regulation of wound healing by growth factors and cytokines. *Physiol. Rev.* **2003**, *83*, 835–870. [CrossRef] [PubMed]
5. Qing, C. The molecular biology in wound healing & non-healing wound. *Chin. J. Traumatol. Zhonghua Chuang Shang Za Zhi* **2017**, *20*, 189–193. [PubMed]
6. Demidova-Rice, T.N.; Hamblin, M.R.; Herman, I.M. Acute and impaired wound healing: Pathophysiology and current methods for drug delivery, part 2: Role of growth factors in normal and pathological wound healing: Therapeutic potential and methods of delivery. *Adv. Skin Wound Care* **2012**, *25*, 349–370. [CrossRef] [PubMed]
7. Turner, N.J.; Badylak, S.F. The use of biologic scaffolds in the treatment of chronic nonhealing wounds. *Adv. Wound Care* **2015**, *4*, 490–500. [CrossRef] [PubMed]
8. Ojeh, N.; Pastar, I.; Tomic-Canic, M.; Stojadinovic, O. Stem cells in skin regeneration, wound healing, and their clinical applications. *Int. J. Mol. Sci.* **2015**, *16*, 25476–25501. [CrossRef] [PubMed]
9. Gurtner, G.C.; Werner, S.; Barrandon, Y.; Longaker, M.T. Wound repair and regeneration. *Nature* **2008**, *453*, 314–321. [CrossRef] [PubMed]
10. Sen, C.K.; Gordillo, G.M.; Roy, S.; Kirsner, R.; Lambert, L.; Hunt, T.K.; Gottrup, F.; Gurtner, G.C.; Longaker, M.T. Human skin wounds: A major and snowballing threat to public health and the economy. *Wound Repair Regen.* **2009**, *17*, 763–771. [CrossRef] [PubMed]
11. Dickinson, L.E.; Gerecht, S. Engineered biopolymeric scaffolds for chronic wound healing. *Front. Physiol.* **2016**, *7*, 341. [CrossRef] [PubMed]
12. Duscher, D.; Barrera, J.; Wong, V.W.; Maan, Z.N.; Whittam, A.J.; Januszyk, M.; Gurtner, G.C. Stem cells in wound healing: The future of regenerative medicine? A mini-review. *Gerontology* **2016**, *62*, 216–225. [CrossRef] [PubMed]
13. Rosemann, A. Why regenerative stem cell medicine progresses slower than expected. *J. Cell. Biochem.* **2014**, *115*, 2073–2076. [CrossRef] [PubMed]
14. Kirby, G.T.; Mills, S.J.; Cowin, A.J.; Smith, L.E. Stem cells for cutaneous wound healing. *BioMed Res. Int.* **2015**, *2015*, 285869. [CrossRef] [PubMed]
15. Cerqueira, M.T.; Pirraco, R.P.; Marques, A.P. Stem cells in skin wound healing: Are we there yet? *Adv. Wound Care* **2016**, *5*, 164–175. [CrossRef] [PubMed]
16. Major, M.R.; Wong, V.W.; Nelson, E.R.; Longaker, M.T.; Gurtner, G.C. The foreign body response: At the interface of surgery and bioengineering. *Plast. Reconstr. Surg.* **2015**, *135*, 1489–1498. [CrossRef] [PubMed]
17. Morais, J.M.; Papadimitrakopoulos, F.; Burgess, D.J. Biomaterials/tissue interactions: Possible solutions to overcome foreign body response. *AAPS J.* **2010**, *12*, 188–196. [CrossRef] [PubMed]
18. Baraniak, P.R.; McDevitt, T.C. Stem cell paracrine actions and tissue regeneration. *Regen. Med.* **2010**, *5*, 121–143. [CrossRef] [PubMed]
19. Liang, X.; Ding, Y.; Zhang, Y.; Tse, H.F.; Lian, Q. Paracrine mechanisms of mesenchymal stem cell-based therapy: Current status and perspectives. *Cell Transplant.* **2014**, *23*, 1045–1059. [CrossRef] [PubMed]
20. Quan, R.; Zheng, X.; Xu, S.; Zhang, L.; Yang, D. Gelatin-chondroitin-6-sulfate-hyaluronic acid scaffold seeded with vascular endothelial growth factor 165 modified hair follicle stem cells as a three-dimensional skin substitute. *Stem Cell Res. Ther.* **2014**, *5*, 118. [CrossRef] [PubMed]

21. Shen, Y.; Dai, L.; Li, X.; Liang, R.; Guan, G.; Zhang, Z.; Cao, W.; Liu, Z.; Mei, S.; Liang, W.; et al. Epidermal stem cells cultured on collagen-modified chitin membrane induce in situ tissue regeneration of full-thickness skin defects in mice. *PLoS ONE* **2014**, *9*, e87557. [CrossRef] [PubMed]

22. Bahrami, H.; Keshel, S.H.; Chari, A.J.; Biazar, E. Human unrestricted somatic stem cells loaded in nanofibrous pcl scaffold and their healing effect on skin defects. *Artif. Cells Nanomed. Biotechnol.* **2016**, *44*, 1556–1560. [CrossRef] [PubMed]

23. Metallo, C.M.; Azarin, S.M.; Moses, L.E.; Ji, L.; de Pablo, J.J.; Palecek, S.P. Human embryonic stem cell-derived keratinocytes exhibit an epidermal transcription program and undergo epithelial morphogenesis in engineered tissue constructs. *Tissue Eng. Part A* **2010**, *16*, 213–223. [CrossRef] [PubMed]

24. Cherbuin, T.; Movahednia, M.M.; Toh, W.S.; Cao, T. Investigation of human embryonic stem cell-derived keratinocytes as an in vitro research model for mechanical stress dynamic response. *Stem Cell Rev.* **2015**, *11*, 460–473. [CrossRef] [PubMed]

25. Guenou, H.; Nissan, X.; Larcher, F.; Feteira, J.; Lemaitre, G.; Saidani, M.; Del Rio, M.; Barrault, C.C.; Bernard, F.X.; Peschanski, M.; et al. Human embryonic stem-cell derivatives for full reconstruction of the pluristratified epidermis: A preclinical study. *Lancet* **2009**, *374*, 1745–1753. [CrossRef]

26. Wu, D.C.; Boyd, A.S.; Wood, K.J. Embryonic stem cell transplantation: Potential applicability in cell replacement therapy and regenerative medicine. *Front. Biosci.* **2007**, *12*, 4525–4535. [CrossRef] [PubMed]

27. Takahashi, K.; Yamanaka, S. Induction of pluripotent stem cells from mouse embryonic and adult fibroblast cultures by defined factors. *Cell* **2006**, *126*, 663–676. [CrossRef] [PubMed]

28. Avior, Y.; Sagi, I.; Benvenisty, N. Pluripotent stem cells in disease modelling and drug discovery. *Nat. Rev. Mol. Cell Biol.* **2016**, *17*, 170–182. [CrossRef] [PubMed]

29. Sebastiano, V.; Zhen, H.H.; Haddad, B.; Bashkirova, E.; Melo, S.P.; Wang, P.; Leung, T.L.; Siprashvili, Z.; Tichy, A.; Li, J.; et al. Human col7a1-corrected induced pluripotent stem cells for the treatment of recessive dystrophic epidermolysis bullosa. *Sci. Transl. Med.* **2014**, *6*, 264ra163. [CrossRef] [PubMed]

30. Umegaki-Arao, N.; Pasmooij, A.M.; Itoh, M.; Cerise, J.E.; Guo, Z.; Levy, B.; Gostynski, A.; Rothman, L.R.; Jonkman, M.F.; Christiano, A.M. Induced pluripotent stem cells from human revertant keratinocytes for the treatment of epidermolysis bullosa. *Sci. Transl. Med.* **2014**, *6*, 264ra164. [CrossRef] [PubMed]

31. Yang, R.; Zheng, Y.; Burrows, M.; Liu, S.; Wei, Z.; Nace, A.; Guo, W.; Kumar, S.; Cotsarelis, G.; Xu, X. Generation of folliculogenic human epithelial stem cells from induced pluripotent stem cells. *Nat. Commun.* **2014**, *5*, 3071. [CrossRef] [PubMed]

32. Itoh, M.; Umegaki-Arao, N.; Guo, Z.; Liu, L.; Higgins, C.A.; Christiano, A.M. Generation of 3d skin equivalents fully reconstituted from human induced pluripotent stem cells (ipscs). *PLoS ONE* **2013**, *8*, e77673. [CrossRef] [PubMed]

33. Asti, A.; Gioglio, L. Natural and synthetic biodegradable polymers: Different scaffolds for cell expansion and tissue formation. *Int. J. Artif. Organs* **2014**, *37*, 187–205. [PubMed]

34. Hinderer, S.; Layland, S.L.; Schenke-Layland, K. Ecm and ecm-like materials—Biomaterials for applications in regenerative medicine and cancer therapy. *Adv. Drug Deliv. Rev.* **2016**, *97*, 260–269. [CrossRef] [PubMed]

35. Kohane, D.S.; Langer, R. Polymeric biomaterials in tissue engineering. *Pediatr. Res.* **2008**, *63*, 487–491. [CrossRef] [PubMed]

36. Stoppel, W.L.; Ghezzi, C.E.; McNamara, S.L.; Black, L.D., 3rd; Kaplan, D.L. Clinical applications of naturally derived biopolymer-based scaffolds for regenerative medicine. *Ann. Biomed. Eng.* **2015**, *43*, 657–680. [CrossRef] [PubMed]

37. Rabotyagova, O.S.; Cebe, P.; Kaplan, D.L. Protein-based block copolymers. *Biomacromolecules* **2011**, *12*, 269–289. [CrossRef] [PubMed]

38. Chow, D.; Nunalee, M.L.; Lim, D.W.; Simnick, A.J.; Chilkoti, A. Peptide-based biopolymers in biomedicine and biotechnology. *Mater. Sci. Eng. R Rep.* **2008**, *62*, 125–155. [CrossRef] [PubMed]

39. Dawson, E.; Mapili, G.; Erickson, K.; Taqvi, S.; Roy, K. Biomaterials for stem cell differentiation. *Adv. Drug Deliv. Rev.* **2008**, *60*, 215–228. [CrossRef] [PubMed]

40. Ghasemi-Mobarakeh, L.; Prabhakaran, M.P.; Tian, L.; Shamirzaei-Jeshvaghani, E.; Dehghani, L.; Ramakrishna, S. Structural properties of scaffolds: Crucial parameters towards stem cells differentiation. *World J. Stem Cells* **2015**, *7*, 728–744. [CrossRef] [PubMed]

41. Willerth, S.M.; Sakiyama-Elbert, S.E. Combining stem cells and biomaterial scaffolds for constructing tissues and cell delivery. In *StemBook*; Harvard Stem Cell Institute: Cambridge, MA, USA, 2008.

42. Formigli, L.; Paternostro, F.; Tani, A.; Mirabella, C.; Quattrini Li, A.; Nosi, D.; D'Asta, F.; Saccardi, R.; Mazzanti, B.; Lo Russo, G.; et al. Mscs seeded on bioengineered scaffolds improve skin wound healing in rats. *Wound Repair Regen.* **2015**, *23*, 115–123. [CrossRef] [PubMed]

43. Hsu, S.H.; Hsieh, P.S. Self-assembled adult adipose-derived stem cell spheroids combined with biomaterials promote wound healing in a rat skin repair model. *Wound Repair Regen.* **2015**, *23*, 57–64. [CrossRef] [PubMed]

44. Kosaraju, R.; Rennert, R.C.; Maan, Z.N.; Duscher, D.; Barrera, J.; Whittam, A.J.; Januszyk, M.; Rajadas, J.; Rodrigues, M.; Gurtner, G.C. Adipose-derived stem cell-seeded hydrogels increase endogenous progenitor cell recruitment and neovascularization in wounds. *Tissue Eng. Part A* **2016**, *22*, 295–305. [CrossRef] [PubMed]

45. Zeng, Y.; Zhu, L.; Han, Q.; Liu, W.; Mao, X.; Li, Y.; Yu, N.; Feng, S.; Fu, Q.; Wang, X.; et al. Preformed gelatin microcryogels as injectable cell carriers for enhanced skin wound healing. *Acta Biomater.* **2015**, *25*, 291–303. [CrossRef] [PubMed]

46. Bhowmick, S.; Scharnweber, D.; Koul, V. Co-cultivation of keratinocyte-human mesenchymal stem cell (hmsc) on sericin loaded electrospun nanofibrous composite scaffold (cationic gelatin/hyaluronan/chondroitin sulfate) stimulates epithelial differentiation in hmscs: In vitro study. *Biomaterials* **2016**, *88*, 83–96. [CrossRef] [PubMed]

47. Fu, Y.; Guan, J.; Guo, S.; Guo, F.; Niu, X.; Liu, Q.; Zhang, C.; Nie, H.; Wang, Y. Human urine-derived stem cells in combination with polycaprolactone/gelatin nanofibrous membranes enhance wound healing by promoting angiogenesis. *J. Transl. Med.* **2014**, *12*, 274. [CrossRef] [PubMed]

48. Machula, H.; Ensley, B.; Kellar, R. Electrospun tropoelastin for delivery of therapeutic adipose-derived stem cells to full-thickness dermal wounds. *Adv. Wound Care* **2014**, *3*, 367–375. [CrossRef] [PubMed]

49. Ozpur, M.A.; Guneren, E.; Canter, H.I.; Karaaltin, M.V.; Ovali, E.; Yogun, F.N.; Baygol, E.G.; Kaplan, S. Generation of skin tissue using adipose tissue-derived stem cells. *Plast. Reconstr. Surg.* **2016**, *137*, 134–143. [CrossRef] [PubMed]

50. Caiado, F.; Carvalho, T.; Silva, F.; Castro, C.; Clode, N.; Dye, J.F.; Dias, S. The role of fibrin e on the modulation of endothelial progenitors adhesion, differentiation and angiogenic growth factor production and the promotion of wound healing. *Biomaterials* **2011**, *32*, 7096–7105. [CrossRef] [PubMed]

51. Assi, R.; Foster, T.R.; He, H.; Stamati, K.; Bai, H.; Huang, Y.; Hyder, F.; Rothman, D.; Shu, C.; Homer-Vanniasinkam, S.; et al. Delivery of mesenchymal stem cells in biomimetic engineered scaffolds promotes healing of diabetic ulcers. *Regen. Med.* **2016**, *11*, 245–260. [CrossRef] [PubMed]

52. Navone, S.E.; Pascucci, L.; Dossena, M.; Ferri, A.; Invernici, G.; Acerbi, F.; Cristini, S.; Bedini, G.; Tosetti, V.; Ceserani, V.; et al. Decellularized silk fibroin scaffold primed with adipose mesenchymal stromal cells improves wound healing in diabetic mice. *Stem Cell Res. Ther.* **2014**, *5*, 7. [CrossRef] [PubMed]

53. Chlapanidas, T.; Tosca, M.C.; Farago, S.; Perteghella, S.; Galuzzi, M.; Lucconi, G.; Antonioli, B.; Ciancio, F.; Rapisarda, V.; Vigo, D.; et al. Formulation and characterization of silk fibroin films as a scaffold for adipose-derived stem cells in skin tissue engineering. *Int. J. Immunopathol. Pharmacol.* **2013**, *26*, 43–49. [CrossRef] [PubMed]

54. He, S.; Shen, L.; Wu, Y.; Li, L.; Chen, W.; Hou, C.; Yang, M.; Zeng, W.; Zhu, C. Effect of brain-derived neurotrophic factor on mesenchymal stem cell-seeded electrospinning biomaterial for treating ischemic diabetic ulcers via milieu-dependent differentiation mechanism. *Tissue Eng. Part A* **2015**, *21*, 928–938. [CrossRef] [PubMed]

55. Liu, S.; Zhang, H.; Zhang, X.; Lu, W.; Huang, X.; Xie, H.; Zhou, J.; Wang, W.; Zhang, Y.; Liu, Y.; et al. Synergistic angiogenesis promoting effects of extracellular matrix scaffolds and adipose-derived stem cells during wound repair. *Tissue Eng. Part A* **2011**, *17*, 725–739. [CrossRef] [PubMed]

56. Schmitt, A.; Rodel, P.; Anamur, C.; Seeliger, C.; Imhoff, A.B.; Herbst, E.; Vogt, S.; van Griensven, M.; Winter, G.; Engert, J. Calcium alginate gels as stem cell matrix-making paracrine stem cell activity available for enhanced healing after surgery. *PLoS ONE* **2015**, *10*, e0118937. [CrossRef] [PubMed]

57. Bussche, L.; Harman, R.M.; Syracuse, B.A.; Plante, E.L.; Lu, Y.C.; Curtis, T.M.; Ma, M.; Van de Walle, G.R. Microencapsulated equine mesenchymal stromal cells promote cutaneous wound healing in vitro. *Stem Cell Res. Ther.* **2015**, *6*, 66. [CrossRef] [PubMed]

58. Bellini, M.Z.; Caliari-Oliveira, C.; Mizukami, A.; Swiech, K.; Covas, D.T.; Donadi, E.A.; Oliva-Neto, P.; Moraes, A.M. Combining xanthan and chitosan membranes to multipotent mesenchymal stromal cells as bioactive dressings for dermo-epidermal wounds. *J. Biomater. Appl.* **2015**, *29*, 1155–1166. [CrossRef] [PubMed]

59. Zeinali, R.; Biazar, E.; Keshel, S.H.; Tavirani, M.R.; Asadipour, K. Regeneration of full-thickness skin defects using umbilical cord blood stem cells loaded into modified porous scaffolds. *ASAIO J.* **2014**, *60*, 106–114. [CrossRef] [PubMed]

60. Chen, D.; Hao, H.; Tong, C.; Liu, J.; Dong, L.; Ti, D.; Hou, Q.; Liu, H.; Han, W.; Fu, X. Transdifferentiation of umbilical cord-derived mesenchymal stem cells into epidermal-like cells by the mimicking skin microenvironment. *Int. J. Low Extrem. Wounds* **2015**, *14*, 136–145. [CrossRef] [PubMed]

61. Garg, R.K.; Rennert, R.C.; Duscher, D.; Sorkin, M.; Kosaraju, R.; Auerbach, L.J.; Lennon, J.; Chung, M.T.; Paik, K.; Nimpf, J.; et al. Capillary force seeding of hydrogels for adipose-derived stem cell delivery in wounds. *Stem Cells Transl. Med.* **2014**, *3*, 1079–1089. [CrossRef] [PubMed]

62. Cerqueira, M.T.; da Silva, L.P.; Santos, T.C.; Pirraco, R.P.; Correlo, V.M.; Reis, R.L.; Marques, A.P. Gellan gum-hyaluronic acid spongy-like hydrogels and cells from adipose tissue synergize promoting neoskin vascularization. *ACS Appl. Mater. Interfaces* **2014**, *6*, 19668–19679. [CrossRef] [PubMed]

63. O'Loughlin, A.; Kulkarni, M.; Creane, M.; Vaughan, E.E.; Mooney, E.; Shaw, G.; Murphy, M.; Dockery, P.; Pandit, A.; O'Brien, T. Topical administration of allogeneic mesenchymal stromal cells seeded in a collagen scaffold augments wound healing and increases angiogenesis in the diabetic rabbit ulcer. *Diabetes* **2013**, *62*, 2588–2594. [CrossRef] [PubMed]

64. Kim, C.H.; Lee, J.H.; Won, J.H.; Cho, M.K. Mesenchymal stem cells improve wound healing in vivo via early activation of matrix metalloproteinase-9 and vascular endothelial growth factor. *J. Korean Med. Sci.* **2011**, *26*, 726–733. [CrossRef] [PubMed]

65. Lequeux, C.; Oni, G.; Wong, C.; Damour, O.; Rohrich, R.; Mojallal, A.; Brown, S.A. Subcutaneous fat tissue engineering using autologous adipose-derived stem cells seeded onto a collagen scaffold. *Plast. Reconstr. Surg.* **2012**, *130*, 1208–1217. [CrossRef] [PubMed]

66. Lough, D.M.; Wetter, N.; Madsen, C.; Reichensperger, J.; Cosenza, N.; Cox, L.; Harrison, C.; Neumeister, M.W. Transplantation of an lgr6+ epithelial stem cell-enriched scaffold for repair of full-thickness soft-tissue defects: The in vitro development of polarized hair-bearing skin. *Plast. Reconstr. Surg.* **2016**, *137*, 495–507. [CrossRef] [PubMed]

67. Falanga, V.; Iwamoto, S.; Chartier, M.; Yufit, T.; Butmarc, J.; Kouttab, N.; Shrayer, D.; Carson, P. Autologous bone marrow-derived cultured mesenchymal stem cells delivered in a fibrin spray accelerate healing in murine and human cutaneous wounds. *Tissue Eng.* **2007**, *13*, 1299–1312. [CrossRef] [PubMed]

68. Almine, J.F.; Bax, D.V.; Mithieux, S.M.; Nivison-Smith, L.; Rnjak, J.; Waterhouse, A.; Wise, S.G.; Weiss, A.S. Elastin-based materials. *Chem. Soc. Rev.* **2010**, *39*, 3371–3379. [CrossRef] [PubMed]

69. Wise, S.G.; Mithieux, S.M.; Weiss, A.S. Engineered tropoelastin and elastin-based biomaterials. *Adv. Protein Chem. Struct. Biol.* **2009**, *78*, 1–24. [PubMed]

70. Shen, Y.I.; Cho, H.; Papa, A.E.; Burke, J.A.; Chan, X.Y.; Duh, E.J.; Gerecht, S. Engineered human vascularized constructs accelerate diabetic wound healing. *Biomaterials* **2016**, *102*, 107–119. [CrossRef] [PubMed]

71. Rodrigues, C.; de Assis, A.M.; Moura, D.J.; Halmenschlager, G.; Saffi, J.; Xavier, L.L.; Fernandes Mda, C.; Wink, M.R. New therapy of skin repair combining adipose-derived mesenchymal stem cells with sodium carboxymethylcellulose scaffold in a pre-clinical rat model. *PLoS ONE* **2014**, *9*, e96241. [CrossRef] [PubMed]

72. Kim, K.L.; Han, D.K.; Park, K.; Song, S.H.; Kim, J.Y.; Kim, J.M.; Ki, H.Y.; Yie, S.W.; Roh, C.R.; Jeon, E.S.; et al. Enhanced dermal wound neovascularization by targeted delivery of endothelial progenitor cells using an RGD-*g*-PLLA scaffold. *Biomaterials* **2009**, *30*, 3742–3748. [CrossRef] [PubMed]

73. Geesala, R.; Bar, N.; Dhoke, N.R.; Basak, P.; Das, A. Porous polymer scaffold for on-site delivery of stem cells–protects from oxidative stress and potentiates wound tissue repair. *Biomaterials* **2016**, *77*, 1–13. [CrossRef] [PubMed]

74. Lee, P.Y.; Cobain, E.; Huard, J.; Huang, L. Thermosensitive hydrogel peg-plga-peg enhances engraftment of muscle-derived stem cells and promotes healing in diabetic wound. *Mol. Ther.* **2007**, *15*, 1189–1194. [CrossRef] [PubMed]

75.   Natesan, S.; Zamora, D.O.; Wrice, N.L.; Baer, D.G.; Christy, R.J. Bilayer hydrogel with autologous stem cells derived from debrided human burn skin for improved skin regeneration. *J. Burn Care Res.* **2013**, *34*, 18–30. [CrossRef] [PubMed]

76.   Yang, F.; Cho, S.W.; Son, S.M.; Bogatyrev, S.R.; Singh, D.; Green, J.J.; Mei, Y.; Park, S.; Bhang, S.H.; Kim, B.S.; et al. Genetic engineering of human stem cells for enhanced angiogenesis using biodegradable polymeric nanoparticles. *Proc. Natl. Acad. Sci. USA* **2010**, *107*, 3317–3322. [CrossRef] [PubMed]

77.   Chen, S.; Shi, J.; Zhang, M.; Chen, Y.; Wang, X.; Zhang, L.; Tian, Z.; Yan, Y.; Li, Q.; Zhong, W.; et al. Mesenchymal stem cell-laden anti-inflammatory hydrogel enhances diabetic wound healing. *Sci. Rep.* **2015**, *5*, 18104. [CrossRef] [PubMed]

78.   Gu, J.; Liu, N.; Yang, X.; Feng, Z.; Qi, F. Adiposed-derived stem cells seeded on plcl/p123 eletrospun nanofibrous scaffold enhance wound healing. *Biomed. Mater.* **2014**, *9*, 035012. [CrossRef] [PubMed]

79.   Baker, M.I.; Walsh, S.P.; Schwartz, Z.; Boyan, B.D. A review of polyvinyl alcohol and its uses in cartilage and orthopedic applications. *J. Biomed. Mater. Res. Part B Appl. Biomater.* **2012**, *100*, 1451–1457. [CrossRef] [PubMed]

80.   Ribeiro, J.; Pereira, T.; Amorim, I.; Caseiro, A.R.; Lopes, M.A.; Lima, J.; Gartner, A.; Santos, J.D.; Bartolo, P.J.; Rodrigues, J.M.; et al. Cell therapy with human mscs isolated from the umbilical cord wharton jelly associated to a pva membrane in the treatment of chronic skin wounds. *Int. J. Med. Sci.* **2014**, *11*, 979–987. [CrossRef] [PubMed]

81.   Dong, Y.; Hassan, W.U.; Kennedy, R.; Greiser, U.; Pandit, A.; Garcia, Y.; Wang, W. Performance of an in situ formed bioactive hydrogel dressing from a peg-based hyperbranched multifunctional copolymer. *Acta Biomater.* **2014**, *10*, 2076–2085. [CrossRef] [PubMed]

82.   Hassan, W.; Dong, Y.; Wang, W. Encapsulation and 3d culture of human adipose-derived stem cells in an in-situ crosslinked hybrid hydrogel composed of peg-based hyperbranched copolymer and hyaluronic acid. *Stem Cell Res. Ther.* **2013**, *4*, 32. [CrossRef] [PubMed]

83.   Martino, S.; D'Angelo, F.; Armentano, I.; Kenny, J.M.; Orlacchio, A. Stem cell-biomaterial interactions for regenerative medicine. *Biotechnol. Adv.* **2012**, *30*, 338–351. [CrossRef] [PubMed]

84.   Liu, P.; Deng, Z.; Han, S.; Liu, T.; Wen, N.; Lu, W.; Geng, X.; Huang, S.; Jin, Y. Tissue-engineered skin containing mesenchymal stem cells improves burn wounds. *Artif. Organs* **2008**, *32*, 925–931. [CrossRef] [PubMed]

85.   Catanzano, O.; D'Esposito, V.; Acierno, S.; Ambrosio, M.R.; De Caro, C.; Avagliano, C.; Russo, P.; Russo, R.; Miro, A.; Ungaro, F.; et al. Alginate-hyaluronan composite hydrogels accelerate wound healing process. *Carbohydr. Polym.* **2015**, *131*, 407–414. [CrossRef] [PubMed]

86.   Dash, B.C.; Mandal, B.B.; Kundu, S.C. Silk gland sericin protein membranes: Fabrication and characterization for potential biotechnological applications. *J. Biotechnol.* **2009**, *144*, 321–329. [CrossRef] [PubMed]

87.   Dinescu, S.; Galateanu, B.; Albu, M.; Cimpean, A.; Dinischiotu, A.; Costache, M. Sericin enhances the bioperformance of collagen-based matrices preseeded with human-adipose derived stem cells (hadscs). *Int. J. Mol. Sci.* **2013**, *14*, 1870–1889. [CrossRef] [PubMed]

88.   Kim, K.O.; Lee, Y.; Hwang, J.W.; Kim, H.; Kim, S.M.; Chang, S.W.; Lee, H.S.; Choi, Y.S. Wound healing properties of a 3-D scaffold comprising soluble silkworm gland hydrolysate and human collagen. *Colloids Surf. B* **2014**, *116*, 318–326. [CrossRef] [PubMed]

89.   Xu, K.; Cantu, D.A.; Fu, Y.; Kim, J.; Zheng, X.; Hematti, P.; Kao, W.J. Thiol-ene michael-type formation of gelatin/poly(ethylene glycol) biomatrices for three-dimensional mesenchymal stromal/stem cell administration to cutaneous wounds. *Acta Biomater.* **2013**, *9*, 8802–8814. [CrossRef] [PubMed]

9

Dynamic Cultivation of Mesenchymal Stem Cell Aggregates

Dominik Egger [1] (ID), Carla Tripisciano [2], Viktoria Weber [2], Massimo Dominici [3,4] and Cornelia Kasper [1,*] (ID)

[1]   Department of Biotechnology, University of Natural Resources and Life Sciences, Muthgasse 18, 1190 Vienna, Austria; dominik.egger@boku.ac.at
[2]   Christian Doppler Laboratory for Innovative Therapy Approaches in Sepsis, Danube University Krems, Dr.-Karl-Dorrek-Straße 30, 3500 Krems, Austria; carla.tripisciano@donau-uni.ac.at (C.T.); viktoria.weber@donau-uni.ac.at (V.W.)
[3]   Division of Oncology, Department of Medical and Surgical Sciences for Children & Adults, University-Hospital of Modena and Reggio Emilia, Via Università 4, 41121 Modena, Italy; massimo.dominici@unimore.it
[4]   Technopole of Mirandola TPM, 41037 Mirandola, Modena, Italy
*   Correspondence: cornelia.kasper@boku.ac.at

Abstract: Mesenchymal stem cells (MSCs) are considered as primary candidates for cell-based therapies due to their multiple effects in regenerative medicine. Pre-conditioning of MSCs under physiological conditions—such as hypoxia, three-dimensional environments, and dynamic cultivation—prior to transplantation proved to optimize their therapeutic efficiency. When cultivated as three-dimensional aggregates or spheroids, MSCs display increased angiogenic, anti-inflammatory, and immunomodulatory effects as well as improved stemness and survival rates after transplantation, and cultivation under dynamic conditions can increase their viability, proliferation, and paracrine effects, alike. Only few studies reported to date, however, have utilized dynamic conditions for three-dimensional aggregate cultivation of MSCs. Still, the integration of dynamic bioreactor systems, such as spinner flasks or stirred tank reactors might pave the way for a robust, scalable bulk expansion of MSC aggregates or MSC-derived extracellular vesicles. This review summarizes recent insights into the therapeutic potential of MSC aggregate cultivation and focuses on dynamic generation and cultivation techniques of MSC aggregates.

Keywords: mesenchymal stem cells; aggregates; spheroids; dynamic cultivation; bioreactor cultivation; extracellular vesicles; therapeutic potential; scaffold-free

---

1. Therapeutic Relevance of Mesenchymal Stem Cells

In the field of regenerative medicine, mesenchymal stem cells (MSCs) are considered primary candidates for cellular therapies and tissue engineering. They can be harvested from a variety of tissues, such as bone marrow, adipose tissue, or umbilical cords. Minimal criteria for the characterization of human MSCs defined in a position paper by the International Society for Cellular Therapies (ISCT) comprise plastic adherence, trilineage differentiation (adipogenic, chondrogenic, osteogenic), as well as a specific surface marker expression profile ($CD105^+$, $CD73^+$, $CD90^+$, $CD14^-$, $CD19^-$, $CD34^-$, $CD45^-$ and $HLADR^-$) [1]. Although MSCs might display similar properties across different species this review considers only results from research on human MSCs.

The regenerative potential of MSCs is not limited to their ability to differentiate into adipocytes, chondrocytes, and osteoblasts, as indicated by a number of studies reporting MSC differentiation into neurons [2], cardiomyocytes [3], and corneal epithelial cells [4] along with effects related to injury repair, such as migration to injury sites [5,6], angiogenesis [7] and anti-scarring effects [8]. MSCs display immunomodulatory and anti-inflammatory properties mediated by cellular cross talk [9] or by secretion of trophic factors, such as transforming growth factor-β (TGF-β), IL-6, prostaglandin E2 (PGE2), platelet-derived growth factor (PDGF), insulin-like growth factor (IGF), fibroblast growth factor (FGF), epidermal growth factor (EGF), stromal cell-derived factor 1 (SCDF-1), and vascular endothelial growth factor (VEGF) [10,11]. MSCs have been applied in a number of clinical trials with promising results for the treatment of graft-versus-host disease (GvHD), myocardial injuries, as well as bone and cartilage defects [12], and further investigations were conducted in the context of pulmonary disease, ischemic stroke, liver disease, and diabetes [13,14].

Stem Cell-Derived Extracellular Vesicles

It is generally recognized that MSCs exert their therapeutic effects via the secretion of paracrine factors and stimulation of host cells rather than via direct engraftment and cell replacement, and there is increasing evidence for the significance of MSC-derived extracellular vesicles (EVs) in this context [15,16]. EVs are small phospholipid vesicles released from a wide variety of cell types, which are commonly classified into exosomes (30–100 nm; intraluminal vesicles originating from multivesicular bodies), microvesicles (100–1000 nm, released from the plasma membrane), and apoptotic bodies (1–5 μm) according to their biogenesis and size. Exosomes preferentially expose molecules related to endosomal trafficking, such as tetraspanins (CD9, CD63, and CD81) or Alix, while microvesicles are enriched with surface markers derived from their parent cells, such as CD73 and CD90 for MSC-derived EVs. However, as there are considerable overlaps in both, size and marker profiles of exosomes and microvesicles, and their precise separation is not yet technically feasible, the use of the collective term "extracellular vesicles" has been recommended by the International Society for Extracellular Vesicles (ISEV) [17].

EVs are central mediators in a number of physiological processes, including intercellular communication, cell signaling, and maintenance of tissue homeostasis, but also in pathological settings, such as inflammation and cancer. They can be internalized by a variety of cell types and transfer bioactive molecules (e.g., cytokines, growth factors, as well as coding and regulatory genomic material, such as mRNA, miRNA, siRNA, piRNA) to their recipient cells [18]. The structure of EVs protects their cargo from enzymatic degradation, and the presence of membrane proteins enables tailored delivery to their target cells [19]. Additionally, the solubility, local availability, and bioactivity of specific factors can be enhanced by their association with EV membranes [20].

The functional consequences of EV-mediated transfer of bioactive molecules include the induction, amplification, and regulation of immune responses, which has sparked considerable interest in the application of EVs as therapeutic agents [21]. In fact, MSC-derived EVs have been shown to recapitulate the ability of their parent cells to deliver signals related to immune regulation [22,23]. In vitro data indicate that (1) MSC-derived EVs can target a range of adaptive and innate immune cells [24], that (2) EVs from different MSC sources employ different immunomodulatory mechanisms [25] and can have different effects on their target cells [15,26], and that (3) MSC-derived EVs may mediate both, immunosuppressive properties and enhanced immune responses [27]. While these findings show that MSC-derived EVs can recapitulate the ability of their parent cells to deliver signals related to immune regulation [22,23], several studies have provided evidence that MSC-derived EVs do not fully reflect the effects exerted by their parent cells [28–30]. These diverging findings may at least partially result from different experimental approaches, such as different culture conditions of MSCs, isolation, and standardization of MSC-derived EV populations, as well as variable in vitro co-culture conditions, highlighting the requirement for standardized protocols [17]. The tissue source, the isolation, as well as the culture conditions can indeed influence the biological activity of MSC-derived EVs, as recently reviewed [14]. As an example, proteomic analysis of EVs from bone marrow-derived MSCs revealed significantly increased expression of proteins associated with angiogenic signaling under ischemic conditions [31] and hypoxic preconditioning enhanced the release of EVs enriched in miRNAs involved in wound healing [32].

In addition to the in vitro data, a number of studies have investigated the therapeutic effects of MSC-derived EVs in vivo using animal disease models of myocardial infarction, stroke, kidney failure, and liver fibrosis, as summarized in [33]. A recent review of controlled trials using MSC-derived EVs concluded that their administration to animals was safe and could contribute to improved organ function following injury [34]. Two clinical studies in humans using MSC-derived EVs have been reported so far. In the first case, a patient suffering from steroid refractory GvHD was successfully treated with MSC-derived EVs. Although it remains unclear which fraction or components of the EV preparation were responsible for the anti-inflammatory effects, the study suggests that MSC-derived EVs modulated the response of patients' immune cells [35]. In the second study, patients suffering from chronic kidney disease and administered twice with cord-blood MSC-derived EVs showed improved kidney function and beneficial modulation of inflammatory markers, i.e., increase of TGF-β1 and IL-10 and decrease of TNF-α levels in response to treatment [36].

This suggests that MSC-derived EVs could represent an alternative to whole cell therapies. They may have a superior safety profile as compared to whole cells, and due to their size in the nanometer range, injected EVs can circulate through capillaries without entrapment by filter organs [37] and can cross biological barriers [38]. Their perceived capacity to survive and retain their activity during storage further supports MSC-derived EVs as a promising alternative tool for cell-free therapies [39–41]. As products of viable active stem cells, MSC-derived EVs are classified as "biological medicinal products", but aspects such as information on the active substances and the mechanism of action are still not fully elucidated. Based on the criteria for "high risk medicinal products" (HRMPs) classification, such as lack of knowledge on the mechanisms of action, no clear understanding of the target, and limited relevance of animal models, EV-based therapeutics might be categorized as such, leading to the demand of strict pre-clinical safety tests. To this purpose, the ISEV has discussed aspects concerning safety and regulatory matters to be taken into account for clinical application or medicinal manufacturing, pointing out that, as allogenic EVs are routinely transfused within blood products and there is little evidence on adverse effects and as the response to previous treatments with autologous and allogenic MSCs has been positive, MSC-EVs should not be considered as HRMPs [42].

2. 3D Aggregate Cultivation of MSCs

While biological, chemical, physical, and mechanical cues can profoundly influence cellular characteristics, commonly used MSC cultivation conditions, such as 2D cultivation on plastic surfaces under static conditions are far from representing the physiological environment of these cells. To reflect physiological conditions in vitro, cells can either be cultivated on 3D matrices or in a scaffold-free manner as cellular aggregates, often referred to as spheroids. While the general term 'aggregate' describes any multicellular entity of condensed cells, the term 'spheroid' refers to spherical cellular aggregates. In embryonic stem cell research, where aggregate cultivation has been used since decades, it is also referred to as organoid culture [43].

The dynamics of spheroid formation comprises three stages. Cadherin-cadherin interaction and integrin binding to extracellular matrix (ECM) proteins mediate first cell-cell contacts to form loose cellular aggregates. This is followed by a delay period of reorganization in which cell aggregates pause in compaction. In the third stage, strong interaction of cadherins is a major factor for the morphological transition from loose cellular aggregates to compact spheroids [44] (Figure 1).

While MSC aggregate cultivation has mainly been conducted in the context of chondrogenic differentiation, it is increasingly used to study cellular behavior under 3D conditions to more closely resemble a physiological setting [45]. Cellular behavior in tissues is determined by diffusive mass transfer, causing gradients of oxygen, nutrients, metabolic waste products, and paracrine mediators [46], which is not appropriately reflected in 2D cultivation. Moreover, MSCs cultivated as aggregates experience a different strain and rigidity during cultivation and adapt their adhesion behavior and phenotype [47] accordingly, potentially resulting in increased immunomodulatory, anti-inflammatory, and angiogenic effects. Due to this increased angiogenic and vasculogenic potential, MSC aggregates might be used as vascularization units and can be considered as building blocks for tissue engineering [48,49].

While there are a number of reviews on the characteristics of MSC aggregates [47,50–53], the impact of dynamic cultivation on both, MSC aggregates and EVs released from these aggregates remains to be summarized. As physiological pre-conditioning strategies including a 3D environment and dynamic cultivation under hypoxic conditions have been shown to optimize the therapeutic potential of MSCs [50], they may enhance the effects of MSC-derived EVs, but this hypothesis remains to be tested.

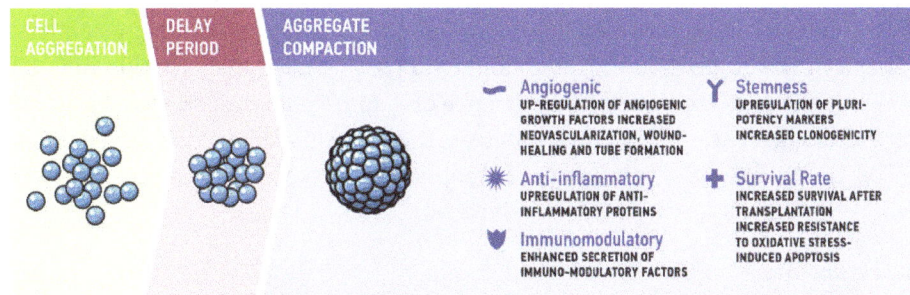

Figure 1. Three-step aggregation process: during the first phase of cell aggregation, cadherin–cadherin interactions and integrin binding to extracellular matrix proteins mediate the first cell–cell contacts. After a delay period of reorganization, aggregate compaction is mediated by cadherins.

3. Therapeutic Potential of Aggregate Cultivation

Compared to 2D monolayer cultivation, scaffold-free aggregate cultivation of cells improves their biological properties, resulting in increased cell viability, proliferation, and differentiation, as well as in physiologically relevant metabolism, phenotype, and genotype [20]. For MSC aggregates, in particular, enhanced anti-inflammatory [54], angiogenic, and tissue regenerative effects [55] as well as enhanced differentiation [56], maintenance of stem cell properties and delayed replicative senescence were observed [55,57]. This shift to a more physiological cellular behavior is not only relevant for clinical application, but can also enhance the significance of in vitro models. The following section highlights the therapeutic potential of MSC aggregate cultivation with respect to (1) angiogenic properties, (2) anti-inflammatory properties, (3) immunomodulatory characteristics, (4) stemness, and (5) cell survival and anti-apoptotic properties (Figure 1).

3.1. Angiogenic Properties

MSCs cultivated as aggregates exhibit increased angiogenic properties. Human MSC aggregates showed improved therapeutic efficacy for ischemia treatment through increased angiogenic factor secretion which has been attributed to the hypoxic environment inside the aggregates [58]. HGF, VEGF, and FGF-2 levels were 20- to 145-fold higher in medium conditioned with MSC aggregates, as compared to medium from MSCs cultivated in 2D monolayers. Recent studies not only report an improvement in terms of paracrine effects, but also functional improvement through increased neovascularization [59,60], wound healing [60], tube formation, and migration of fibroblasts into a wounded area [61].

3.2. Anti-Inflammatory and Immunomodulatory Effects

Furthermore, MSC aggregates display immunosuppressive effects. Secretion of TNF-α from macrophages decreased in co-culture with MSC aggregates as compared to co-cultivation with a 2D MSC monolayer [54,62]. The secretion of PGE2 [63], HGF [54,63–65], and TGF-β [62] which are known to suppress pro-inflammatory markers and direct stimulated macrophages towards an anti-inflammatory phenotype increased upon MSC aggregate cultivation. Likewise, anti-inflammatory factors, such as TNF-α-stimulated gene protein TSG-6 which is known to counteract TNF-α and IL-1 inflammation, were elevated [54]. Furthermore, MSC aggregates suppressed inflammation in a mouse model of zymosan-induced peritonitis [54] and reduced acute kidney injury in a rat ischemia–reperfusion model [66]. The expression of anti-inflammatory and immunomodulatory factors, however, can be further increased by optimizing the microenvironment via the spheroid size, oxygen tension, and inflammatory stimulus [67].

3.3. Stemness

During 2D monolayer cultivation, MSCs may undergo aging, loss of clonogenicity, or spontaneous differentiation [68,69]. For clinical application however, it is crucial to maintain the stemness of MSCs during in vitro cultivation. Compared to 2D cultivation, MSC aggregates display increased expression of the pluripotency marker genes *Nanog*, *Sox2*, and *Oct4* [70,71]. miRNAs—namely miR-489, miR-370, and miR-433—which are related to the maintenance of a quiescent adult stem cell state, were highly expressed in MSC aggregates [70], and an increased clonogenicity was observed after aggregate cultivation [70,71]. In a following study, delayed replicative senescence of aggregate-derived MSCs was observed in comparison to monolayer-derived MSCs [55].

3.4. Cell Survival and Anti-Apoptotic Effects

The survival of cells after transplantation plays an important role in the therapeutic outcome. As an example, more than 85% of systemically injected MSCs were found in the precapillaries [37]. MSCs cultivated as aggregates displayed better survival in ischemic conditions [72] and higher resistance to oxidative stress-induced apoptosis [73]. Additionally, the pro-apoptotic molecule Bax was downregulated, while the anti-apoptotic molecule Bcl-2 was upregulated in MSC aggregates [57,72], which might contribute to the overall post-transplantation survival of MSCs.

4. Generation of MSC Aggregates

To generate aggregates, MSC adhesion to tissue culture plates must be avoided. Methods for the generation of aggregates from a single cell suspension can be classified into cluster-based self-assembly and collision-based assembly [74]. Cluster-based self-assembly is a process in a static environment where cells are prevented from attaching to a surface and thus come in contact with each other to form aggregates. In contrast, collision-based assembly takes place in a dynamic environment, where cells collide upon centrifugation or mixing of a single cell suspension (Figure 2).

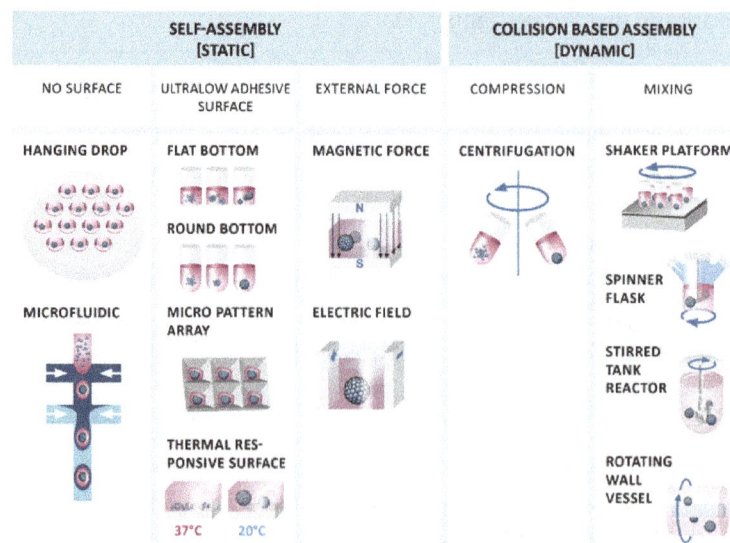

Figure 2. Different techniques for static cluster-based self-assembly and dynamic collision-based assembly of MSC aggregates. Self-assembly of MSCs can be forced using no or ultralow adhesive surfaces or external forces. Collision-based assembly is conducted by compression or mixing.

4.1. Static Cluster-Based Self-Assembly

In cluster-based self-assembly, single cells are separated into compartments and undergo the typical three-step process of aggregate formation as shown in Figure 1. Hanging drop cultivation may be the most common cluster-based self-assembly method [49,75,76]. Specialized cell culture plates allow formation of hanging drops from a single cell suspension with subsequent formation of cell aggregates. Beside its labor intensity, the only drawback of this method is that medium changes are challenging and prone to error or destruction of aggregates or the hanging drops. To overcome this limitation, automated [77], robot assisted [78] and microfluidic based [79] high-throughput hanging drop cultivation systems have been developed recently.

Cell culture plates with ultralow adhesive surfaces can be used to generate aggregates, as well [56,62,75]. This method is also referred to as 'liquid overlay' method. On flat bottom plates, cells form aggregates of heterogeneous size and shape, whereas aggregate shape and size can be very well controlled in round-shaped cavities, such as round bottom multiwell plates. Based on this principle, different kinds of microwell arrays made from micropatterned agarose [80], polydimethylsiloxane (PDMS) [81] or polyethylene glycol (PEG) hydrogels [82] have been developed to generate large quantities of uniformly sized and shaped aggregates in a cost-effective manner. Other modifications, such as thermally responsive surfaces [83] or polycationic chitosan membranes [71,84], have also been applied to form aggregates. These methods yielded viable aggregates, although heterogeneous in shape and size. Microfluidic systems were also used to generate size controlled aggregates [85]. As an example double-emulsion droplets were used to generate picoliter-sized bioreactors for the self-assembly of MSC spheroids [86]. External forces such as magnetic force [87], electric field [88], or ultrasound wave traps [89] to concentrate cells for aggregation are not as common, and only magnetic force has been used for the aggregation of MSCs so far [90,91].

4.2. Dynamic Collision-Based Assembly

Methods for dynamic, collision-based assembly of MSC aggregates include forced aggregation by centrifugation [92] or mixing mediated by shaker platforms [75,93], spinner flasks [56,59], rotating wall vessels (RWVs) [56], and stirred tank reactors (STRs) [94]. Aggregation by centrifugation has mainly been used for chondrogenic differentiation of MSCs [95] and is also known as pellet or micromass culture. Collision-based assembly by mixing was observed with a seeding density of as low as 2×10^4 cells/mL in spinner flasks and RWVs [56], with 1×10^5 cells/mL in a STR [94] and led to randomly sized spheroids, whereas mixing in ultralow adhesive multiwell plates on a shaker platform [75] and compression by centrifugation [92] yielded aggregates with narrower size and homogeneous shape distribution.

5. Dynamic Cultivation of Aggregates

The therapeutic effects of MSCs, MSC conditioned medium, or MSC-derived EVs have been shown to support regeneration after organ and tissue injury. In vitro pre-conditioning strategies can enhance survival, engraftment, and paracrine properties of MSCs and, therefore, optimize their therapeutic potential [50]. Specifically, dynamic cultivation conditions, such as fluid flow have substantial impact on cellular behavior (Figure 3). Increased proliferation, viability, differentiation potential but also paracrine effects were observed in perfusion bioreactors, on horizontal or orbital shaking platforms, or in stirred systems, such as spinner flasks or stirred tank reactors [96–98]. However, only a few studies have harnessed dynamic conditions for the generation and cultivation of MSC aggregates (Table 1), as dynamic cultivation aggregates do not necessarily need to be generated by dynamic collision-based assembly. Studies report formation by centrifugation [92], in hanging drops [75], in ultralow adhesive multiwell plates [75], or in a microwell array [82] followed by cultivation in a dynamic cultivation system, such as shaker platforms [75,82,92,93], in spinner flasks [56,59], RWV [56], or STR [94]. In microwell arrays, all seeded cells are involved in the formation of aggregates, thus the size and cell number per aggregate can be precisely controlled.

Table 1. Comparison of different studies reporting on dynamic cultivation conditions for the cultivation of MSC aggregates.

Ref.	Cultivation System	Cells Per Spheroid	Initial Cell Density (c/mL)	Rotation (rpm)	Duration (days)	Surface Marker	Differentiation Capacity	Effects	Aggregate Size (μm)
[56]	SF and cultivation in spinner flask	random	2×10^4	30	7	~	A↑, O↑ (compared to 2D static)	Hypoxia-linked genes ↑, changes in ECM organization, IL24↑	56–135 (avrg. 99)
[56]	SF and cultivation in rotating wall vessel bioreactor	random	2×10^4	15	7	~	A↑, O↑ (compared to 2D static)	-	18–44 (avrg. 32)
[75]	SF and cultivation on orbital shaker	random	5×10^4	95	Aborted after 3 days	-		-	Multi aggregation
[75]	Formation in hanging drop, cultivation in suspension on orbital shaker	5000	2.5×10^5	95	Aborted after 3 days	-			Multi aggregation
[75]	96-well plate on orbital shaker followed by static cultivation	$1–2 \times 10^4$	$0.6–1.3 \times 10^5$	95	2 dynamic followed by 21 static	-	O↑ (compared to control)	Col1, Col3, OPN, BMP-2↑	200
[59]	SF and cultivation in spinner flask	random	6×10^5	70	3	~	-	Anti-apoptotic, angiogenic factors, preservation of ECM, enhanced survival after transplantation	100–350
[92]	Formation by centrifugation followed by orbital shaker	300/600/1000	$1.8–6 \times 10^6$	45	21	-	A↑, O↑ (compared to 2D static)	Active proliferation in the center of the spheroid, undifferentiated up to 16 days,	157/100/177 (day 7)
[93]	SF and cultivation in shaker flask on horizontal shaker	random	1×10^5	80	7	~	A, C, O (dissociated cells in 2D static after 3D dynamic)	Active proliferation in the center of the spheroid, up to 6-fold expansion	-
[94]	SF and cultivation in stirred tank bioreactor	random	1×10^5	600	6	~	A, C, O (dissociated cells in 2D static after 3D dynamic)	Approx. 2-fold expansion	-
[82]	Microwell array on orbital shaker	400	5×10^5 cells/array	30	7	-	-	No proliferation, EV production ↑	150

SF: spontaneous formation of aggregates, A: adipogenic differentiation, C: chondrogenic differentiation, O: osteogenic differentiation, decrease: ↓, increase: ↑, comparable to 2D: ~, not measured: -.

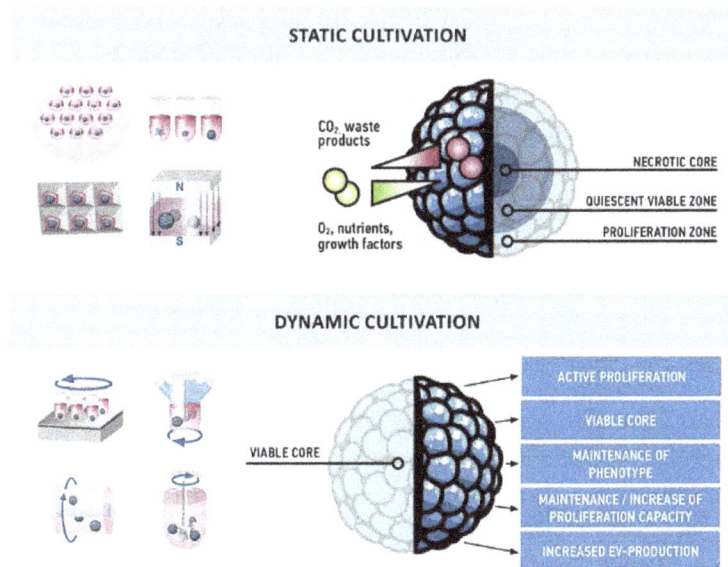

Figure 3. Comparison of effects observed in static and dynamic cultivation of MSC aggregates. Due to gradients of nutrients and waste products, aggregates cultivated under static conditions are usually structured in three layers: a necrotic core in the center, a quiescent viable zone of non-proliferative cells, and an outer layer with proliferating cells. In contrast, dynamic cultivation conditions result in a viable core and active proliferation throughout the aggregate. Cells from these aggregates maintain their phenotype, proliferation capacity, and display an increased production of EVs.

5.1. Proliferation and Viability

Under static conditions, a necrotic core develops with time due to oxygen, nutrition, and waste product gradients along the diameter of an aggregate, and thus aggregates cultivated under static conditions have not been reported to exceed approximately 500 µm in diameter [99,100]. Interestingly, none of the studies on dynamic cultivation of MSC aggregates reported a necrotic core inside the aggregates. In contrast, two studies reported active proliferation at the core of the aggregates [92,93], and an up to six-fold expansion of cells was observed [93,94]. Increased convection in dynamic cultivation seems to improve oxygen and nutrient supply and to inhibit the formation of gradients. In contrast, Cha et al. observed active proliferation only in 3D static conditions whereas in 3D dynamic conditions cells seemed to rest, although they were more active in terms of EV production [82]. The different cellular behavior might not only be owed to different aggregate formation and cultivation techniques but also to different media compositions. High expansion was observed in optimized serum-free [93] or platelet lysate supplemented [94] medium, whereas less or no proliferation was observed using fetal bovine serum-containing medium [82].

5.2. Stemness

Mechanical stimulation by shear forces during dynamic cultivation can trigger spontaneous differentiation of MSCs and thus compromise their stemness. Therefore, surface marker expression and differentiation capacity of MSCs have been evaluated in previous studies after dynamic aggregate cultivation. None of the studies analyzing surface marker expression according to the guidelines of the ISCT found alterations of the phenotype [56,59,93,94]. However, during aggregate formation and cellular reorganization mesenchymal stem cell markers seemed to be altered [56,82]. When cells were cultivated for an extended period as aggregates or were dissociated after 3D dynamic cultivation and cultivated again in 2D models, cells expressed a typical MSC phenotype. Also, MSC aggregates were kept in an undifferentiated state over a period of 16 days on a shaker platform at 45 rpm [92] and for a period of 6 days at 600 rpm in a STR with an average shear stress of 0.2 Pa [94]. Thus, aggregate formation might shield the inner cell mass from shear forces and helps to avoid spontaneous differentiation. Regarding differentiation capacity all studies that either differentiated MSC aggregates [56,75,92] or differentiated MSCs after 3D dynamic cultivation [93,94] observed robust trilineage differentiation as analyzed by histological stainings and/or gene expression. The studies testing differentiation of MSC aggregates against 2D static observed increased adipogenic and osteogenic differentiation [56,75,93]. However, it remains unclear if the cultivation under 3D dynamic conditions increases the differentiation potential in comparison to 3D static cultivation.

5.3. Therapeutic Potential

In the context of therapeutic potential, a direct comparison between dynamic and static cultivation of 3D aggregates would be of interest since until now the gene regulation of 3D dynamic cultivated cells was only compared to 2D static cultivation. The study observed 710 genes that were differently expressed (277 downregulated and 433 upregulated genes). The most differently expressed genes were classified under (1) biological adhesion, structural molecule activity, and ECM which pointed to changes in the cytoskeleton; (2) developmental process which affected numerous secreted factors like IL-24; and (3) hypoxia related genes [56]. Also, the secretion of angiogenic factors like VEGF, HGF, and HGF-2 was significantly increased in an ischemic limb model after dynamic cultivation when cells were grafted as spheroids compared to dissociated cells [59]. The same study observed a higher survival rate after transplantation. Furthermore, the production of EVs was strongly increased during dynamic aggregate cultivation compared to static aggregates [82] and cytokine levels in these EVs was significantly higher. These findings suggest that dynamic cultivation of MSC aggregates might increase the therapeutic potential of MSCs or of MSC-derived EVs.

6. Concluding Remarks

Only few studies so far have addressed dynamic cultivation of MSC aggregates and since different techniques for the generation and cultivation were used, results are diverse and not directly comparable. Moreover, until now only Cha et al. specifically compared 3D static to 3D dynamic cultivation of MSC aggregates [82]. However, existing reports on the dynamic cultivation of MSC aggregates highlight the maintenance of stemness, improved differentiation capacity, and to some extent active proliferation of cells (Figure 3).

The dynamic cultivation of MSC aggregates might be a suitable strategy to develop a passage-free expansion system. Since convection reduces oxygen, nutrient and waste gradients along the diameter of aggregates, larger viable aggregates without necrotic cells in their core can be generated under dynamic conditions. Currently, the growth rate is not competitive to microcarrier-based expansion systems [101,102]. Up to now, however, no study has investigated the cellular growth in dynamically cultivated MSC aggregates for more than seven days, thus after optimization MSC aggregates might be a viable option for in vitro expansion. Interestingly, up to six-fold expansion of MSCs was observed within seven days with a medium optimized for dynamic aggregate cultivation of MSCs [93] indicating further optimization potential.

Also, defined starting conditions in terms of uniform aggregate shape, size, and cell number are needed in future studies. As more manufacturers offer ready-to-use micropatterned multiwell plates that do not need additional centrifugation steps or rinsing agents, this might be the best option to generate lager numbers of uniform spheroids at low cost and a minimum of time.

Next to relevant biological implications and advantages of the 3D culture, the large number of clinical studies (>700 from www.clinicaltrials.gov) based on MSC demand for novel conditions where cells may be cultured in a more cost-effective manner. This could allow the reduction of MSC manufacturing costs, contributing to the progress towards larger phase III studies and, ultimately, to a wider diffusion of their therapeutic potential.

Although more studies on the direct comparison of 3D static to 3D dynamic cultivation of MSC aggregates are needed, the upregulation of hypoxic genes which results in increased angiogenesis, the upregulation of the cancer suppressing cytokine IL-24, and the increased EV production found in dynamically cultivated MSC aggregates may be promising for future clinical application. Due to this high therapeutic potential, the large-scale production of MSC-EVs and their standardization is becoming a crucial issue for clinical translation [103–105].

Author Contributions: D.E. designed the concept of the review, performed the literature research, wrote the manuscript and designed the figures. C.T. and V.W. contributed to parts about extracellular vesicles and critically reviewed the manuscript. M.D. contributed to parts on the therapeutic potential of MSCs and critically reviewed the manuscript. C.K. designed the concept of the review and reviewed and approved the manuscript.

Funding: C.T. and V.W. received funding from the Christian Doppler Society (Christian Doppler Laboratory for Innovative Therapy Approaches in Sepsis).

Acknowledgments: The authors would like to thank Maria Egger (mooi design) for preparing the figures.

Conflicts of Interest: The authors declare no conflict of interest.

Abbreviations

2D	Two-dimensional
3D	Three-dimensional
ECM	Extracellular matrix
EGF	Epidermal growth factor
EVs	Extracellular vesicles
FGF	Fibroblast growth factor
GvHD	Graft-versus-host disease
HRMPs	High risk medicinal products
IGF	Insulin-like growth factor
ISCT	International Society for Cellular Therapies
MSCs	Mesenchymal stem cells
PDGF	Platelet-derived growth factor
PDMS	Polydimethylsiloxane
PEG	Polyethylene glycol
PGE2	Prostaglandin E2
RWV	Rotating wall vessel
SCDF-1	Stromal cell-derived factor 1
STR	Stirred tank reactor
TGF-β	Transforming growth factor-β
VEGF	Vascular endothelial growth factor

References

1. Dominici, M.; Le Blanc, K.; Mueller, I.; Slaper-Cortenbach, I.; Marini, F.; Krause, D.; Deans, R.; Keating, A.; Prockop, D.; Horwitz, E. Minimal criteria for defining multipotent mesenchymal stromal cells. The International Society for Cellular Therapy position statement. *Cytotherapy* **2006**, *8*, 315–317. [CrossRef] [PubMed]

2. Kim, J.; Park, S.; Kim, Y.J.; Jeon, C.S.; Lim, K.T.; Seonwoo, H.; Cho, S.P.; Chung, T.D.; Choung, P.H.; Choung, Y.H.; et al. Monolayer Graphene-Directed Growth and Neuronal Differentiation of Mesenchymal Stem Cells. *J. Biomed. Nanotechnol.* **2015**, *11*, 2024–2033. [CrossRef] [PubMed]

3. Shen, H.; Wang, Y.; Zhang, Z.; Yang, J.; Hu, S.; Shen, Z. Mesenchymal Stem Cells for Cardiac Regenerative Therapy: Optimization of Cell Differentiation Strategy. *Stem Cells Int.* **2015**, *2015*, 524756. [CrossRef] [PubMed]

4. Harkin, D.G.; Foyn, L.; Bray, L.J.; Sutherland, A.J.; Li, F.J.; Cronin, B.G. Concise Reviews: Can Mesenchymal Stromal Cells Differentiate into Corneal Cells? A Systematic Review of Published Data. *Stem Cells* **2015**, *33*, 785–791. [CrossRef] [PubMed]

5. Oh, J.Y.; Kim, M.K.; Shin, M.S.; Lee, H.J.; Ko, J.H.; Wee, W.R.; Lee, J.H. The anti-inflammatory and anti-angiogenic role of mesenchymal stem cells in corneal wound healing following chemical injury. *Stem Cells* **2008**, *26*, 1047–1055. [CrossRef] [PubMed]

6. Maxson, S.; Lopez, E.A.; Yoo, D.; Danilkovitch-Miagkova, A.; Leroux, M.A. Concise Review: Role of Mesenchymal Stem Cells in Wound Repair. *Stem Cells Transl. Med.* **2012**, *1*, 142–149. [CrossRef] [PubMed]

7. Sorrell, J.M.; Baber, M.A.; Caplan, A.I. Influence of adult mesenchymal stem cells on in vitro vascular formation. *Tissue Eng. Part A* **2009**, *15*, 1751–1761. [CrossRef] [PubMed]

8. Meirelles Lda, S.; Fontes, A.M.; Covas, D.T.; Caplan, A.I. Mechanisms involved in the therapeutic properties of mesenchymal stem cells. *Cytokine Growth Factor Rev.* **2009**, *20*, 419–427. [CrossRef] [PubMed]

9. Murphy, M.B.; Moncivais, K.; Caplan, A.I. Mesenchymal stem cells: Environmentally responsive therapeutics for regenerative medicine. *Exp. Mol. Med.* **2013**, *45*, e54. [CrossRef] [PubMed]

10. Wang, Y.; Chen, X.; Cao, W.; Shi, Y. Plasticity of mesenchymal stem cells in immunomodulation: Pathological and therapeutic implications. *Nat. Immunol.* **2014**, *15*, 1009–1016. [CrossRef] [PubMed]

11. Gnecchi, M.; Zhang, Z.P.; Ni, A.G.; Dzau, V.J. Paracrine Mechanisms in Adult Stem Cell Signaling and Therapy. *Circ. Res.* **2008**, *103*, 1204–1219. [CrossRef] [PubMed]

12. Trounson, A.; McDonald, C. Stem Cell Therapies in Clinical Trials: Progress and Challenges. *Cell Stem Cell* **2015**, *17*, 11–22. [CrossRef] [PubMed]

13. Squillaro, T.; Peluso, G.; Galderisi, U. Clinical Trials With Mesenchymal Stem Cells: An Update. *Cell Transp.* **2016**, *25*, 829–848. [CrossRef] [PubMed]

14. Linero, I.; Chaparro, O. Paracrine Effect of Mesenchymal Stem Cells Derived from Human Adipose Tissue in Bone Regeneration. *PLoS ONE* **2014**, *9*, e0119262. [CrossRef] [PubMed]

15. Stephen, J.; Bravo, E.L.; Colligan, D.; Fraser, A.R.; Petrik, J.; Campbell, J.D.M. Mesenchymal stromal cells as multifunctional cellular therapeutics—A potential role for extracellular vesicles. *Transfus. Apher. Sci.* **2016**, *55*, 62–69. [CrossRef] [PubMed]

16. Mushahary, D.; Spittler, A.; Kasper, C.; Weber, V.; Charwat, V. Isolation, cultivation, and characterization of human mesenchymal stem cells. *Cytom. Part A* **2018**, *93*, 19–31. [CrossRef] [PubMed]

17. Lötvall, J.; Hill, A.F.; Hochberg, F.; Buzas, E.I.; Di Vizio, D.; Gardiner, C.; Gho, Y.S.; Kurochkin, I.V.; Mathivanan, S.; Quesenberry, P.; et al. Minimal experimental requirements for definition of extracellular vesicles and their functions: A position statement from the International Society for Extracellular Vesicles. *J. Extracell. Vesicles* **2014**, *3*, 26913. [CrossRef] [PubMed]

18. Mulcahy, L.A.; Pink, R.C.; Carter, D.R. Routes and mechanisms of extracellular vesicle uptake. *J. Extracell. Vesicles* **2014**, *3*, 24641. [CrossRef] [PubMed]

19. Biancone, L.; Bruno, S.; Deregibus, M.C.; Tetta, C.; Camussi, G. Therapeutic potential of mesenchymal stem cell-derived microvesicles. *Nephrol. Dial. Transp.* **2012**, *27*, 3037–3042. [CrossRef] [PubMed]

20. Sun, D.M.; Zhuang, X.Y.; Xiang, X.Y.; Liu, Y.L.; Zhang, S.Y.; Liu, C.R.; Barnes, S.; Grizzle, W.; Miller, D.; Zhang, H.G. A Novel Nanoparticle Drug Delivery System: The Anti-inflammatory Activity of Curcumin Is Enhanced When Encapsulated in Exosomes. *Mol. Ther.* **2010**, *18*, 1606–1614. [CrossRef] [PubMed]

21. Robbins, P.D.; Morelli, A.E. Regulation of immune responses by extracellular vesicles. *Nat. Rev. Immunol.* **2014**, *14*, 195–208. [CrossRef] [PubMed]

22. Mokarizadeh, A.; Delirezh, N.; Morshedi, A.; Mosayebi, G.; Farshid, A.A.; Mardani, K. Microvesicles derived from mesenchymal stem cells: Potent organelles for induction of tolerogenic signaling. *Immunol. Lett.* **2012**, *147*, 47–54. [CrossRef] [PubMed]

23. Budoni, M.; Fierabracci, A.; Luciano, R.; Petrini, S.; Di Ciommo, V.; Muraca, M. The Immunosuppressive Effect of Mesenchymal Stromal Cells on B Lymphocytes Is Mediated by Membrane Vesicles. *Cell Transp.* **2013**, *22*, 369–379. [CrossRef]

24. Di Trapani, M.; Bassi, G.; Midolo, M.; Gatti, A.; Kamga, P.T.; Cassaro, A.; Carusone, R.; Adamo, A.; Krampera, M. Differential and transferable modulatory effects of mesenchymal stromal cell-derived extracellular vesicles on T, B and NK cell functions. *Sci. Rep.* **2016**, *6*, 24120. [CrossRef] [PubMed]

25. Del Fattore, A.; Luciano, R.; Pascucci, L.; Goffredo, B.M.; Giorda, E.; Scapaticci, M.; Fierabracci, A.; Muraca, M. Immunoregulatory Effects of Mesenchymal Stem Cell-Derived Extracellular Vesicles on T Lymphocytes. *Cell Transp.* **2015**, *24*, 2615–2627. [CrossRef] [PubMed]

26. Del Fattore, A.; Luciano, R.; Saracino, R.; Battafarano, G.; Rizzo, C.; Pascucci, L.; Alessandri, G.; Pessina, A.; Perrotta, A.; Fierabracci, A.; et al. Differential effects of extracellular vesicles secreted by mesenchymal stem cells from different sources on glioblastoma cells. *Expert Opin. Biol. Ther.* **2015**, *15*, 495–504. [CrossRef] [PubMed]

27. Burrello, J.; Monticone, S.; Gai, C.; Gomez, Y.; Kholia, S.; Camussi, G. Stem Cell-Derived Extracellular Vesicles and Immune-Modulation. *Front. Cell Dev. Biol.* **2016**, *4*, 83. [CrossRef] [PubMed]

28. Favaro, E.; Carpanetto, A.; Caorsi, C.; Giovarelli, M.; Angelini, C.; Cavallo-Perin, P.; Tetta, C.; Camussi, G.; Zanone, M.M. Human mesenchymal stem cells and derived extracellular vesicles induce regulatory dendritic cells in type 1 diabetic patients. *Diabetologia* **2016**, *59*, 325–333. [CrossRef] [PubMed]

29. Conforti, A.; Scarsella, M.; Starc, N.; Giorda, E.; Biagini, S.; Proia, A.; Carsetti, R.; Locatelli, F.; Bernardo, M.E. Microvescicles derived from mesenchymal stromal cells are not as effective as their cellular counterpart in the ability to modulate immune responses in vitro. *Stem Cells Dev.* **2014**, *23*, 2591–2599. [CrossRef] [PubMed]

30. de Andrade, A.V.G.; Bertolino, G.; Riewaldt, J.; Bieback, K.; Karbanova, J.; Odendahl, M.; Bornhauser, M.; Schmitz, M.; Corbeil, D.; Tonn, T. Extracellular Vesicles Secreted by Bone Marrow- and Adipose Tissue-Derived Mesenchymal Stromal Cells Fail to Suppress Lymphocyte Proliferation. *Stem Cells Dev.* **2015**, *24*, 1374–1376. [CrossRef] [PubMed]

31. Anderson, J.D.; Johansson, H.J.; Graham, C.S.; Vesterlund, M.; Pham, M.T.; Bramlett, C.S.; Montgomery, E.N.; Mellema, M.S.; Bardini, R.L.; Contreras, Z. Comprehensive proteomic analysis of mesenchymal stem cell exosomes reveals modulation of angiogenesis via nuclear factor-kappaB signaling. *Stem Cells* **2016**, *34*, 601–613. [CrossRef] [PubMed]

32. Lo Sicco, C.; Reverberi, D.; Balbi, C.; Ulivi, V.; Principi, E.; Pascucci, L.; Becherini, P.; Bosco, M.C.; Varesio, L.; Franzin, C. Mesenchymal stem cell-derived extracellular vesicles as mediators of anti-inflammatory effects: Endorsement of macrophage polarization. *Stem Cells Transl. Med.* **2017**, *6*, 1018–1028. [CrossRef] [PubMed]

33. Börger, V.; Bremer, M.; Ferrer-Tur, R.; Gockeln, L.; Stambouli, O.; Becic, A.; Giebel, B. Mesenchymal Stem/Stromal Cell-Derived Extracellular Vesicles and Their Potential as Novel Immunomodulatory Therapeutic Agents. *Int. J. Mol. Sci.* **2017**, *18*, 1450. [CrossRef] [PubMed]

34. Akyurekli, C.; Le, Y.; Richardson, R.B.; Fergusson, D.; Tay, J.; Allan, D.S. A Systematic Review of Preclinical Studies on the Therapeutic Potential of Mesenchymal Stromal Cell-Derived Microvesicles. *Stem Cell Rev. Rep.* **2015**, *11*, 150–160. [CrossRef] [PubMed]

35. Kordelas, L.; Rebmann, V.; Ludwig, A.K.; Radtke, S.; Ruesing, J.; Doeppner, T.R.; Epple, M.; Horn, P.A.; Beelen, D.W.; Giebel, B. MSC-derived exosomes: A novel tool to treat therapy-refractory graft-versus-host disease. *Leukemia* **2014**, *28*, 970–973. [CrossRef] [PubMed]

36. Nassar, W.; El-Ansary, M.; Sabry, D.; Mostafa, M.A.; Fayad, T.; Kotb, E.; Temraz, M.; Saad, A.N.; Essa, W.; Adel, H. Umbilical cord mesenchymal stem cells derived extracellular vesicles can safely ameliorate the progression of chronic kidney diseases. *Biomater. Res.* **2016**, *20*, 21. [CrossRef] [PubMed]

37. Toma, C.; Wagner, W.R.; Bowry, S.; Schwartz, A.; Villanueva, F. Fate Of Culture-Expanded Mesenchymal Stem Cells in The Microvasculature In Vivo Observations of Cell Kinetics. *Circ. Res.* **2009**, *104*, 398–402. [CrossRef] [PubMed]

38. Yang, T.Z.; Martin, P.; Fogarty, B.; Brown, A.; Schurman, K.; Phipps, R.; Yin, V.P.; Lockman, P.; Bai, S.H. Exosome Delivered Anticancer Drugs Across the Blood-Brain Barrier for Brain Cancer Therapy in Danio Rerio. *Pharm. Res.-Dordr.* **2015**, *32*, 2003–2014. [CrossRef] [PubMed]

39. Otsuru, S.; Desbourdes, L.; Guess, A.J.; Hofmann, T.J.; Relation, T.; Kaito, T.; Dominici, M.; Iwamoto, M.; Horwitz, E.M. Extracellular vesicles released from mesenchymal stromal cells stimulate bone growth in osteogenesis imperfecta. *Cytotherapy* **2017**, *20*, 72–73. [CrossRef] [PubMed]

40. Kumeda, N.; Ogawa, Y.; Akimoto, Y.; Kawakami, H.; Tsujimoto, M.; Yanoshita, R. Characterization of Membrane Integrity and Morphological Stability of Human Salivary Exosomes. *Biol. Pharm. Bull.* **2017**, *40*, 1183–1191. [CrossRef] [PubMed]

41. Jeyaram, A.; Jay, S.M. Preservation and Storage Stability of Extracellular Vesicles for Therapeutic Applications. *AAPS J.* **2018**, *20*, 1. [CrossRef] [PubMed]

42. Lener, T.; Gimona, M.; Aigner, L.; Börger, V.; Buzas, E.; Camussi, G.; Chaput, N.; Chatterjee, D.; Court, F.A.; Portillo, H.A.d. Applying extracellular vesicles based therapeutics in clinical trials–an ISEV position paper. *J. Extracell. Vesicles* **2015**, *4*, 30087. [CrossRef] [PubMed]

43. Clevers, H. Modeling Development and Disease with Organoids. *Cell* **2016**, *165*, 1586–1597. [CrossRef] [PubMed]

44. Lin, R.Z.; Chang, H.Y. Recent advances in three-dimensional multicellular spheroid culture for biomedical research. *Biotechnol. J.* **2008**, *3*, 1172–1184. [CrossRef] [PubMed]

45. Chimenti, I.; Massai, D.; Morbiducci, U.; Beltrami, A.P.; Pesce, M.; Messina, E. Stem Cell Spheroids and Ex Vivo Niche Modeling: Rationalization and Scaling-Up. *J. Cardiovasc. Transl.* **2017**, *10*, 150–166. [CrossRef] [PubMed]

46. Kinney, M.A.; Hookway, T.A.; Wang, Y.; McDevitt, T.C. Engineering three-dimensional stem cell morphogenesis for the development of tissue models and scalable regenerative therapeutics. *Ann. Biomed. Eng.* **2014**, *42*, 352–367. [CrossRef] [PubMed]

47. Follin, B.; Juhl, M.; Cohen, S.; Perdersen, A.E.; Kastrup, J.; Ekblond, A. Increased Paracrine Immunomodulatory Potential of Mesenchymal Stromal Cells in Three-Dimensional Culture. *Tissue Eng. Part B-Rev.* **2016**, *22*, 322–329. [CrossRef] [PubMed]

48. Laschke, M.W.; Menger, M.D. Life is 3D: Boosting Spheroid Function for Tissue Engineering. *Trends Biotechnol.* **2017**, *35*, 133–144. [CrossRef] [PubMed]

49. Yoon, H.H.; Bhang, S.H.; Shin, J.Y.; Shin, J.; Kim, B.S. Enhanced Cartilage Formation via Three-Dimensional Cell Engineering of Human Adipose-Derived Stem Cells. *Tissue Eng. Part A* **2012**, *18*, 1949–1956. [CrossRef] [PubMed]

50. Schafer, R.; Spohn, G.; Baer, P.C. Mesenchymal Stem/Stromal Cells in Regenerative Medicine: Can Preconditioning Strategies Improve Therapeutic Efficacy? *Transfus. Med. Hemother.* **2016**, *43*, 256–267. [CrossRef] [PubMed]

51. Petrenko, Y.; Sykova, E.; Kubinova, S. The therapeutic potential of three-dimensional multipotent mesenchymal stromal cell spheroids. *Stem Cell Res. Ther.* **2017**, *8*, 94. [CrossRef] [PubMed]

52. Sart, S.; Tsai, A.C.; Li, Y.; Ma, T. Three-dimensional aggregates of mesenchymal stem cells: Cellular mechanisms, biological properties, and applications. *Tissue Eng. Part B Rev.* **2014**, *20*, 365–380. [CrossRef] [PubMed]

53. Xie, L.; Mao, M.; Zhou, L.; Jiang, B. Spheroid Mesenchymal Stem Cells and Mesenchymal Stem Cell-Derived Microvesicles: Two Potential Therapeutic Strategies. *Stem Cells Dev.* **2016**, *25*, 203–213. [CrossRef] [PubMed]

54. Bartosh, T.J.; Ylostalo, J.H.; Mohammadipoor, A.; Bazhanov, N.; Coble, K.; Claypool, K.; Lee, R.H.; Choi, H.; Prockop, D.J. Aggregation of human mesenchymal stromal cells (MSCs) into 3D spheroids enhances their antiinflammatory properties. *Proc. Natl. Acad. Sci. USA* **2010**, *107*, 13724–13729. [CrossRef] [PubMed]

55. Cheng, N.C.; Chen, S.Y.; Li, J.R.; Young, T.H. Short-term spheroid formation enhances the regenerative capacity of adipose-derived stem cells by promoting stemness, angiogenesis, and chemotaxis. *Stem Cells Transl. Med.* **2013**, *2*, 584–594. [CrossRef] [PubMed]

56. Frith, J.E.; Thomson, B.; Genever, P.G. Dynamic three-dimensional culture methods enhance mesenchymal stem cell properties and increase therapeutic potential. *Tissue Eng. Part C Methods* **2010**, *16*, 735–749. [CrossRef] [PubMed]

57. Cesarz, Z.; Tamama, K. Spheroid Culture of Mesenchymal Stem Cells. *Stem Cells Int.* **2016**, *2016*, 9176357. [CrossRef] [PubMed]

58. Kwon, S.H.; Bhang, S.H.; Jang, H.K.; Rhim, T.; Kim, B.S. Conditioned medium of adipose-derived stromal cell culture in three-dimensional bioreactors for enhanced wound healing. *J. Surg. Res.* **2015**, *194*, 8–17. [CrossRef] [PubMed]

59. Bhang, S.H.; Cho, S.W.; La, W.G.; Lee, T.J.; Yang, H.S.; Sun, A.Y.; Baek, S.H.; Rhie, J.W.; Kim, B.S. Angiogenesis in ischemic tissue produced by spheroid grafting of human adipose-derived stromal cells. *Biomaterials* **2011**, *32*, 2734–2747. [CrossRef] [PubMed]

60. Hsu, S.H.; Hsieh, P.S. Self-assembled adult adipose-derived stem cell spheroids combined with biomaterials promote wound healing in a rat skin repair model. *Wound Repair Regen.* **2015**, *23*, 57–64. [CrossRef] [PubMed]

61. Costa, M.H.G.; McDevitt, T.C.; Cabral, J.M.S.; da Silva, C.L.; Ferreira, F.C. Tridimensional configurations of human mesenchymal stem/stromal cells to enhance cell paracrine potential towards wound healing processes. *J. Biotechnol.* **2017**, *262*, 28–39. [CrossRef] [PubMed]

62. Zimmermann, J.A.; McDevitt, T.C. Pre-conditioning mesenchymal stromal cell spheroids for immunomodulatory paracrine factor secretion. *Cytotherapy* **2014**, *16*, 331–345. [CrossRef] [PubMed]

63. Ylostalo, J.H.; Bartosh, T.J.; Coble, K.; Prockop, D.J. Human Mesenchymal Stem/Stromal Cells Cultured as Spheroids are Self-activated to Produce Prostaglandin E2 that Directs Stimulated Macrophages into an Anti-inflammatory Phenotype. *Stem Cells* **2012**, *30*, 2283–2296. [CrossRef] [PubMed]

64. Park, I.S.; Rhie, J.W.; Kim, S.H. A novel three-dimensional adipose-derived stem cell cluster for vascular regeneration in ischemic tissue. *Cytotherapy* **2014**, *16*, 508–522. [CrossRef] [PubMed]

65. Amos, P.J.; Kapur, S.K.; Stapor, P.C.; Shang, H.L.; Bekiranov, S.; Khurgel, M.; Rodeheaver, G.T.; Peirce, S.M.; Katz, A.J. Human Adipose-Derived Stromal Cells Accelerate Diabetic Wound Healing: Impact of Cell Formulation and Delivery. *Tissue Eng. Part A* **2010**, *16*, 1595–1606. [CrossRef] [PubMed]

66. Xu, Y.; Shi, T.P.; Xu, A.X.; Zhang, L. 3D spheroid culture enhances survival and therapeutic capacities of MSCs injected into ischemic kidney. *J. Cell. Mol. Med.* **2016**, *20*, 1203–1213. [CrossRef] [PubMed]

67. Murphy, K.C.; Whitehead, J.; Falahee, P.C.; Zhou, D.; Simon, S.I.; Leach, J.K. Multifactorial Experimental Design to Optimize the Anti-Inflammatory and Proangiogenic Potential of Mesenchymal Stem Cell Spheroids. *Stem Cells* **2017**, *35*, 1493–1504. [CrossRef] [PubMed]

68. Baer, P.C.; Griesche, N.; Luttmann, W.; Schubert, R.; Luttmann, A.; Geiger, H. Human adipose-derived mesenchymal stem cells in vitro: Evaluation of an optimal expansion medium preserving stemness. *Cytotherapy* **2010**, *12*, 96–106. [CrossRef] [PubMed]

69. Park, E.; Patel, A.N. Changes in the expression pattern of mesenchymal and pluripotent markers in human adipose-derived stem cells. *Cell Biol. Int.* **2010**, *34*, 979–984. [CrossRef] [PubMed]

70. Guo, L.; Zhou, Y.; Wang, S.; Wu, Y.J. Epigenetic changes of mesenchymal stem cells in three-dimensional (3D) spheroids. *J. Cell. Mol. Med.* **2014**, *18*, 2009–2019. [CrossRef] [PubMed]

71. Cheng, N.C.; Wang, S.; Young, T.H. The influence of spheroid formation of human adipose-derived stem cells on chitosan films on stemness and differentiation capabilities. *Biomaterials* **2012**, *33*, 1748–1758. [CrossRef] [PubMed]

72. Bhang, S.H.; Lee, S.; Shin, J.Y.; Lee, T.J.; Kim, B.S. Transplantation of Cord Blood Mesenchymal Stem Cells as Spheroids Enhances Vascularization. *Tissue Eng. Part A* **2012**, *18*, 2138–2147. [CrossRef] [PubMed]

73. Zhang, Q.; Nguyen, A.L.; Shi, S.; Hill, C.; Wilder-Smith, P.; Krasieva, T.B.; Le, A.D. Three-dimensional spheroid culture of human gingiva-derived mesenchymal stem cells enhances mitigation of chemotherapy-induced oral mucositis. *Stem Cells Dev.* **2012**, *21*, 937–947. [CrossRef] [PubMed]

74. Achilli, T.M.; Meyer, J.; Morgan, J.R. Advances in the formation, use and understanding of multi-cellular spheroids. *Expert Opin. Biol. Ther.* **2012**, *12*, 1347–1360. [CrossRef] [PubMed]

75. Hildebrandt, C.; Buth, H.; Thielecke, H. A scaffold-free in vitro model for osteogenesis of human mesenchymal stem cells. *Tissue Cell* **2011**, *43*, 91–100. [CrossRef] [PubMed]

76. Foty, R. A Simple Hanging Drop Cell Culture Protocol for Generation of 3D Spheroids. *JoVE-J. Vis. Exp.* **2011**, *51*. [CrossRef] [PubMed]

77. Neto, A.I.; Correia, C.R.; Oliveira, M.B.; Rial-Hermida, M.I.; Alvarez-Lorenzo, C.; Reis, R.L.; Mano, J.F. A novel hanging spherical drop system for the generation of cellular spheroids and high throughput combinatorial drug screening. *Biomater. Sci.* **2015**, *3*, 581–585. [CrossRef] [PubMed]

78. Tung, Y.C.; Hsiao, A.Y.; Allen, S.G.; Torisawa, Y.S.; Ho, M.; Takayama, S. High-throughput 3D spheroid culture and drug testing using a 384 hanging drop array. *Analyst* **2011**, *136*, 473–478. [CrossRef] [PubMed]

79. Aijian, A.P.; Garrell, R.L. Digital Microfluidics for Automated Hanging Drop Cell Spheroid Culture. *Jala-J. Lab. Autom.* **2015**, *20*, 283–295. [CrossRef] [PubMed]

80. Napolitano, A.P.; Chai, P.; Dean, D.M.; Morgan, J.R. Dynamics of the self-assembly of complex cellular aggregates on micromolded nonadhesive hydrogels. *Tissue Eng.* **2007**, *13*, 2087–2094. [CrossRef] [PubMed]

81. Lopa, S.; Piraino, F.; Kemp, R.J.; Di Caro, C.; Lovati, A.B.; Di Giancamillo, A.; Moroni, L.; Peretti, G.M.; Rasponi, M.; Moretti, M. Fabrication of Multi-well PDMS Chips for Spheroid Cultures and Implantable Fibrin Constructs with High Cell Density Regions through Low-cost Rapid Prototyping Techniques. *Tissue Eng. Part A* **2015**, *21*, 3503.

82. Cha, J.M.; Shin, E.K.; Sung, J.H.; Moon, G.J.; Kim, E.H.; Cho, Y.H.; Park, H.D.; Bae, H.; Kim, J.; Bang, O.Y. Efficient scalable production of therapeutic microvesicles derived from human mesenchymal stem cells. *Sci. Rep.* **2018**, *8*, 1171. [CrossRef] [PubMed]

83. Kim, J.; Ma, T. Endogenous extracellular matrices enhance human mesenchymal stem cell aggregate formation and survival. *Biotechnol. Prog.* **2013**, *29*, 441–451. [CrossRef] [PubMed]

84. Huang, G.S.; Dai, L.G.; Yen, B.L.; Hsu, S.H. Spheroid formation of mesenchymal stem cells on chitosan and chitosan-hyaluronan membranes. *Biomaterials* **2011**, *32*, 6929–6945. [CrossRef] [PubMed]

85. Ota, H.; Kodama, T.; Miki, N. Rapid formation of size-controlled three dimensional hetero-cell aggregates using micro-rotation flow for spheroid study. *Biomicrofluidics* **2011**, *5*. [CrossRef] [PubMed]

86. Chan, H.F.; Zhang, Y.; Ho, Y.P.; Chiu, Y.L.; Jung, Y.; Leong, K.W. Rapid formation of multicellular spheroids in double-emulsion droplets with controllable microenvironment. *Sci. Rep.* **2013**, *3*, 3462. [CrossRef] [PubMed]

87.   Ino, K.; Okochi, M.; Honda, H. Application of Magnetic Force-Based Cell Patterning for Controlling Cell–cell Interactions in Angiogenesis. *Biotechnol. Bioeng.* **2009**, *102*, 882–890. [CrossRef] [PubMed]

88.   Sebastian, A.; Buckle, A.M.; Markx, G.H. Tissue engineering with electric fields: Immobilization of mammalian cells in multilayer aggregates using dielectrophoresis. *Biotechnol. Bioeng.* **2007**, *98*, 694–700. [CrossRef] [PubMed]

89.   Wang, T.; Green, R.; Nair, R.R.; Howell, M.; Mohapatra, S.; Guldiken, R.; Mohapatra, S.S. Surface Acoustic Waves (SAW)-Based Biosensing for Quantification of Cell Growth in 2D and 3D Cultures. *Sensors* **2015**, *15*, 32045–32055. [CrossRef] [PubMed]

90.   Lewis, E.E.L.; Wheadon, H.; Lewis, N.; Yang, J.L.; Mulling, M.; Hursthouse, A.; Stirling, D.; Dalby, M.J.; Berry, C.C. A Quiescent, Regeneration-Responsive Tissue Engineered Mesenchymal Stem Cell Bone Marrow Niche Model via Magnetic Levitation. *ACS Nano* **2016**, *10*, 8346–8354. [CrossRef] [PubMed]

91.   Du, V.; Fayol, D.; Reffay, M.; Luciani, N.; Bacri, J.C.; Gay, C.; Wilhelm, C. Magnetic engineering of stable rod-shaped stem cell aggregates: Circumventing the pitfall of self-bending. *Integr. Biol.* **2015**, *7*, 170–177. [CrossRef] [PubMed]

92.   Baraniak, P.R.; McDevitt, T.C. Scaffold-free culture of mesenchymal stem cell spheroids in suspension preserves multilineage potential. *Cell Tissue Res.* **2012**, *347*, 701–711. [CrossRef] [PubMed]

93.   Alimperti, S.; Lei, P.; Wen, Y.; Tian, J.; Campbell, A.M.; Andreadis, S.T. Serum-free spheroid suspension culture maintains mesenchymal stem cell proliferation and differentiation potential. *Biotechnol. Prog.* **2014**, *30*, 974–983. [CrossRef] [PubMed]

94.   Egger, D.; Schwedhelm, I.; Hansmann, J.; Kasper, C. Hypoxic Three-Dimensional Scaffold-Free Aggregate Cultivation of Mesenchymal Stem Cells in a Stirred Tank Reactor. *Bioengineering* **2017**, *4*, 47. [CrossRef] [PubMed]

95.   Muraglia, A.; Corsi, A.; Riminucci, M.; Mastrogiacomo, M.; Cancedda, R.; Bianco, P.; Quarto, R. Formation of a chondro-osseous rudiment in micromass cultures of human bone-marrow stromal cells. *J. Cell Sci.* **2003**, *116*, 2949–2955. [CrossRef] [PubMed]

96.   Martin, I.; Wendt, D.; Heberer, M. The role of bioreactors in tissue engineering. *Trends Biotechnol.* **2004**, *22*, 80–86. [CrossRef] [PubMed]

97.   Zhao, J.J.; Griffin, M.; Cai, J.; Li, S.X.; Bulter, P.E.M.; Kalaskar, D.M. Bioreactors for tissue engineering: An update. *Biochem. Eng. J.* **2016**, *109*, 268–281. [CrossRef]

98.   King, J.A.; Miller, W.M. Bioreactor development for stem cell expansion and controlled differentiation. *Curr. Opin. Chem. Biol.* **2007**, *11*, 394–398. [CrossRef] [PubMed]

99.   Groebe, K.; MuellerKlieser, W. On the relation between size of necrosis and diameter of tumor spheroids. *Int. J. Radiat. Oncol.* **1996**, *34*, 395–401. [CrossRef]

100. Alvarez-Perez, J.; Ballesteros, P.; Cerdan, S. Microscopic images of intraspheroidal pH by 1H magnetic resonance chemical shift imaging of pH sensitive indicators. *Magma* **2005**, *18*, 293–301. [CrossRef] [PubMed]

101. Chen, A.K.L.; Reuveny, S.; Oh, S.K.W. Application of human mesenchymal and pluripotent stem cell microcarrier cultures in cellular therapy: Achievements and future direction. *Biotechnol. Adv.* **2013**, *31*, 1032–1046. [CrossRef] [PubMed]

102. Timmins, N.E.; Kiel, M.; Gunther, M.; Heazlewood, C.; Doran, M.R.; Brooke, G.; Atkinson, K. Closed system isolation and scalable expansion of human placental mesenchymal stem cells. *Biotechnol. Bioeng.* **2012**, *109*, 1817–1826. [CrossRef] [PubMed]

103. Yu, B.; Zhang, X.M.; Li, X.R. Exosomes Derived from Mesenchymal Stem Cells. *Int. J. Mol. Sci.* **2014**, *15*, 4142–4157. [CrossRef] [PubMed]

104. Rani, S.; Ryan, A.E.; Griffin, M.D.; Ritter, T. Mesenchymal Stem Cell-derived Extracellular Vesicles: Toward Cell-free Therapeutic Applications. *Mol. Ther.* **2015**, *23*, 812–823. [CrossRef] [PubMed]

105. Yeo, R.W.Y.; Lai, R.C.; Zhang, B.; Tan, S.S.; Yin, Y.J.; Teh, B.J.; Lim, S.K. Mesenchymal stem cell: An efficient mass producer of exosomes for drug delivery. *Adv. Drug Deliver. Rev.* **2013**, *65*, 336–341. [CrossRef] [PubMed]

10

Evaluation of Peripheral Blood and Cord Blood Platelet Lysates in Isolation and Expansion of Multipotent Mesenchymal Stromal Cells

Ioanna Christou [1,†], Panagiotis Mallis [1,†] ⓘ, Efstathios Michalopoulos [1,*],
Theofanis Chatzistamatiou [1] ⓘ, George Mermelekas [2], Jerome Zoidakis [2] ⓘ,
Antonia Vlahou [2] and Catherine Stavropoulos-Giokas [1]

[1] Hellenic Cord Blood Bank, Biomedical Research Foundation Academy of Athens,
   4 Soranou Ephessiou Street, 115 27 Athens, Greece; x.janna@hotmail.com (I.C.);
   pmallis@bioacademy.gr (P.M.); tchatzistamatiou@dha.gov.ae (T.C.); cstavrop@bioacademy.gr (C.S.-G.)
[2] Biotechnology division, Biomedical Research Foundation Academy of Athens, 4 Soranou Ephessiou Street,
   115 27 Athens, Greece; gmermelekas@yahoo.com (G.M.); izoidakis@bioacademy.gr (J.Z.);
   vlahoua@bioacademy.gr (A.V.)
*   Correspondence: smichal@bioacademy.gr
†   These authors contributed equally to this work as first authors.

Abstract: Background: Multipotent Mesenchymal Stromal Cells (MSCs) are used in tissue engineering and regenerative medicine. The in vitro isolation and expansion of MSCs involve the use of foetal bovine serum (FBS). However, many concerns have been raised regarding the safety of this product. In this study, alternative additives derived either from peripheral or cord blood were tested as an FBS replacement. **Methods:** Platelet lysates (PL) from peripheral and cord blood were used for the expansion of MSCs. The levels of growth factors in peripheral blood (PB) and cord blood (CB) PLs were determined using the Multiple Reaction Monitoring (MRM). Finally, the cell doubling time (CDT), tri-lineage differentiation and phenotypic characterization of the MSCs expanded with FBS and PLs were determined. **Results:** MSCs treated with culture media containing FBS and PB-PL, were successfully isolated and expanded, whereas MSCs treated with CB-PL could not be maintained in culture. Furthermore, the MRM analysis yielded differences in growth factor levels between PB-PL and CB-PL. In addition, the MSCs were successfully expanded with FBS and PB-PL and exhibited tri-lineage differentiation and stable phenotypic characteristics. **Conclusion:** PB-PL could be used as an alternative additive for the production of MSCs culture medium applied to xenogeneic-free expansion and maintenance of MSCs in large scale clinical studies.

Keywords: Cord blood; Multiple Reaction Monitoring; multipotent Mesenchymal Stem Cells; peripheral blood; platelet lysate

1. Introduction

The field of tissue engineering and regenerative medicine is rapidly evolving and involves the use of specified and unspecified cellular populations in combination with various types of scaffolds [1,2]. Currently, multipotent Mesenchymal Stromal Cells (MSCs) are clinically used in approaches of regenerative medicine and cellular therapies [3–5]. However, these applications demand a significant number of in vitro expanded MSCs [6,7]. Common culture methods for the expansion of MSCs involve the use of foetal bovine serum (FBS) as a supplement, in combination with basal culture medium [8,9]. A satisfactory number of these cells can be obtained easily from different sources including bone marrow, umbilical cords, Wharton jelly, umbilical cord blood, amniotic fluid and lipoaspirates from adipose tissue [10–13]. According to the International Society for Cellular Therapy (ISCT), MSCs are defined as plastic adherent cells, positive for specific surface antigens including CD105, CD73, CD90 and negative for hematopoietic markers such as CD45, CD34, CD14, CD19 and HLA class II and trilineage mesodermal differentiation to adipocytes, chondroblasts and osteoblasts [14–17]. Recently, it was determined that the source of MSCs affects significantly their characteristics such as their heterogeneity in morphology, proliferative activity, differentiation and therapeutic potentials [17,18].

Furthermore, MSCs have the ability to secrete a variety of trophic factors that contribute to tissue remodelling and immunomodulation and can be applied as first or second line treatment for various diseases [19–21].Moreover, combining them with chitosan scaffolds, could be a useful tool for osteochondral tissue regeneration [22,23].Additionally, due to their immunomodulatory properties, MSCs represent an attractive cell source for treatment of autoimmune disorders such as multiple sclerosis (MS), amyotrophic lateral sclerosis (ALS), type I diabetes mellitus, Crohn's disease and systemic lupus erythematosus [24–27]. It has been proven that MSCs can adjust the immunoreaction directly or indirectly [28]. In a direct manner, MSCs can induce apoptosis of T cells through Fas/Fas ligand and TNF receptor signalling pathways. Alternatively, apoptosis in T cells can be induced via the secretion of IL-6, IL-10, nitric oxide (NO), idoleamine 2,3 dioxygenase (IDO) and prostaglandin E_2 (PGE_2). In this way, the autoreactive T cell population can be adjusted, thus providing enough time to the damaged tissues to remodel and regenerate [28]. Currently, clinical trials using MSCs for ischemic stroke, myocardial infraction and graft versus host disease have also been performed worldwide [29,30].

In order to support, the wide use of MSCs in regenerative medicine and tissue engineering approaches, in vitro culturing and expansion conditions must be developed. In addition, large scale clinical translation trials in accordance with good manufacturing practices (GMP) requires the use of a well-defined culture medium in order to maintain the cellular quality, while avoiding adverse patient reactions [30]. Nowadays, FBS, derived from the whole blood of bovine foetuses, is the most widely used supplement for cell culture medium preparation. FBS is a rich source of growth factors like transforming growth factor- beta 1 (TGF-β1), fibroblast growth factor (FGF), epidermal growth factor (EGF), vascular endothelial growth factor (VEGF), platelet derived growth factor (PDGF), insulin-like growth factor (IGF), growth hormones and albumin. Thus, it is the optimum additive in culture for the expansion of various types of cells [31,32]. However, many concerns are arising regarding the safety of this product are arising. FBS could contain prions (causing mad cow disease), xenogeneic antigens, bovine proteins or transfer zoonotic infections to the cultured cells. These cause significant complications to patients receiving cultured MSC therapies [33,34]. Another disadvantage is the different concentration in the amount of growth factors between different lots of FBS. Annually, it is estimated that 600,000 L of FBS are demanded for cell culturing but only 1/3 is suitable for GMP use and clinical grade cell expansion. However, more than 200 phase I/II clinical trials, report the use of FBS as the primary supplement for the in vitro expansion of MSCs (according to www.clinicaltrials.gov as of 25 March 2013).

To address these issues, alternative strategies for the culture and expansion of cells are currently being developed and focused in the production of culture medium free of any animal derivatives. The substitution of FBS with human serum has provided contradictory results in the expansion, proliferation and differentiation capacities of MSCs [35–37]. Human platelet lysate (hPL) from pooled expired plasma apheresis showed promising results, when used in MSC culture. Given that, hPL contains significant amounts of TGF-β1, FGF, VEGF, PDGF and IGF it can be used efficiently for MSC applications [38,39]. One serious drawback regarding the use of hPL is the availability of expired pooled plasma apheresis. Umbilical cord blood, could possibly address this problem, thus providing an alternative source for the production of platelet lysate, since it contains a similar number of platelets with the peripheral blood. In addition, a growing number of publications reported the use of umbilical cord blood as the primary source for the production of platelet rich plasma (PRP) and fibrin, which are applied in clinical practices [40–42].

In this study, human peripheral blood platelet lysate (PB-PL) and human umbilical cord blood platelet lysate (CB-PL) were evaluated as possible substitutions to FBS in culture medium for human MSCs (hMSCs) culture. In order to limit the biological variability of platelet concentrations between the human donors that could result in batch to batch variation of platelet lysate pooling of the peripheral blood and cord blood units was applied for the production of PB-PL and CB-PL respectively. Thus, the probability for variations on MSCs isolation, expansion and trilineage differentiation was minimized.

In addition, an evaluation of the growth factor levels with targeted proteomic methods including MRM was performed. Finally, we tested three different additives in culture media including PB-PL, CB-PL and FBS in order to compare the isolation, expansion and tri-lineage differentiation capacities of hMSCs.

2. Materials and Methods

2.1. Preparation of Human Platelet Lysate

2.1.1. Peripheral Blood Platelet Lysate

Peripheral blood units ($n = 50$) with an average volume of 450 ± 45 mL were collected from healthy donors at the Evagelismos Hospital, following the Greek regulatory procedures for blood donation. The blood units that were used for the production of hPB-PL were expired and considered as not valid for transfusion. Total platelets (PLTs) in each blood unit were determined by a haematological analyser (Nihon Khoden, MEK-6400C, Tokyo, Japan). The blood units were centrifuged at $1050 \times g$ for 15 min. Then, the supernatant, containing plasma and platelets, was isolated and centrifuged again at $3972 \, g \times$ for 15 min to comprise 1 unit of PRP and finally stored overnight at -80 °C. After at least 12 h of storage at -80 °C, the PRP units were thawed at 4 °C for 12 h and centrifuged at $3972 \times g$ for 30 min. Five PRP units were used for the production of 1 PRP pool. Each PRP pool contained about 655×10^6 PLTs/mL (Table 1). The supernatants were passed through 0.65 μm filter, reducing in this way the membrane fragments, resulting in the production of platelet lysate. Finally, the PB-PLs were stored in 20-mL PL bags (Macopharma SA, Mouvaux, France) at -80 °C until further processing.

Table 1. Cord and peripheral PRP pools features.

Total Platelet Concentration ($\times 10^6$/mL)		*p*-Value
CB PRP Pools	PB PRP Pools	
698 ± 23	655 ± 21	0.17

Overview of the average PLT concentration in the cord and peripheral PRP pools. No statistically significant differences in Total Platelet concentration. Statistical significant is considered when $p < 0.05$.

2.1.2. Cord Blood Platelet Lysate

Umbilical cord blood units ($n = 100$) with an average volume of 90 ± 7 mL were collected after informed consent from the mothers by experienced midwives and immediately distributed to HCBB. The collections were performed in accordance with the ethical standards of the Greek National Ethical Committee and were approved by our Institution's ethical board. The cord blood units were collected from end term normal and caesarean deliveries (gestational ages 36–40 weeks)—which had been processed within 24 h after collection and which did not fulfil the criteria outlined by the Hellenic Cord Blood Bank (HCBB)—for processing and storage in liquid nitrogen. A detailed description of HCBB criteria is available in supplementary Table S1. According to their blood type, cord blood units were pooled in a final volume of 400 ± 50 mL. Automated cell counting in each cord blood unit with haematological analyser was performed for determination of total cell concentration. Then, centrifugation was performed at $324 \times g$ for 9 min (at 22 °C optional). The supernatant was isolated and centrifuged again at $3972 \times g$ for 15 min to comprise 1 unit of PRP. PRP units were pooled (270 ± 30 mL) and frozen at -80 °C. Each PRP pool contained about 698×10^6 PLTs/mL. Cord blood platelet lysate was prepared in accordance with the protocol described in peripheral blood platelet lysate preparation and stored at $-80\,°C$ until further processing.

2.2. Protein Determination and Quantification Using Multiple Reaction Monitoring

All platelet lysate samples either derived from peripheral ($n = 4$) or cord blood ($n = 4$) were centrifuged in order to remove insoluble material, prior to processing. The total protein content for each sample was determined by the Bradford assay. An appropriate volume (~2 μL) of each mixture, corresponding to 10 μg of total protein, was diluted to a volume of 20 μL with urea buffer (8 M urea, 50 mM NH_4HCO_3) followed by reduction (10 mM DTE) and alkylation (40 mM Iodoacetamide). The samples were then diluted to a final volume of 90 μL with 50 mM NH_4HCO_3 in order to obtain a final concentration of 1.5 M for urea. Trypsin was added at an enzyme protein ratio of 1:100 and the solution was incubated overnight. After trypsinization, samples were acidified with 0.1% formic acid, desalted by zip-tip and dried (speedVac). Subsequently, the samples were reconstituted in appropriate volume of mobile phase A (water, 0.1% formic acid) to a final protein concentration of 0.5 μg/μL. These samples were analysed by LC/MRM. In total, 16 growth factors were quantified with the above method. The human spectral library was searched using the Skyline software and Peptide Atlas repository to identify proteotypic peptides for the growth factors of peripheral and cord blood platelet lysate. Data analysis was performed using Skyline software and all chromatograms were manually inspected to ensure the quality and accuracy of peak picking. The sum of peak areas of two to four transitions per peptide was used to calculate the signal intensity for the selected growth factors. A detailed list of MRM transitions is available in supplementary Table S2.

2.3. Liquid Chromatography-Mass Spectrometry Setup

Liquid chromatography was performed using an Agilent 1200 series nano-pump system (Agilent Technologies Inc., Wilmington, DE, USA), coupled with a C18 nano-column (150 mm × 75 μm, particle size 5 μm) from Agilent. Peptide separation and elution was achieved with a 40 min 5–45% ACN/water 0.1% FA gradient at a flow rate of 300 nL/min. Six microliters of each sample (corresponding to 3 μg of total protein content) were injected.

Tryptic peptides were analysed on an AB/MDS Sciex 4000 QTRAP (AB SCIEX Pte Ltd., Orlando, FL, USA), with a nanoelectrospray ionization source controlled by Analyst 1.5 software (Sciex). The mass spectrometer was operated in MRM mode, with the first (Q1) and third quadrupole (Q3) at 0.7 unit mass resolution. Two to four transitions were recorded for each peptide. Optimum collision energies for each transition were automatically calculated by the Skyline software (v4.1, ProteoWizard, Washington, DC, USA).

2.4. Collection of Human Umbilical Cords

Fresh human umbilical cords (5 to 10 cm) were collected from normal deliveries (gestational ages 36–40 weeks) after informed consent form the mothers by experienced midwives trained in cord blood collection. The umbilical cords ($n = 10$) were stored into Phosphate Buffer Saline 1× (PBS 1×, Gibco, Life Technologies, Grand Island, NY, USA) supplemented with 10 U/mL Penicillin & 10 μg/mL Streptomycin (Gibco, Life Technologies, Grand Island, NY, USA) at 4 °C and processed within 24 h from reception at the HCBB. The collections were performed in accordance with the ethical standards of the Greek National Ethical Committee and were approved by our Institution's ethical board.

2.5. Isolation and Culture of Wharton's Jelly MSCs

After removing the umbilical arteries and vein, the Wharton Jelly tissue was cut with scissors into small pieces (1–3 mm^3), placed into 6-well plates (Costar, Corning Life Sciences, Canton, MA, USA) and cultured with growth media in a humidified atmosphere with 5% CO_2 at 37 °C. Upon reaching the sufficient number of adherent cells in the 6-well plates, cells were detached using 0.25% trypsin EDTA solution (Gibco), washed with PBS 1× and re-plated into the 75 cm^2 flasks (Costar) with the appropriate culture medium. On reaching 80% of confluency, the cells were trypsinized, washed and resuspended into the 175 cm^2 flasks. The same procedure was repeated until the cells reached passage 5 (P5) of culture. The growth media that were used for the expansion of Wharton's Jelly-Mesenchymal Stromal Cells (WJ-MSCs) was α-Minimum Essentials Medium (α-MEM, Gibco) supplemented either with 15% Foetal Bovine Serum (FBS, Gibco) or 10% PB-PL or 10% CB-PL. Each growth medium was supplemented with 10 U/mL penicillin (Gibco) and 10 μg/mL streptomycin (Gibco) and 2 mM L-glutamine (Gibco). The growth medium was changed twice every week and the cultures were maintained in a humidified atmosphere with 5% CO_2 at 37 °C.

2.6. Cell Viability and Growth Rate

Cell viability of the FBS and PB-PL expanded WJ-MSCs was determined using Trypan blue. The comparison of cell doubling time (CDT) between the three different culturing conditions until reaching P5 was estimated, by plating at P1 1000 cells/cm^2 in flasks. The number of population doubling was calculated by the classical formula:

$$CDT = \frac{\log_{10}(N/N_0)}{\log_{10}(2)} x\ (T) \tag{1}$$

where N is the number of cells at the end of the culture, N_0 is the number of cells seeded and T is the culture duration in hours.

2.7. Differentiation Capacity of MSCs

The capacity of the FBS and PB-PL expanded WJ-MSCs to differentiate into the osteogenic, chondrogenic and adipogenic lineages was determined. For this purpose, the cells were seeded in 6-well plates. WJ-MSCs were differentiated into osteogenic lineage using basal medium (Mesencult, StemCell Technologies, Vancouver, BC, Canada) supplemented with 15% osteogenic stimulatory supplements (StemCell Technologies), 0.01 mM dexamethasone (StemCell Technologies) and 50 μg/mL ascorbic acid (StemCell Technologies). Osteogenic differentiation was assessed after 25 days with Alizarin Red S (Sigma-Aldrich, Darmstadt, Germany) staining. Chondrogenic differentiation was induced in a spheroid culture using high glucose D-MEM supplemented with 0.01 mM dexamethasone (StemCell Technologies), 35 μg/mL ascorbic acid-2-phospate (StemCell Technologies), 10 ng/mL transforming growth factor- β1 (Sigma-Aldrich), liquid medium supplement (ITS+ premix, Sigma-Aldrich) for 30 days. The pellets were fixed with 10% formalin (Sigma-Aldrich), paraffin embedded and cut into 5 μm sections. Chondrogenic differentiation was assessed with Alcian blue (Fluka, Sigma-Aldrich) staining. Finally, the adipogenic differentiation of WJ-MSCs was committed with the use of basal medium (Mesencult, StemCell Technologies) supplemented with 10% of adipogenic stimulatory supplements (StemCell Technologies) for 25 days and assessed by staining of lipid vacuoles with Oil Red-O (Sigma-Aldrich) staining.

2.8. Colony-Forming Unit-Fibroblast (CFU-F) Assay

The CFU-F assay performed in MSCs expanded either with FBS ($n = 3$) or PB-PL ($n = 3$) at passage 2, 3, 4 and 5. The MSCs were trypsinzed, counted and seeded at a density of 500 cells/well on 6-well tissue culture plates with MSC growth medium without the addition of FBS or PB-PL and cultured for a time period of 15 days in humidified atmosphere at 37 °C. The medium was changed biweekly. After 15 days of cultivation, cells were fixed with formalin (Sigma-Aldrich) 10% for 5 min and stained with Giemsa (Sigma-Aldrich). The stained colonies were counted manually by two independent observers.

2.9. Phenotypic Characterization of WJ-MSCs

Expanded WJ-MSCs with FBS ($n = 3$) and PB-PL ($n = 3$) were analysed for cell surface antigen phenotyping using flow cytometry. Each sample was measured in triplicate. Cells were labelled with fluorescein isothiocyanate-conjugated anti-CD90 (Immunotech, Beckman Coulter, Marseille, France), HLA-ABC (Immunotech), CD29 (Immunotech), CD19 (Immunotech), CD31 (Immunotech), CD45 (Immunotech). Epitopes CD105 (Immunotech), CD73 (Immunotech), CD44 (Immunotech), CD34 (Immunotech) CD3 (Immunotech) and CD14 (Immunotech), HLA-DR (Immunotech) were assessed with phycoerythrin-conjugated and PC5-conjugated mouse anti-human monoclonal antibodies respectively. The WJ-MSCs phenotypes were analysed in Cytomics FC 500 (Beckman Coulter, Marseille, France) flow cytometer with the CXP Analysis software (Beckman Coulter).

2.10. Growth Promotion Study and Media Validation

All peripheral blood and cord blood units used for the production of platelet lysate were tested for bacterial and viral contamination. Specifically, all blood and cord blood units were tested for aerobic and anaerobic bacteria with the BacT/Alert system for a time period of 14 days (BACTEC 9240, Becton Dickinson, Franklin Lakes, NJ, USA) by direct inoculation of at least 1% of the unit or 16 mL of the pooled platelet lysate. Further confirmation of the BacT/Alert system was performed by the use of blood and Sabouraud agar. For viral contamination, all blood and cord blood units were evaluated for HIV I/II, HBV, HGV, HTLV-I/II, CMV, HCV, HAV, WNV and for T. Pallidum and T. Cruzi, with serologic testing. In addition, final MSC culture expanded either with FBS or platelet lysates were tested for bacterial contamination, endotoxin content and mycoplasma contamination. Briefly, for sterility evaluation, 16 mL of the final cell product (MSCs cultured with FBS or platelet lysate) were tested for aerobic and anaerobic bacteria using the previously described BacT/Alert system for a time period of 14 days. The endotoxin content evaluation was performed by the Limulus amebocyte lysate (LAL) test according to European Pharmacopeia (PBI S.p.A., Milano, Italy). Finally, the MycoA-lert test (Cambrex Corporation, Verviers, Belgium) was used for the Mycoplasma contamination.

2.11. Statistical Analysis

Statistical analysis was performed by using Graph Pad Prism v 6.01 (GraphPad Software, San Diego, CA, USA). Comparisons in CDT and between the two experimental conditions (FBS and PB-PL) were performed with the unpaired nonparametric Mann–Whitney U-Test. Statistical significant difference between group values was considered when p value was less than 0.05. Indicated values are mean ± standard deviation.

3. Results

3.1. Preparation of Human Platelet Lysate

3.1.1. Peripheral Blood

The PRP pools (n = 10) with an average volume of 400 ± 45 mL, were processed in order to isolate the platelet lysate. Cell counting with haematological analyser was performed prior to the freezing process. The average platelet concentration was 655 ± 21 × 10^6/mL (Table 1). The mean platelet concentration of each peripheral blood unit is presented in supplementary Table S3.

3.1.2. Cord Blood

The cord blood-derived PRP pools (n = 10) with an average volume of 268 ± 31 mL, were processed in order to isolate the platelet lysate. Automated cell counting in each pool was performed. The average platelet concentration was 698 ± 23 × 10^6/mL (Table 1). The mean platelet concentration of each cord blood unit is presented in supplementary Table S3.

3.2. Protein Determination and Quantification in PB-PL and CB-PL

The total protein content of PB-PL and CB-PL was 88.2 ± 2.7 µg and 31.1 ± 4.3 µg respectively. The identification and quantification of growth factors in PB-PL and CB-PL was accomplished with the MRM technology. In order to obtain reliable results of the growth factor content in PB-PL and CB-PL, normalization based on the initial total protein amount of each sample was performed (PB-PL, CB-PL). The relative signal intensity of each growth factor in PB-PL and CB-PL samples is presented in supplementary figure (Figure S4). The ratio of the amount for growth factors in CB-PL in comparison to PB-PL is presented in Table 2. These results indicated that the PB-PL contained significantly elevated levels of each growth factor when compared to CB-PL.

Table 2. Ratio of growth factors in CB-PL and PB-PL.

Protein Identification	Accession Number	Ratio PB-PL/CB-PL
Interferon gamma receptor 1 precursor	INGR1_HUMAN	6.8 ± 1.2
Interleukin 1A	IL1A_HUMAN	7.0 ± 2.1
Interferon gamma precursor	IFNG_HUMAN	5.6 ± 1.0
Interleukin 1B	IL1B_HUMAN	5.2 ± 0.8
Tumour necrosis factor receptor type 1-associated DEATH domain protein	TRADD_HUMAN	5.4 ± 1.9
Intercellular adhesion molecule 1 precursor	ICAM1_HUMAN	4.4 ± 1.5
Tumour Necrosis Factor A	TNFA_HUMAN	4.3 ± 1.7
Interleukin 6	IL6_HUMAN	3.6 ± 0.6
Vascular Endothelial Growth Factor A	VEGFA_HUMAN	6.2 ± 4.1
Fibroblast Growth Factor 2	FGF2_HUMAN	3.8 ± 0.8
Platelet Derived Growth Factor A	PDGFA_HUMAN	3.9 ± 1.6
Interleukin 8	IL8_HUMAN	3.2 ± 0.6
C-C motif chemokine 3 precursor	CCL3_HUMAN	3.2 ± 0.7
Transforming Growth Factor B1 precursor	TGFB1_HUMAN	2.9 ± 0.3
C-C motif chemokine 5 precursor	CCL5_HUMAN	2.7 ± 0.3
Vascular Cell Adhesion protein 1 precursor	VCAM1_HUMAN	2.4 ± 0.3

3.3. Isolation and Culture Characteristics of WJ-MSCs

WJ-MSCs were successfully isolated from 10 human umbilical cords using the growth media supplemented either with 15% FBS or 10% PB-PL. The first adherent cells appeared after 6 days of culturing using 15% FBS and after 7 days using 10% PB-PL (Figure 1) under standard conditions. The cells were passaged in 75 cm^2 flasks after 18 days. Furthermore, there were no significant morphological differences and exhibited spindle shape morphology. However, PB-PL cells had smaller size as defined by the optical examination with the light microscope. The cells isolated from Wharton-Jelly tissue using both growth media retained their morphology until reaching P10 as shown in Figure 2. In contrast, all attempts to isolate adherent cells from human umbilical cord tissue using the CB-PL growth medium failed, even after 20 days of culture (Figure 1). Thus, only the FBS and PB-PL expanded WJ-MSCs were used for the next set of experiments for this study. The PB-PL expanded WJ-MSCs grew significantly ($p < 0.01$) faster than FBS- expanded WJ-MSCs. To calculate the doubling time of the WJ-MSCs from 10 different human umbilical cords, the cells were grown at maximum of 80% confluence until reaching P5. Our results showed longer doubling time in MSCs (313 ± 49 h) with the FBS containing medium (Table 3). The PB-PL expanded WJ-MSCs showed a higher proliferation rate as the doubling time is significantly lower at 137 ± 21 h (Figure 3).

FBS WJ-MSCs   PB-PL WJ-MSCs   CB-PL WJ-MSCs

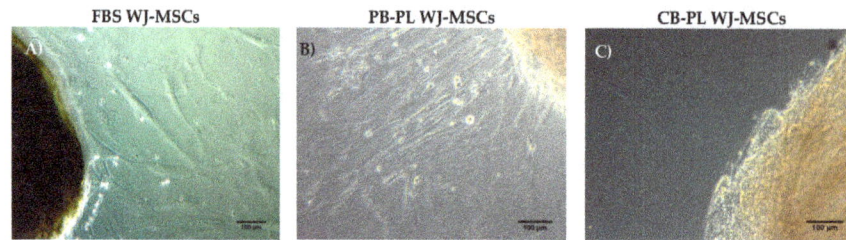

Figure 1. Isolation of MSCs from Wharton Jelly tissue using growth media supplemented with 15% FBS, 10% PB-PL and 10% CB-PL. (**A**) FBS expanded WJ-MSCs after 6 days of culture under standard conditions; (**B**) PB-PL expanded WJ-MSCs after 7 days of culture; (**C**) CB-PL growth medium failed to expand the WJ-MSCs even after 20 days of culture under standard conditions. Original magnification 10×, scale bars 100 µm.

Passage 1   Passage 5   Passage 10

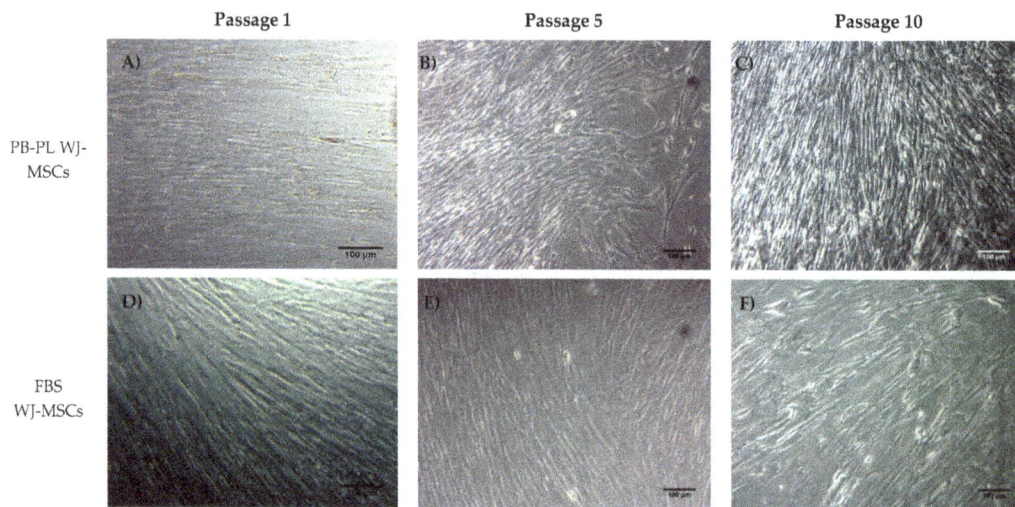

Figure 2. WJ-MSCs in culture at passage 1, 5 and 10. MSCs either with FBS or PB-PL growth medium achieved to retain their morphology until reaching passage 10. PB-PL expanded WJ-MSCs at passage 1 (**A**), 5 (**B**) and 10 (**C**). FBS expanded WJ-MSCs at passage 1 (**D**), 5 (**E**) and 10 (**F**). Original magnification 10×, scale bars 100 µm.

Table 3. Cell culture kinetics of FBS and PB-PL expanded WJ-MSCs.

	FBS Expanded WJ-MSCs			
	$n = 10$	$n = 10$	$n = 10$	$n = 10$
Passage	2	3	4	5
Mean cell Viability (%)	83 ± 1 [†]	88 ± 2	87 ± 3	85 ± 4
Cell Doubling Time (hours)	55 ± 11	60 ± 20	97 ± 30	313 ± 49 [‡]
	PB-PL Expanded WJ-MSCs			
	$n = 10$	$n = 10$	$n = 10$	$n = 10$
Passage	2	3	4	5
Mean cell Viability (%)	88 ± 2 [†]	86 ± 1	88 ± 2	86 ± 3
Cell Doubling Time (hours)	50 ± 8	53 ± 12	77 ± 29	167 ± 33 [‡]

Overview of the culture kinetics of FBS and PB-PL expanded WJ-MSCs. There was no statistically significant difference between two groups, with the exception of FBS expanded WJ-MSCs CDT at P5 that was calculated higher in respect to PB-PL expanded WJ-MSCs. Additionally, the Mean Cell Viability (%) FBS expanded WJ-MSCs at P2 was lower than PB PL expanded WJ-MSCs. [†] $p = 0.0008$; [‡] $p = 0.0001$, $p < 0.05$ indicates statistical significance.

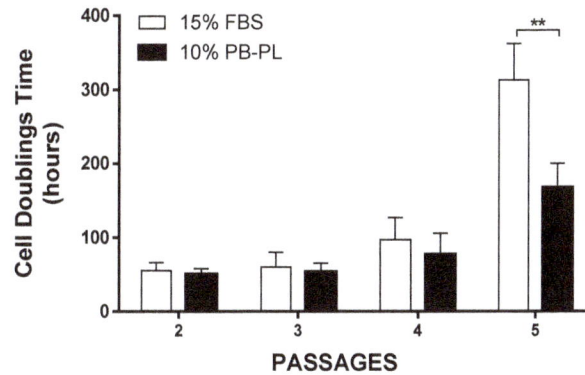

Figure 3. Cell doubling time of WJ-MSCs treated with FBS and PB-PL growth medium. At passage 5, the WJ-MSCs treated with PB-PL growth medium showed statistical significant reduction in CDT, compared to FBS expanded WJ-MSCs. The significance of the difference between the FBS and PB-PL treated WJ-MSCs at passage 5 is represented: ** $p < 0.01$.

3.4. Differentiation of WJ-MSCs

The ability of WJ-MSCs to differentiate to osteogenic, chondrogenic and adipogenic lineages was analysed under particular culture conditions that favour each specific differentiation pattern. MSCs obtained at P2 were exposed to osteogenic, adipogenic and chondrogenic medium for up to 3 weeks. Both WJ-MSCs successfully exhibited calcium deposition and stained positively with Alizarin Red S stain, which is specific for calcium mineralization (Figure 4). However, the stain was more intense in PB-PL expanded cells comparing to FBS expanded cells, indicating a more robust deposition of calcium at the same time point. On the other hand, after 21 days of culturing in adipogenic conditions, both types of WJ-MSCs exhibited limited number of lipidic inclusions visualized with Oil Red-O stain, indicating an immature adipocyte phenotype. Finally, when the WJ-MSCs were induced to differentiate into chondrogenic lineage, there was a visible difference in glycosaminoglycans production between the FBS and PB-PL cultured MSCs. The glycosaminoglycan content of the PB-PL MSCs assessed was higher with Alcian blue stain comparing to the FBS expanded MSCs (Figure 4).

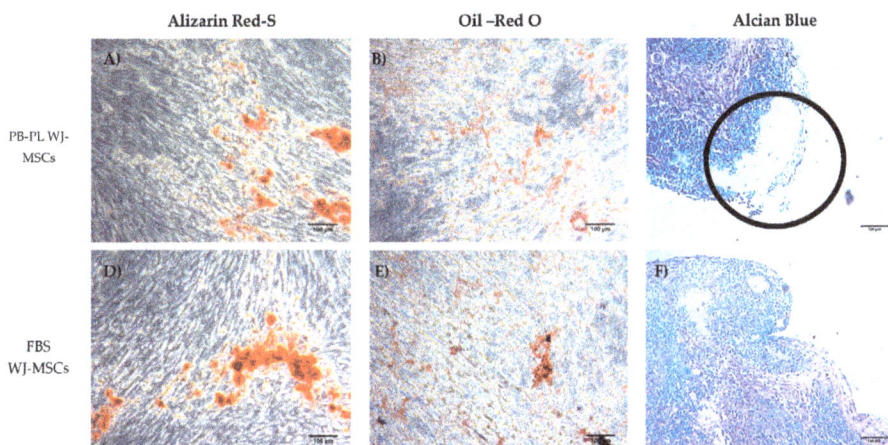

Figure 4. Histological analysis of the induced PB-PL and FBS treated WJ-MSCs into osteogenic, adipogenic and chondrogenic lineages. (**A,D**) Staining of the PB-PL and FBS expanded WJ-MSCs after induction into osteogenic lineage with Alizarin Red S stain. (**B,E**) Staining of the PB-PL and FBS expanded WJ-MSCs after induction into adipogenic lineage with Oil Red O stain. (**C,F**) Staining of the PB-PL and FBS expanded WJ-MSCs after induction into chondrogenic lineage with Alcian Blue stain. (**C**) The high amount of glycosaminoglycan content that was produced in PB-PL expanded WJ-MSCs is indicated by the black circle. Original magnification 10x, scale bares 100 μm.

3.5. CFU–F of WJ-MSCs Cultured with FBS or PB-PL

The clonogenic potential of WJ-MSCs were evaluated with the CFU-F assay. More specifically, after 15 days of cultivation at 37 °C the WJ-MSCs were fixed and stained with Giemsa. The WJ-MSCs that initially isolated and expanded with culture medium containing PB-PL characterized by higher CFU-F number than the WJ-MSCs that isolated and cultured with medium containing FBS at passage 2,3,4 and 5 but this increase was not statistical significant (Figure 5). The highest CFU-F number of FBS expanded WJ-MSCs was at passage 4 (22 ± 2 CFUs) and for PB-PL was at passage 5 (24 ± 2 CFUs).

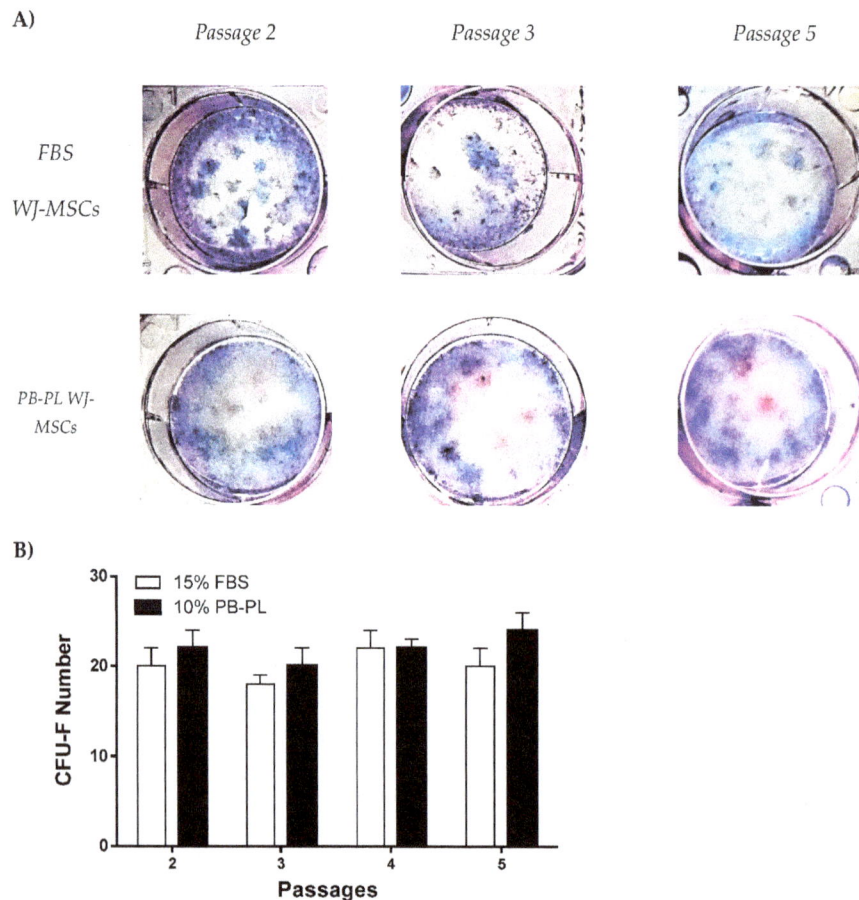

Figure 5. Colony Forming Unit-Fibroblast assay of FBS and PB-PL expanded WJ-MSCs. (**A**) Representable images of CFU-F at passages 2,3 and 5stained with Giemsa (**B**) The CFU-F assay was performed at passages 2,3,4 and 5 where no statistical significant difference was observed between FBS and PB-PL expanded WJ-MSCs.

3.6. Phenotypic Characterization

The phenotypic characterization of WJ-MSCs was carried out with a routinely-used panel for cell surface markers as indicated by the International Society for Cellular Therapy. Three FBS expanded MSCs and three PB-PL expanded MSCs samples were analysed at P3. Both WJ-MSCs were negative to hematopoietic markers CD3, CD19, CD34 and CD45 and CD31, HLA DR. In addition, the MSCs samples were positive for β1 integrin subunit CD29 and matrix receptors CD90, CD105, CD73 and HLA-ABC. Specifically, the comparison in cell surface markers showed some minor variability in their expression between FBS and PB-PL expanded MSCs. Finally, statistical significant difference ($p < 0.05$) was noticed in positive markers CD105, CD73, CD44 and in the hematopoietic negative market CD3 between the two groups. Detailed information on the expression of cell surface markers is described in Table 4.

Table 4. Cell surface markers expression (%) at FBS and PB-PL expanded WJ-MSCs.

Cell Surface Markers	FBS Expanded WJ-MSCs	PB-Expanded WJ-MSCs	p Value
CD90	96.2 ± 0.6	96.7 ± 0.4	0.3707
CD105	96.8 ± 0.1	98.6 ± 0.5	0.0053
HLA-ABC	94.0 ± 0.1	94.7 ± 0.5	0.1927
CD73	96.4 ± 0.7	98.7 ± 0.6	0.0300
CD29	95.6 ± 0.7	94.6 ± 0.5	0.1784
CD44	96.0 ± 0.6	93.7 ± 0.3	0.0176
CD19	1.2 ± 0.1	1.4 ± 0.1	0.1295
CD3	1.8 ± 0.1	1.6 ± 0.1	0.0066
CD31	1.6 ± 0.2	1.6 ± 0.1	0.8620
CD14	1.3 ± 0.1	1.7 ± 0.2	0.1159
HLA-DR	1.1 ± 0.1	1.4 ± 0.2	0.1561
CD45	1.4 ± 0.3	1.3 ± 0.1	0.7445
CD34	1.5 ± 0.1	1.6 ± 0.2	0.4423

Percentage of all WJ-MSCs expressing surface markers as determined by flow cytometry. The percentage of expression is indicated as the mean of all WJ-MSCs ($n = 3$) of each group.

3.7. Growth Promotion Study and Media Validation Test Results

The blood and cord blood units tested for bacterial and viral contamination, were found to be negative for these pathogens. Additionally, at the end of the culture, all WJ-MSCs expanded either with FBS or PB-PL were found to be within acceptable ranges in all performed tests. Specifically, cultures were found to be negative for bacterial contamination with 14 days of cultivation and even after the use of blood and sabouraud agar no microorganism growth was observed. In regard to viral contamination, cell products were found to be within the acceptable values. Furthermore, endotoxin content of the MSC final cultures were below 2.5 EU/μ as defined by European Pharmacopoeia and mycoplasma testing results confirmed that no contamination was detected. A detailed description of the above results is presented in supplementary Tables S5 and S6.

4. Discussion

The aim of this study was the evaluation of peripheral blood and cord blood platelet lysates on the isolation, expansion and differentiation of WJ-MSCs. Currently, the most widely used supplement for the in vitro expansion of MSCs is FBS [43–45]. Despite its great benefits such as rapid cell expansion and maintenance of tri-lineage differentiation capability, the use of FBS in cell cultures is associated with safety concerns [46,47]. Under this scope, the use of platelet lysate from blood units not valid for transfusion has been proposed and used successfully by several groups [46–48]. However, the availability of ready to use expired blood units is limited. On the other hand, cord blood could possibly be used for the production of platelet lysate [40–42]. On a daily basis, a significant number of cord blood units are rejected by the cord blood banks due to stringent selection criteria for hematopoietic stem cell isolation and cryopreservation. It is estimated that only 10–20% of cord blood units fulfilled the criteria for transplantation to patients while the remaining 80% could be used as a source of PL production. The culture media used for the isolation and expansion of WJ-MSCs in the current study, were supplemented with 15% FBS or 10% platelet lysate derived from peripheral or cord blood units.

The platelet lysate was produced from PRP pools either from peripheral blood units ($n = 50$) or cord blood units ($n = 100$) with a mean platelet concentration at $655 \pm 21 \times 10^6/\mu$ and $698 \pm 22 \times 10^6/mL$ respectively. The production of PRP pools was performed in order to avoid batch to batch variations of peripheral blood or cord blood units. In addition, patients with severe disease conditions may be unable to donate large volumes of peripheral blood, in order to be used for autologous platelet lysate preparation. Despite this fact, the cell number that can be obtained from a single patient is very low and can be restricted further after processing steps of platelet lysate. Moreover, large scale expansion of MSCs is required for regenerative medicine applications. This huge number of MSCs can be achieved only by the use of 10–50 conventional tissue culture flasks, thus approximately 150 mL of platelet lysate it is needed for the preparation of culture media. Under this scope and in case of routinely used platelet lysate for clinical-grade expansion of MSCs under GMP, pooling of initial peripheral blood and cord blood units must be performed. As a consequence of the biological variability of platelet concentration in peripheral blood units, measurement of the platelet number in the haematological analyser of initial units has been performed. The obtained results did not show huge discrepancies between each peripheral blood or cord blood unit. Finally, after pooling and production of the PL, the platelet number measured again and found no statistical significant difference between PB and CB-PL.

Our data showed that only 15% FBS and 10% PB-PL successfully achieved the in vitro isolation of MSCs from human umbilical cord tissue after 6 days of culturing under standard conditions. In contrast, the isolation of MSCs from the Wharton Jelly tissue with the use of CB-PL growth medium was unsuccessful. Even after 20 days of culturing with bi-weekly change of the media containing 10% CB-PL, no MSCs were obtained. In this way, further evaluation of only PB-PL as replacement of FBS was performed. In addition, a targeted proteomic approach, MRM, was used for quantification of the growth factors in PB-PL and CB-PL, in order to correlate the growth factor levels with the proliferation and differentiation of the WJ-MSCs. WJ-MSCs that were expanded either with 15% FBS or 10% PB-PL growth media, retained their spindle shape morphology up to passage 10. The WJ-MSCs treated with 10% PB-PL exhibited smaller size as has been previously reported by Chevallier et al. [49], possibly due to their increased proliferative activity compared to 15% FBS treated WJ-MSCs. Furthermore, the WJ-MSCs treated with both growth media successfully differentiated to osteogenic and chondrogenic lineages as confirmed by Alizarin Red S and Alcian blue stains respectively. Moreover, when we tried to compare our study with the study of Chevallier et al. [49], it was noticed an immature adipogenic phenotype of differentiated MSCs with the presence of few lipidic inclusions observed after Oil Red O staining. The fact that these two studies did not succeed in exhibiting a mature phenotype under adipogenic differentiation conditions, might be due to the foetal origin of MSCs and the differentiation protocol that was applied [49]. Alcian blue stain revealed a higher content of glycosaminoglycans in 10% PB-PL treated WJ-MSCs rather than in 15% FBS treated WJ-MSCs. These findings were in accordance with the study of Ranzato et al. [50], who reported an increased production of collagen content after platelet lysate treatment in human non tumorigenic keratinocytes [50]. It is known that glycosaminoglycans are forming large polymers with a core protein—called proteoglycans—thus holding the collagen fibres in a specific orientation. The observed increased glycosaminoglycan and collagen amount seemed to be relevant to our study and Ranazato's et al. [50] study, due to their common biological function and resulted by the high levels of growth factors in PB-PL. On the other hand, in regard to the osteogenic induction of MSCs, we did not notice any increased calcium deposition in 10% PB-PL treated WJ-MSCs, when compared to the previously mentioned study of Chevallier et al. [49]. In both studies, PB-PL growth medium was used from the beginning of the isolation and expansion of MSCs. A possible explanation for the different outcome in mineralization levels after osteogenic induction could be due to the different origin of MSCs that were used [49]. The origin of MSCs is of paramount importance, for the establishment of the epigenetic landmarks in their genome, suggesting that bone marrow MSCs can be driven towards the osteogenic lineage, whereas chondrocyte-like cells can be obtained easier from WJ-MSCs under differentiation conditions. Additionally, small differences might occur at the osteogenic induction protocol of the two studies, thus contributing to the final outcome [49].

As the final step for the completion of this evaluation, we investigated the proliferation, clonogenic activity and phenotypical characteristics between the two groups in this study. The use of a relatively small amount of PB-PL in the culture medium, resulted in accelerated cell growth and an increased number of cells in comparison with the FBS culture medium. This was confirmed by calculating the CDT of MSCs between passages 1 and 5. Specifically, until passage 5 the mean CDT of the PB-PL treated MSCs was 167 ± 33 h, while the mean CDT of FBS treated MSCs was 313 ± 49 h, thus strongly indicated an accelerating proliferation phenotype that was adopted by the WJ-MSCs treated with PB-PL growth medium. The cell viability in the two groups did not show any differences at passage 5 with a mean value of $85 \pm 4\%$ for the FBS expanded WJ-MSCs and $86 \pm 3\%$ for the PB-PL expanded WJ-MSCs. These positive effects on MSCs proliferation and the maintenance of tri-lineage differentiation by PB-PL were previously reported by other groups by using a lower initial percentage of PB-PL in the culture medium [47–49]. Regarding the clonogenic potential of WJ-MSCs, the CFU-F assay was performed. Interestingly, the MSCs cultured with PB-PL presented higher number of CFU-F when compared to FBS expanded WJ-MSCs but this increase was not statistically significant. The successful production and maintaining of CFU-F number from passage 2 to passage 5 by WJ-MSCs indicated further the preservation of their self-renewal and clonogenical properties. Furthermore, flow cytometry analysis of MSCs yielded similar results with previous studies for the expression of surface antigens [7,17]. High expression was observed for CD90, CD105 and CD73, whereas the expression of CD45, CD34, CD19, CD3, CD31, CD14 and HLA-DR was less than 2% of the isolated MSCs. Moreover, statistical significant differences were observed in CD105 and CD73 expression between FBS expanded WJ-MSCs with $97 \pm 1\%$ for CD105, $96 \pm 1\%$ for CD73 and PB-PL expanded WJ-MSCs with $99 \pm 1\%$ for CD105 and $98 \pm 1\%$ for CD73. These differences might reflect a different phenotype acquired by the WJ-MSCs upon expansion with 10% PB-PL, capable for a more robust proliferation activity and at least bilinear differentiation to osteocyte and chondrocyte- like cells.

Our results indicated altered phenotype and functionality of PB-PL treated WJ-MSCs in comparison to cells treated with the regular medium. In addition, despite our efforts no cells were isolated from Wharton Jelly tissue by using 10% CB-PL growth medium. This discrepancy could be due to the amount of growth factors presented in cord blood and peripheral blood. A number of studies have previously aimed to the identification of the exact amount of growth factors contained in platelet lysates [50–53]. The majority of these studies used ELISA assays for the quantification of platelet lysate growth factors. However, ELISA has limited multiplexing capabilities and, in order to ensure that the antibody has satisfactory specificity a Western blot assay is often required. On the other hand, MRM is currently used in preclinical and clinical studies for biomarker discovery, development and validation, thus offering a more feasible method for quantification of a panel of candidate proteins in a large number of samples [54]. In this study, MRM was the optimum method used for the quantification of growth factors, indicating the novelty of this study regarding to previous reports [50–53]. We were able to quantify 23 proteins both in peripheral blood and cord blood platelet lysate. Based on the MRM results and CB-PL/PB-PL ratio calculation, PB-PL contained higher amounts of the growth factors than CB-PL. Among them, FGF, PDGF-A, VEGF-A, TGF-b1, TNF-α, IL1α, IL-1β, IL6, IL8, the key players in proliferation stimulation and preservation of stemness identity in MSCs, were successfully quantified in PB-PL and CB-PL. The unsuccessful isolation of MSCs from Wharton Jelly tissue even after of 20 days of culturing, could be due to low levels of the above growth factors in CB-PL compared to PB-PL. The accelerated growth rate of PB-PL expanded WJ-MSCs could also be related with the presence of proinflammatory cytokines. As already described by others, we confirmed that PB-PL and CB-PL contained chemokines including CCL3, CCL4, CCL5 and the adhesion molecules ICAM1 and VCAM1 [39]. In addition, using the MRM technology, we were able to identify specific receptors for cytokines like IL1R, IL6R, IL10R1/2, for growth factors VGFR1/2, TGFR1/2 and for chemokines CCR1 as a result of platelet lysis. The adhesion molecule ICAM1 in combination with the cytokines TNF-α, IL-1α IL-1β and INF-γ play important roles in innate and adaptive immune reaction, involving transendothelial migration and T-cell mediated host defence [55]. Additionally, it has been shown that PDGF can upregulate the expression of ICAM1, thus acting as a co-stimulatory molecule and activating the HLA class II in antigen presenting cells [49,55]. Based on the current literature, IFN-γ can enhance the immunosuppressive behaviour of MSCs by up regulating the co-stimulatory molecules B7-H1 and IDO [56]. This effect could be further amplified with the combination of IL-1β and TNF-α [57]. The well characterized-immunosuppressive secreted molecule PGE2 seems to be up-regulated upon stimulation of MSCs with IL-6, TNF-αand IFN-γ. Also, CCL2, ICAM1 and VCAM1 have a positive effect on MSCs adhesiveness, survival and proliferation in patients with acute lymphoblastic leukaemia [39,58].

The above results clearly indicated the successful use of PB-PL on the isolation and expansion of WJ-MSCs. However, concerns regarding the safely use of platelet lysates as an alternative supplement to FBS for cell culture medium still remain. Pathogen contamination of platelet lysates is still an important issue. More specifically, platelet lysates units are at particular risk of viral and bacterial contamination mostly by adventitious pathogens at the site of venipuncture or from donor blood transmitting agents. For this purpose and following the Greek regulatory procedures for blood donation, peripheral blood and cord blood units were tested for viral contamination serologically. In addition, sterility testing (BacT/Alert) for aerobic and anaerobic bacteria was performed, thus limiting the contamination possibility. Moreover, the MSC's final culture from all groups was tested for bacteria contamination, evaluation of endotoxin content and mycoplasma. After all these tests, peripheral blood, cord blood units and MSCs final culture product were free of any pathogens, indicating that these producing cells under GMP conditions could theoretically be releasable for clinical use. Other pathogen reduction strategies such as the use of solvent detergent treatments, methylene blue/light, riboflavin/ultraviolet light and amotosalen/ultraviolet light (Intercept) can be applied but may affect the quality of the platelet lysate product and this was the primary reason that these approaches were not selected for the current study.

5. Conclusions

In conclusion, the WJ-MSCs were successfully isolated and expanded by the addition of PB-PL as supplement in growth medium. On the contrary, our efforts on isolation and expansion of WJ-MSCs with the use of CB-PL containing culture medium did not have any successful outcome. In this way, further analysis, using different protocols for the production CB-PL and its use in different types of cellular populations must be performed, in order to conclude safely if the CB-PL could serve as an alternative supplement for cell culture medium preparation. In addition, when we tried to compare the characteristics of PB-PL and FBS treated WJ-MSCs, the PB-PL treated MSCs exhibited increased growth rate, tri-lineage differentiation and preservation of the stemness identity. Comparing the growth promoting effects of WJ-MSCs under PB-PL or FBS culture media treatment, the efficacy of PB-PL in terms of MSCs viability and differentiation did not differ from those with the regular medium. Despite these facts, FBS, the most widely used supplement for the production of culture medium, currently contains xenogeneic antigens, so its use as a medium additive in cellular therapies could lead to patient adverse reactions [52].

Additionally, different Lots of FBS are characterized by variations in levels of growth factors, indicating that Lots with low growth factor concentration cannot be used as an effective supplement for maintaining the MSCs quality under good manufacturing practices. Finally, peripheral blood platelet lysate could be used as alternative additive for the production of MSCs culture medium, while the use of CB-PL is under debate, thus requiring further analysis. Both platelet lysates are devoid of any animal sera risks and hypothetically could efficiently be used for the maintenance and expansion of MSCs in large scale clinical translation studies.

Author Contributions: Ioanna Christou (first author) carried out the experiments regarding the production of peripheral blood and cord blood platelet lysate and the isolation, expansion and enumeration of WJ-MSCs. Panagiotis Mallis (equally contributed first author) contributed in the experiments regarding the tri-lineage differentiation of MSCs and growth factor quantification with the MRM assay, collected the data and performed the statistical analysis of the overall study. Efstathios Michalopoulos and Theofanis Chatzistamatiou designed and supervised the overall study. George Mermelekas and Jerome Zoidakis performed the quantification of growth factors with MRM technology. Antonia Vlahou supervised the experiments regarding the MRM technology. Catherine Stavropoulos Giokas supervised and approved the overall study.

Conflicts of Interest: The authors declare no conflict of interest.

References

1. Koshiro, S.; Kunihiro, Y.; Hiroaki, K.; Kaoru, Y.; Kei, S.; Xiangmei, Z.; Masahiro, K.; Yukichi, Z.; Ken, S.; Shingo, N.; et al. Spontaneous differentiation of human mesenchymal stem cells on poly-lactic-coglycolic acid nano-fiber scaffold. *PLoS ONE* **2016**, *11*, 1–15.

2. Zhang, X.; Yamaoka, K.; Sonomoto, K.; Kaneko, H.; Satake, M.; Yamamoto, Y.; Kondo, M.; Zhao, J.; Miyagawa, I.; Yamagata, K.; et al. Local delivery of mesenchymal stem cells with poly-lactic-co-glycolic acid nano-fiber scaffold suppress arthritis in rats. *PLoS ONE* **2014**, *9*, 1–19. [CrossRef] [PubMed]

3. Firestein, G.S. Evolving concepts of rheumatoid arthritis. *Nature* **2003**, *423*, 356–361. [CrossRef] [PubMed]

4. Horwitz, E.M.; Gordon, P.L.; Koo, W.K.; Marx, J.C.; Neel, M.D.; McNall, R.Y.; Muul, L.; Hofmann, T. Isolated allogeneic bone marrow-derived mesenchymal cells engraft and stimulate growth in children with osteogenesis imperfecta: Implications for cell therapy of bone. *Proc. Natl. Acad. Sci. USA* **2002**, *99*, 8932–8937. [CrossRef] [PubMed]

5. Le Blanc, K.; Mougiakakos, D. Multipotent mesenchymal stromal cells and the innate immune system. *Nat. Rev. Immun.* **2012**, *12*, 383–396. [CrossRef] [PubMed]

6. Yen, B.L.; Huang, H.I.; Chien, C.C.; Jui, H.Y.; Ko, B.S.; Yao, M.; Shun, C.T.; Yen, M.L.; Lee, M.C.; Chen, Y.C. Isolation of multipotent cells from human term placenta. *Stem Cell* **2005**, *23*, 3–9. [CrossRef] [PubMed]

7. Kern, S.; Eichler, H.; Stoeve, J.; Kluter, H.; Bieback, K. Comparative analysis of mesenchymal stem cells from bone marrow, umbilical cord blood, or adipose tissue. *Stem Cell* **2006**, *24*, 1294–1301. [CrossRef] [PubMed]

8. Battula, V.L.; Bareiss, P.M.; Treml, S.; Conrad, S.; Albert, I.; Hojak, S.; Abele, H.; Schewe, B.; Just, L.; Skutella, T.; et al. Human placenta and bone marrow derived MSC cultured in serum-free, b-FGF-containing medium express cell surface frizzled-9 and SSEA-4 and give rise to multilineage differentiation. *Differentiation* **2007**, *75*, 279–291. [CrossRef] [PubMed]

9. Pereira, R.F.; Halford, K.W.; O'Hara, M.D.; Leeper, D.B.; Sokolov, B.P.; Pollard, M.D.; Bagasra, O.; Prockop, D.J. Cultured adherent cells from marrow can serve as long-lasting precursor cells for bone, cartilage and lung in irradiated mice. *Proc. Natl. Acad. Sci. USA* **1995**, *92*, 4857–4861. [CrossRef] [PubMed]

10. Friedenstein, A.J.; Gorskaja, J.F.; Kulagina, N.N. Fibroblast precursors in normal and irradiated mouse hematopoietic organs. *Exp. Hematol.* **1976**, *4*, 267–274. [PubMed]

11. Tuli, R.; Seghatoleslami, M.R.; Tuli, S.; Wang, M.L.; Hozack, W.J.; Manner, P.A.; Danielson, K.G.; Tuan, R.S. A simple, high-yield method for obtaining multi-potential mesenchymal progenitor cells from trabecular bone. *Mol. Biotechnol.* **2003**, *23*, 37–49. [CrossRef]

12. Bakhshi, T.; Zabriskie, R.C.; Bodie, S.; Kidd, S.; Ramin, S.; Paganessi, L.A.; Gregory, S.A.; Fung, H.C.; Christopherson, K.W. Mesenchymal stem cells from the Wharton's jelly of umbilical cord segments provide stromal support for the maintenance of cord blood hematopoietic stem cells during long-term ex vivo culture. *Transfusion* **2008**, *48*, 2638–2644. [CrossRef] [PubMed]

13. Caplan, A.I. Mesenchymal stem cells. *J. Orthop. Res.* **1991**, *9*, 641–650. [CrossRef] [PubMed]

14. Prockop, D.J. Marrow stromal cells as stem cells for nonhematopoietic tissues. *Science* **1997**, *276*, 71–74. [CrossRef] [PubMed]

15. Pittenger, M.F.; Mackay, A.M.; Beck, S.C.; Jaiswal, R.K.; Douglas, R.; Mosca, J.D.; Moorman, M.A.; Simonetti, D.W.; Craig, S.; Marshak, D.R. Multilineage potential of adult human mesenchymal stem cells. *Science* **1999**, *284*, 143–147. [CrossRef] [PubMed]

16. Dennis, J.E.; Merriam, A.; Awadallah, A.; Yoo, J.U.; Johnstone, B.; Caplan, A.I. A quadripotential mesenchymal progenitor cell isolated from the marrow of an adult mouse. *J. Bone Miner. Res.* **1999**, *14*, 700–709. [CrossRef] [PubMed]

17. Dominici, M.; Le Blanc, K.; Mueller, I.; Slaper-Cortenbach, I.; Marini, F.C.; Krause, D.S.; Deans, R.J.; Keating, A.; Prockop, D.J.; Horwitz, E.M. Minimal criteria for defining multipotent mesenchymal stromal cells. The international society for cellular therapy position statement. *Cytotherapy* **2006**, *8*, 315–317. [CrossRef] [PubMed]

18. Wagner, W.; Wein, F.; Seckinger, A.; Frankhauser, M.; Wirkner, U.; Krause, U.; Blake, J.; Schwager, C.; Eckstein, V.; Ansorge, W.; et al. Comparative characteristics of mesenchymal stem cells from human bone marrow, adipose tissue and umbilical cord blood. *Exp. Hematol.* **2005**, *33*, 1402–1416. [CrossRef] [PubMed]

19. Weiss, M.L.; Anderson, C.; Medicetty, S.; Seshareddy, K.B.; Weiss, R.J.; Van der Werff, I.; Troyer, D.; McIntosh, K.R. Immune properties of human umbilical cord Wharton's jelly-derived cells. *Stem Cell* **2008**, *26*, 2865. [CrossRef] [PubMed]

20. Prasanna, S.J.; Gopalakrishnan, D.; Shankar, S.R.; Vasandan, A.B. Proinflammatory cytokines, IFNgamma and TNFalpha, influence immune properties of human bone marrow and Wharton jelly mesenchymal stem cells differentially. *PLoS ONE* **2010**, *5*, 9016. [CrossRef] [PubMed]

21. Yoon, J.H.; Roh, E.Y.; Shin, S.; Jung, N.H.; Song, E.Y.; Chang, J.Y.; Kim, B.J.; Jeon, H.W. Comparison of explant derived and enzymatic digestion-derived MSCs and the growth factors from Wharton's jelly. *Biomed. Res. Int.* **2013**, *2013*, 428726. [CrossRef] [PubMed]

22. Costa-Pinto, A.R.; Salgado, A.J.; Correlo, V.M.; Sol, P.; Bhattacharya, M.; Charbord, P.; Reis, R.L.; Neves, N.M. Adhesion, proliferation and osteogenic differentiation of a mouse mesenchymal stem cell line (BMC9) seeded on novel melt-based chitosan/polyester 3D porous scaffolds. *Tissue Eng. Part A* **2008**, *14*, 1049–1057. [CrossRef] [PubMed]

23. Costa-Pinto, A.R.; Correlo, V.M.; Sol, P.C.; Bhattacharya, M.; Charbord, P.; Delorme, B.; Reis, R.L.; Neves, N.M. Osteogenic differentiation of human bone marrow mesenchymal stem cells seeded on melt based chitosan scaffolds for bone tissue engineering applications. *Biomacromolecules* **2009**, *10*, 2067–2073. [CrossRef] [PubMed]

24. Hajivalili, M.; Pourgholi, F.; Kafil, H.S.; Jadidi-Niaragh, F.; Yousefi, M. Mesenchymal stem cells in the treatment of amyotrophic lateral sclerosis. *Curr. Stem Cell Res. Ther.* **2016**, *11*, 41–50. [CrossRef] [PubMed]

25. Connick, P.; Kolappan, M.; Patani, R.; Scott, M.A.; Crawley, C.; He, X.L.; Richardson, K.; Barber, K.; Webber, D.J.; Wheeler-Kingshott, C.A.; et al. The mesenchymal stem cells in multiple sclerosis (MSCIMS) trial protocol and baseline cohort characteristics: An open-label pre-test: post-test study with blinded outcome assessments. *Trials* **2011**, *12*, 62. [CrossRef] [PubMed]

26. Figueroa, F.E.; Carrión, F.; Villanueva, S.; Khoury, M. Mesenchymal stem cell treatment for autoimmune diseases: A critical review. *Biol. Res.* **2012**, *45*, 269–277. [CrossRef] [PubMed]

27. Dazzi, F.; Krampera, M. Mesenchymal stem cells and autoimmune diseases. *Best. Pract. Res. Clin. Haematol.* **2011**, *24*, 49–57. [CrossRef] [PubMed]

28. De Miguel, M.P.; Fuentes-Julián, S.; Blázquez-Martínez, A.; Pascual, C.Y.; Aller, M.A.; Arias, J.; Arnalich-Montiel, F. Immunosuppressive properties of mesenchymal stem cells: Advances and applications. *Curr. Mol. Med.* **2012**, *12*, 574–591. [CrossRef] [PubMed]

29. Sensebe, L.; Bourin, P.; Tarte, K. Good manufacturing practices production of mesenchymal stem/stromal cells. *Hum. Gene Ther.* **2011**, *22*, 19–26. [CrossRef] [PubMed]

30. Bernardo, M.E.; Cometa, A.M.; Pagliara, D.; Vinti, L.; Rossi, F.; Cristantielli, R.; Palumbo, G.; Locatelli, F. Ex vivo expansion of mesenchymal stromal cells. *Best. Pract. Res. Clin. Haematol.* **2011**, *24*, 73. [CrossRef] [PubMed]

31. Tuschong, L.; Soenen, S.L.; Blaese, R.M.; Candotti, F.; Muul, L.M. Immune response to foetal calf serum by two adenosine deaminase-deficient patients after T cell gene therapy. *Hum. Gene Ther.* **2002**, *13*, 1605–1610. [CrossRef] [PubMed]

32. Jung, S.H.; Panchalingam, K.M.; Rosenberg, L.; Behie, L.A. Ex vivo expansion of human mesenchymal stem cells in defined serum-free media. *Stem Cells Int.* **2012**, *2012*, 1–12. [CrossRef] [PubMed]

33. Schallmoser, K.; Bartmann, C.; Rohde, E.; Reinisch, A.; Kashofer, K.; Stadelmeyer, E.; Drexler, C.; Lanzer, G.; Linkesch, W.; Strunk, D. Human platelet lysate can replace foetal bovine serum for clinical-scale expansion of functional mesenchymal stromal cells. *Transfusion* **2007**, *47*, 1436–1446. [CrossRef] [PubMed]

34. Bieback, K.; Hecker, A.; Kocaomer, A.; Lannert, H.; Schallmoser, K.; Strunk, D.; Ter, H. Human alternatives to foetal bovine serum for the expansion of mesenchymal stromal cells from bone marrow. *Stem Cell* **2009**, *27*, 2331–2341. [CrossRef] [PubMed]

35. Muller, I.; Kordowich, S.; Holzwarth, C.; Spano, C.; Isensee, G.; Staiber, A.; Viebahn, S.; Gieseke, F.; Langer, H.; Gawaz, M.P.; et al. Animal serum-free culture conditions for isolation and expansion of multipotent mesenchymal stromal cells from human BM. *Cytotherapy* **2006**, *8*, 437–444. [CrossRef] [PubMed]

36. Bernardo, M.E.; Avanzini, M.A.; Perotti, C.; Cometa, A.M.; Moretta, A.; Lenta, E.; Del Fante, C.; Novara, F.; de Silvestri, A.; Amendola, G.; et al. Optimization of in vitro expansion of human multipotent mesenchymal stromal cells for cell-therapy approaches: Further insights in the search for a foetal calf serum substitute. *J. Cell. Physiol.* **2007**, *211*, 121–130. [CrossRef] [PubMed]

37. Lange, C.; Cakiroglu, F.; Spiess, A.N.; Cappallo-Obermann, H.; Dierlamm, J.; Zander, A.R. Accelerated and safe expansion of human mesenchymal stromal cells in animal serum-free medium for transplantation and regenerative medicine. *J. Cell. Physiol.* **2007**, *213*, 18–26. [CrossRef] [PubMed]

38. Schallmoser, K.; Strunk, D. Preparation of pooled human platelet lysate (pHPL) as an efficient supplement for animal serum-free human stem cell cultures. *J. Vis. Exp.* **2009**, *32*, 1523. [CrossRef] [PubMed]

39. Fekete, N.; Gadelorge, M.; Fürst, D.; Maurer, C.; Dausend, J.; Fleury-Cappellesso, S.; Mailänder, V.; Lotfi, R.; Ignatius, A.; Sensebé, L.; et al. Platelet lysate from whole blood-derived pooled platelet concentrates and apheresis-derived platelet concentrates for the isolation and expansion of human bone marrow mesenchymal stromal cells: Production process, content and identification of active components. *Cytotherapy* **2011**, *14*, 540–554.

40. Murphy, M.B.; Blashki, D.; Buchanan, R.M.; Yazdi, I.K.; Ferrari, M.; Simmons, P.J.; Tasciotti, E. Adult and umbilical cord blood-derived platelet-rich plasma for mesenchymal stem cell proliferation, chemotaxis and cryo-preservation. *Biomaterials* **2012**, *33*, 5308–5316. [CrossRef] [PubMed]

41. Longo, V.; Rebulla, P.; Pupella, S.; Zolla, L.; Rinalducci, S. Proteomic characterization of platelet gel releasate from adult peripheral and cord blood. *Proteom. Clin. Appl.* **2016**, *10*, 870–882. [CrossRef] [PubMed]

42. Parazzi, V.; Lavazza, C.; Boldrin, V.; Montelatici, E.; Pallotti, F.; Marconi, M.; Lazzari, L. Extensive characterization of platelet gel releasate from cord blood in regenerative medicine. *Cell Transplant.* **2015**, *24*, 2573–2584. [CrossRef] [PubMed]

43. Chatzistamatiou, T.K.; Papassavas, A.C.; Michalopoulos, E.; Gamaloutsos, C.; Mallis, P.; Gontika, I.; Panagouli, E.; Koussoulakos, S.L.; Stavropoulos-Giokas, C. Optimizing isolation culture and freezing methods to preserve Wharton's jelly's mesenchymal stem cell (MSC) properties: An MSC banking protocol validation for the Hellenic Cord Blood Bank. *Transfusion* **2014**, *54*, 3108–3120. [CrossRef] [PubMed]

44. Cardoso, T.C.; Ferrari, H.F.; Garcia, A.F.; Novais, J.B.; Silva-Frade, C.; Ferrarezi, M.C.; Andrade, A.L.; Gameiro, R. Isolation and characterization of Wharton's jelly-derived multipotent mesenchymal stromal cells obtained from bovine umbilical cord and maintained in a defined serum-free three-dimensional system. *BMC Biotechnol.* **2012**, *4*, 12–18. [CrossRef] [PubMed]

45. Fong, C.Y.; Subramanian, A.; Biswas, A.; Bongso, A. Freezing of fresh Wharton's Jelly from human umbilical cords yields high post-thaw mesenchymal stem cell numbers for cell-based therapies. *J. Cell. Biochem.* **2016**, *117*, 815–827. [CrossRef] [PubMed]

46. Ben, A.N.; Jenhani, F.; Regaya, Z.; Berraeis, L.; Ben, O.T.; Ducrocq, E.; Domenech, J. Phenotypical and functional characteristics of mesenchymal stem cells from bone marrow: Comparison of culture using different media supplemented with human platelet lysate or foetal bovine serum. *Stem Cell Res. Ther.* **2012**, *3*, 25043.

47. Bernardi, M.; Albiero, E.; Alghisi, A.; Chieregato, K.; Lievore, C.; Madeo, D.; Rodeghiero, F.; Astori, G. Production of human platelet lysate by use of ultrasound for ex vivo expansion of human bone marrow-derived mesenchymal stromal cells. *Cytotherapy* **2013**, *15*, 920–929. [CrossRef] [PubMed]

48. Griffiths, S.; Baraniak, P.R.; Copland, I.B.; Nerem, R.M.; McDevitt, T.C. Human platelet lysate stimulates high-passage and senescent human multipotent mesenchymal stromal cell growth and rejuvenation in vitro. *Cytotherapy* **2013**, *15*, 1469–1483. [CrossRef] [PubMed]

49. Chevallier, N.; Anagnostou, F.; Zilber, S.; Bodivit, G.; Maurin, S.; Barrault, A.; Bierling, P.; Hernigou, P.; Layrolle, P.; Rouard, H. Osteoblastic differentiation of human mesenchymal stem cells with platelet lysate. *Biomaterials* **2010**, *2010*, 270–278. [CrossRef] [PubMed]

50. Ranzato, E.; Martinotti, S.; Volante, A.; Mazzucco, L.; Burlando, B. Platelet lysate modulates MMP-2 and MMP-9 expression, matrix deposition and cell-to-matrix adhesion in keratinocytes and fibroblasts. *Exp. Dermatol.* **2011**, *20*, 308–313. [CrossRef] [PubMed]

51. Mojica-Henshaw, M.P.; Jacobson, P.; Morris, J.; Kelley, L.; Pierce, J.; Boyer, M.; Reems, J.A. Serum-converted platelet lysate can substitute for fetal bovine serum in human mesenchymal stromal cell cultures. *Cytotherapy* **2013**, *15*, 1458–1468. [CrossRef] [PubMed]

52. Shih, D.T.; Burnouf, T. Preparation, quality criteria and properties of human blood platelet lysate supplements for ex vivo stem cell expansion. *New Biotechnol.* **2015**, *32*, 199–211. [CrossRef] [PubMed]

53. Ebhardt, H.A.; Root, A.; Sander, C.; Aebersold, R. Applications of targeted proteomics in systems biology and translational medicine. *Proteomics* **2015**, *15*, 3193–3208. [CrossRef] [PubMed]

54. Lawson, C.; Wolf, S. ICAM-1 signaling in endothelial cells. *Pharmacol. Rep.* **2009**, *61*, 22–32. [CrossRef]

55. Le Blanc, K.; Tammik, C.; Rosendahl, K.; Zetterberg, E.; Ringdén, O. HLA expression and immunologic properties of differentiated and undifferentiated mesenchymal stem cells. *Exp. Hematol.* **2003**, *31*, 890–896. [CrossRef]

56.   Sheng, H.; Wang, Y.; Jin, Y.; Zhang, Q.; Zhang, Y.; Wang, L.; Shen, B.; Yin, S.; Liu, W.; Cui, L.; et al. A critical role of IFNgamma in priming MSC-mediated suppression of T cell proliferation through up-regulation of B7-H1. *Cell Res.* **2008**, *18*, 846–857. [CrossRef] [PubMed]
57.   Ploederl, K.; Strasser, C.; Hennerbichler, S.; Peterbauer-Scherb, A.; Gabriel, C. Development and validation of a production process of platelet lysate for autologous use. *Platelets* **2010**, *22*, 204–209. [CrossRef] [PubMed]
58.   Su, C.Y.; Kuo, Y.P.; Lin, Y.C.; Huang, C.T.; Tseng, Y.H.; Burnouf, T. A virally inactivated functional growth factor preparation from human platelet concentrates. *Vox Sang.* **2009**, *97*, 119–128. [CrossRef] [PubMed]

11

Development of Self-Assembled Nanoribbon Bound Peptide-Polyaniline Composite Scaffolds and their Interactions with Neural Cortical Cells

Andrew M. Smith, Harrison T. Pajovich and Ipsita A. Banerjee *

Department of Chemistry, Fordham University, 441 East Fordham Road, Bronx, New York, NY 10458, USA;
asmith169@fordham.edu (A.M.S.); hpajovich@fordham.edu (H.T.P.)
* Correspondence: banerjee@fordham.edu

Academic Editor: Gary Chinga Carrasco

Abstract: Degenerative neurological disorders and traumatic brain injuries cause significant damage to quality of life and often impact survival. As a result, novel treatments are necessary that can allow for the regeneration of neural tissue. In this work, a new biomimetic scaffold was designed with potential for applications in neural tissue regeneration. To develop the scaffold, we first prepared a new bolaamphiphile that was capable of undergoing self-assembly into nanoribbons at pH 7. Those nanoribbons were then utilized as templates for conjugation with specific proteins known to play a critical role in neural tissue growth. The template (Ile-TMG-Ile) was prepared by conjugating tetramethyleneglutaric acid with isoleucine and the ability of the bolaamphiphile to self-assemble was probed at a pH range of 4 through 9. The nanoribbons formed under neutral conditions were then functionalized step-wise with the basement membrane protein laminin, the neurotropic factor artemin and Type IV collagen. The conductive polymer polyaniline (PANI) was then incorporated through electrostatic and π–π stacking interactions to the scaffold to impart electrical properties. Distinct morphology changes were observed upon conjugation with each layer, which was also accompanied by an increase in Young's Modulus as well as surface roughness. The Young's Modulus of the dried PANI-bound biocomposite scaffolds was found to be 5.5 GPa, indicating the mechanical strength of the scaffold. Thermal phase changes studied indicated broad endothermic peaks upon incorporation of the proteins which were diminished upon binding with PANI. The scaffolds also exhibited in vitro biodegradable behavior over a period of three weeks. Furthermore, we observed cell proliferation and short neurite outgrowths in the presence of rat neural cortical cells, confirming that the scaffolds may be applicable in neural tissue regeneration. The electrochemical properties of the scaffolds were also studied by generating I-V curves by conducting cyclic voltammetry. Thus, we have developed a new biomimetic composite scaffold that may have potential applications in neural tissue regeneration.

Keywords: self-assembly; templates; tissue regeneration; peptide amphiphiles

1. Introduction

The nervous system consists of a network of interconnected cells that play a critical role in the reception and transmission of electrical signals throughout the body [1,2]. However, damage to the nervous system caused by brain injuries or neurodegenerative disorders such as, Alzheimer's, Parkinson's, epilepsy, multiple sclerosis, or chronic traumatic encephalopathy, can lead to severe impairment in daily function and quality of life [3,4]. The slow growth and fragility of nervous tissue poses a unique challenge for treatment interventions. Current treatments are limited to nerve autographing and the use of nerve conduits [5], as well as development of novel antagonists [6]. These methods are challenged by the lack of donors, tissue rejection, scar tissue growth, implantation decay and lack of sufficient structural and biochemical information at the biomolecular level [7]. Tissue Engineering (TE) poses an alternative treatment option to conventional methods. TE seeks to repair, restore and replace damaged tissues and harbor growth of healthy tissue [8]. This is accomplished by creating a biomimetic three-dimensional matrix that exemplifies properties of the extracellular matrix (ECM), which can eventually aid in re-growing tissue [9]. These scaffolds are tailored to specific tissues to ensure compatibility and alleviate immune response and scar tissue growth and support new tissue by proper adhesion and integration [10,11].

Since the inception of TE, a multitude of materials, both natural and synthetic, have been discovered to promote neural tissue growth [12]. For example, functionalized carbon nanotubes and graphene nanotubes have been successful in promoting cell differentiation and migration, while efficiently maintaining conductive properties within the tissue [13–15]. Polymers [16] such as polyethylene glycol (PEG), poly ε-caprolactone (PCL), poly(lactic-co-glycolic acid) (PLGA) and poly-lactic acid (PLA) are some of the most widely used synthetic polymers [17]. Specifically, they have been used to create neural guidance conduits and cylindrical porous electrospun composites to promote axonal growth and to bridge neural ending defects [18]. For instance, it was reported that composites of poly ε-caprolactone electrospun membranes and gelatin improved cell adhesion, proliferation and differentiation of PC-12 nerve cells and supported neurite outgrowth [19]. In a recent study, it was shown that irradiation of graphitic carbon nitride integrated with graphene oxide (GO) that was bound to electrospun PCL/gelatin fibers resulted in neural stimulation upon irradiation with visible-light and thereby supported neuronal differentiation [20]. Amongst the naturally occurring proteoglycans, hyaluronic acid, chitosan, chondrotin sulfate and heparin sulfate have gained significant prominence [21] in the preparation of composite materials for neural TE. For example, in one study, chondroitin-6-sulfate and neural growth factor (NGF) were fused into PEG gels and promoted neurite extension and viability of cortical cells [22,23]. In a separate study, heparin-mimicking polymers—prepared by combining glucosamine-like 2-methacrylamido glucopyranose monomers with three separate sulfonated units—showed higher cytocompatibility and promoted differentiation of embryonic stem cells to neuronal cells as compared to natural heparin [24]. Forsythe and co-workers developed three-dimensional graphene-heparin-poly-L-lysine polyelectrolytes that promoted neuron cell adhesion, proliferation and neurite outgrowth [25]. It has also been shown that hyaluronic acid bound electrospun PCL scaffolds as well as agarose-chitosan blends enhance mechanical properties and increased proliferation of neural cells [26,27].

In addition to the aforementioned biomaterials, peptide amphiphiles have gained prominence in numerous biomedical applications due to their facile self-assembling properties, biocompatibility and relative ease of functionalization [28]. For example, when peptide nanofibers formed by self-assembly of amphiphilic (palmitoyl-GGGAAAKRK) were utilized for siRNA delivery into the brain, they showed higher intra-cellular uptake after being delivered intra-cranially [29]. Stupp and co-workers recently showed that hybrid DNA-peptide nanotubes that had been prepared by altering the sequence of the DNA strands and incorporating the cell-adhesion motif RGDS displayed selectivity and enhanced cell adhesion and differentiation of neural stem cells into neurons but not astrocytes [30]. Scaffolds formed by utilizing the self-assembling peptide RADA16-I have demonstrated potential in closing neural gaps and regenerating axons and healing spinal cord injuries [31]. Researchers have also developed hybrid matrices by combining Type I collagen and peptide amphiphile based nanofibrous scaffolds functionalized with IKVAV or YIGSR that showed specific responses to cerebellar cortex Granule cells and Purkinje cells. Specifically, the IKVAV hybrid scaffolds showed an increase in granule cell density and growth of Purkinje cell dendrite and axons in the presence of peptide nanofibers over specific concentration ranges compared to collagen [32].

In this work, we have developed a new biomimetic scaffold with potential for neural tissue engineering. We conjugated 3,3-tetramethylene glutaric acid (TMG) with isoleucine (Ile) to form a new bolaamphiphile, wherein TMG was the inner head group while the two isoleucine groups formed the tail groups at each end. The self-assembling ability of the Ile-TMG-Ile conjugate was probed at a pH range of 4–9. We observed that under neutral conditions, the conjugate self-assembled into nanoribbons, which were then utilized as templates for developing the scaffold. TMG has been shown to be biocompatible and is a well-known aldose-reductase inhibitor in vitro. It was once touted for its potential in inhibiting diabetic angiopathy and cataract formation by preventing the formation of sorbitol [33]. However, it was found to be relatively inactive as an aldose-reductase inhibitor in vivo, as significant amounts of TMG were unable to reach the retina or lens. We utilized TMG due to its unique structure containing the glutaric acid back bone functionalized with a cyclopentyl ring system. We conjugated it with isoleucine, as it is a key factor known to enhance the activity of alanine-serine-cysteine transporter (Asc-1), which mediates the release of Gly and Ser from neurons and modulates N-methyl-D-aspartate receptor (NMDAR) synaptic activity [34]. The nanoribbons formed upon self-assembly were utilized as templates for preparing tailored scaffolds for neural tissue regeneration. The template Ile-TMG-Ile nanoribbons were first conjugated with laminin, a major component of the extracellular matrix of vascular tissue in the brain. Laminin has been shown to increase the binding abilities of nanoscaffolds as well as to promote cell migration in newly formed cells [35,36]. A recent study conducted using a laminin functionalized PCl-chitosan scaffold showed increased mechanical properties, cell attachment and proliferation [37]. Another study found that laminin based scaffolds vastly improved neuronal survival in the injured brains of mice, which led to greater performance on spatial learning tasks [38].

We then conjugated the laminin bound assemblies with Artemin, which is a glial cell line derived neurotropic factor. Artemin is known to support signaling and increase growth in both peripheral and central nervous tissue by binding to GFR alpha3–RET, an artemin specific receptor in the MAP kinase pathway [39]. It has also been shown to attenuate neuropathic pain in individuals with spinal cord injuries and plays a protective role against deteriorating motor neurons in ALS patients [40,41]. To form the biocomposite, we then conjugated the laminin-artemin-bound templates with Type IV Collagen. It is well known that Type IV Collagen is a component of the basement membrane of vascular tissues in brain and forms mesh-like structures with advantageous mechanical properties. Furthermore, Type IV collagen promotes cell adhesion and stability [42].

Finally, emeraldine base polyaniline (PANI)—a conductive polymer—was incorporated to impart electric properties to the scaffold. PANI consists of repeat units of benzene rings which are separated by secondary amine groups and a quinoid ring system attached to imine groups [43]. The protonated form of emeraldine is conductive as it can form a semiquinone radical cation [44]. In a study where emeraldine polyaniline was blended with gelatin and then electrospun into nanofibers, the scaffold showed a marked increase in conductivity after incorporation of PANI [45]. Thus, we have created a new composite scaffold that consists of self-assembled nanoribbons, conjugated with key proteinaceous components to enhance growth and proliferation of neuronal cells as well as a conductive polymer, polyaniline to impart electrical properties. The formed scaffold demonstrated biodegradability, enhanced mechanical properties as well as promoted growth and proliferation of cortical cells and promoted axonal outgrowth Thus, these newly formed scaffolds may have potential applications in neural tissue engineering.

2. Materials and Methods

Amino acid isoleucine and 3,3-tetramethylene glutaric acid, dimethylformamide (DMF), N-Hydroxy Succinimide (NHS), 1-ethyl-3-(3-dimethylaminopropyl) carbodiimide (EDAC) and triethylamine (TEA), Bradford reagent, Bovine serum albumin (BSA) were purchased from Sigma Aldrich. Buffer solutions of various pH values were purchased from Fisher Scientific. Mouse laminin (sc-29012) and laminin alpha-2 antibody (B-4) were purchased from Santa Cruz Biotechnology. NHS-rhodamine was purchased from Thermo Scientific. Anti-collagen Type IV (rabbit) antibody was purchased from Rockland. Human artemin (category 4515-20, lot P70215) was purchased from Bio Vision. Type IV collagen (AG 19502) was purchased from Neuromics. Rat cortical cells (E18) and cell culture media (NbActiv4) were purchased from BrainBits. Neuroblast cell culture media and Glutamax were purchased from Gibco. Polyaniline was purchased from Ark Pharmaceuticals. Solvent N-methyl-2-Pyrrolidine was purchased from VWR. The digital multimeter model M-1000D was purchased from Elenco and a 1.62 mm diameter platinum electrode was purchased from Bioanalytical Systems Incorporated.

2.1. Synthesis of Ile-TMG-Ile

To TMG (1M) were added NHS (0.1M) and EDAC (0.1M) in DMF for activating the carboxylic acid groups of TMG. The mixture was stirred for one hour at 4 °C followed by the addition of Ile (2M). Two drops of TEA were added and the reaction mixture was stirred at 4 °C for 24 h. After 24 h, the solution was rotary evaporated to remove the solvent. The product obtained was found to be a white solid. The ESI-MS obtained by HPLC-MS (Agilent 6100 series, Santa Clara, CA, USA) showed a very weak M+ peak at m/z 411.2; peaks were also observed at m/z 952.4 and at m/z 805.5 due to the formation of oligmers. The strongest peak was seen at m/z 393.2 due to loss of hydroxyl radical. The product most likely undergoes McLafferty rearrangement, as expected for amides. Smaller fragments were observed at m/z = 360.2; 230.7, 361.2 and 115.0. Thus, in addition to Ile-TMG-Ile, side products that included oligomers were also formed. The product was recrystallized using methanol and dried under vacuum before further analysis. The yield of the product was found to 57.2%. The formation of the product was confirmed by ^{1}H NMR spectroscopy using a Bruker 400 MHz NMR (Billerica, MA, USA) in deuterated DMSO with TMS as a solvent. ^{1}H NMR (DMSO-d6) spectrum showed peaks at δ 0.9 (t, 6H); δ 1.2 (d, 6H); δ 1.4 (t, 2H); δ 1.7 (m, 4H); δ 2.3 (d, 2H); δ 2.7 (s, 4H); δ 2.9 (s, 4H); 3.4 (s, 4H); δ 8.1 (s, 2H); δ 12.2 (s, 2H). ^{13}C NMR (DMSO-d6) showed peaks at δ 17.3; δ 22.2; 25.3; δ 29.6; δ 33.4; δ 39.5 δ 42.7; δ 45.9 and δ 174.2.

2.2. Self-Assembly of Ile-TMG-Ile Template

The synthesized product was allowed to self-assemble in buffer solutions of varying pH values. In general, the product (45 mM) was allowed to assemble under acidic (pH 4, potassium acid phthalate buffer); neutral (pH 7, potassium phosphate monobasic-sodium hydroxide buffer) and basic (pH 9, boric acid, potassium chloride, sodium hydroxide buffer) conditions over a period of three to four weeks at room temperature. The growth of assemblies was monitored by dynamic light scattering periodically. After four weeks of growth, the assemblies were centrifuged and washed thrice with deionized water to remove the buffer and left in deionized water for further analysis.

2.3. Preparation of Scaffold

The washed Ile-TMG-Ile assemblies grown at pH 7 were utilized for preparation of scaffolds. An aqueous solution of the fibrillar assemblies (2 mM) was treated with EDAC (1 mM) and NHS (1 mM) for one hour at 4 °C to activate the free carboxylic groups in Ile-TMG-Ile. Mouse laminin (0.1 mg/ mL, 200 µL) was then added to the activated template. The mixture was stirred at 4 °C for 24 h to allow for adhesion of laminin. The laminin bound templates were then washed and centrifuged thrice with deionized water to remove any unbound laminin followed by addition of EDAC (1 mM) and NHS (1 mM) for one hour at 4 °C. To the laminin bound templates, artemin (0.1 mg/mL, 200 µL) was then added and shaken at 4 °C for 24 h and washed and centrifuged thrice to remove unbound artemin. The laminin and artemin functionalized construct was once again allowed to react with EDAC (1 mM) and NHS (1 mM) for one hour at 4 °C followed by the addition of Type IV collagen (0.1 mg/mL, 100 µL) and was shaken at 4 °C for 24 h. The biocomposite was washed and centrifuged to remove unbound collagen. The formed biocomposite scaffold was then vacuum dried. To the dried scaffold, polyaniline (PANI) (0.1 mg/mL) in N-methyl-2-Pyrrolidine (5 mL) was added and the mixture was shaken at 4 °C for 24 h and then centrifuged for three hours to remove unbound PANI.

2.4. Binding Efficiency of Laminin, Artemin and Type IV Collagen on the Assemblies

The efficiency of binding of each of the protein components was examined by UV-Vis spectroscopy which was monitored at 272 nm and the absorbance before and after binding to each of the proteins was measured. The absorbance of laminin (0.1 mg/mL, 200 μL) was compared, with that of washed laminin bound assemblies which were prepared as described above. The volumes of the solutions were kept constant. Our results indicated 84.6% binding of laminin to the assemblies. Subsequently the protein concentration of laminin bound assemblies was determined using the Bradford method, based on a standard curve obtained for BSA (1 mg/mL). The concentration of laminin on the assemblies was found to be 2.5 μM. Similarly, upon binding with artemin, the binding efficiency was found to be 92.3% and the protein concentration was found to be 3.19 μM. Finally, for Type IV collagen, the binding efficiency was found to be 88.3% and the protein concentration after binding to Type IV Collagen was found to be 3.65 μM.

2.5. In Vitro Biodegradability Studies

To examine the biodegradability of the formed scaffold, 45 mg of the scaffold was weighed in a petri dish to which 10 mL of simulated body fluid buffer (SBF) was added. The weight of the scaffold was measured every 10 h over a period of 22 days. In general, at each time point, the scaffold was rinsed with deionized water and air dried at room temperature before measurement of weight. The SBF was replaced as necessary each time with the same volume (10 mL). Studies were carried out in triplicate. The results were then analyzed as a function of time. The SBF was prepared according to previously established methods [46]. Briefly, to prepare the simulated body fluid (SBF) for biodegradability studies, 750 mL of distilled water was first brought to a constant temperature of 36.5 °C. The solution was constantly stirred while adding 7.996 g of NaCl, 0.350 g NaHCO$_3$, 0.224 g KCl, 0.228 g K$_2$HPO$_4$ 3H$_2$O, 0.305 g MgCl$_2$ 6H$_2$O, 40 mL HCl (1 M), 0.278 g CaCl$_2$, 0.071 g Na$_2$SO$_4$ and 6.057 g (CH$_2$OH)$_3$CNH$_2$. The pH was adjusted to 7.4 with dropwise addition of 1 M HCl. The total volume was then brought to 1 L using distilled water and then the solution was stored at 4 °C before use.

2.6. Cell Studies

To examine cell viability, rat cortical cells (E18, lot, BrainBits) were cultured in NbActiv4 media (BrainBits) containing 1% 10,000 g/mL amphotericin and 100 units/mL penicillin and streptomycin. The cells were grown to confluence and kept in a humidified atmosphere of 5% CO$_2$ at 37 °C. To examine the effects of the scaffold, cells were plated in 12-well Falcon polystyrene tissue culture plates at a density of 1×10^3 cells per well. After allowing the cells to adhere to the well plates for two hours, scaffolds were added at varying concentration (6 μM, 10 μM and 13 μM). The scaffolds were allowed to interact with the cells for 24, 48 and 72 h. These studies were carried out in triplicate. After each allotted period of time, cell viability and growth was examined by via trypan blue method. Cortical cells in media alone were used as a control for this study. Once stained with trypan blue, live and dead cells were counted using a hemocytometer and averaged. The percent viability was then calculated as follows: (living cortical cells)/(living cortical cells + dead cortical cells) × 100. To observe the interactions of the cortical cells with the scaffolds, cells were plated and allowed to interact with scaffolds as before. The media was changed every 48 h. Images were taken using an AmScope IN200TA-P Inverted Tissue Culture Microscope (Irvine, CA, USA) with a USB camera at various magnifications every 24 h over a period of seven days.

2.7. Electrochemical Studies

We first examined the resistance of control PANI, PANI-bound scaffolds and the biocomposite scaffold in the absence of PANI and determined the conductivity. The resistance of the scaffolds was tested in HCl (1M). A digital multimeter, model M-1000D, from Elenco (Philadelphia, PA, USA) was then used to determine the resistance. To examine the electrochemical properties of the scaffold, we conducted cyclic voltammetry in the presence of PANI, PANI-bound scaffolds and the biocomposite scaffold before it was bound to PANI as control. To obtain I-V curves, the solutions were dried directly onto the 2.01 mm^2 electrode using a vacuum pump for a period of 24 h, adding an addition layer every 6 h. A voltage cell was created using this working electrode, a platinum counter electrode, an Ag/AgCl reference electrode and 1M HCl. Prior to electrochemical measurements, nitrogen was bubbled into the cell for 20 min to remove dissolved oxygen from the solution. PowerSuite by Princeton Applied Research and a Princeton Applied Research potentiostat model 263A (Oakridge, TN, USA) were used to obtain I-V curves from a potential of -0.2 V to 0.9 V at 10 mV/s.

2.8. Characterization

2.8.1. Fourier Transform Infrared (FTIR) Spectroscopy

FTIR spectroscopy was conducted using a Thermo Scientific Nicolet IS50 FTIR (Waltham, MA, USA) with OMNIC software. In general, samples were run at a range of 400 cm^{-1} to 4000 cm^{-1} with 100 scans per sample and the data obtained was averaged.

2.8.2. Scanning Electron Microscopy (SEM)

To examine the morphologies of the assemblies as well as the incorporation of each layer after conjugation, samples were air-dried on to carbon double stick tape and carbon coated to prevent charging. Samples were imaged at various magnifications between of 2 kV to 10 kV utilizing a Zeiss EVO MA10 scanning electron microscope (Thornwood, NY, USA).

2.8.3. Transmission Electron Microscopy (TEM)

In order to further elucidate the morphologies of the assemblies we also conducted TEM analysis using JEOL 1200 EX transmission electron microscope (Peabody, MA, USA). Samples were air-dried on to formvar/carbon 200 mesh copper grids overnight before analysis. Samples were imaged at various magnifications at 80 KeV.

2.8.4. Dynamic Light Scattering (DLS)

To monitor the growth of the assemblies, we conducted DLS using a NICOMP 380 ZLS sizer (Willow Grove, PA, USA). Samples were diluted to appropriate concentrations and each sample was run at least three times and the data obtained was averaged.

2.8.5. Atomic Force Microscopy (AFM)

We examined the nanoscale morphology of the assemblies after conjugation with each layer using a Bruker Multimode 8 AFM (Santa Barbara, CA, USA). Furthermore, the mechanical properties of the scaffolds were also determined by conducting Peak Force Microscopy. The tip was moved to various points (at least five points per sample) on each sample and the values obtained were averaged. Young's Modulus was determined by fitting the data into a Hertzian Model. In general, we used RTESPA-175 Antimony (n) doped Si tip with a spring constant of 40 N/m.

2.8.6. Differential Scanning Calorimetry (DSC)

To explore the thermal phase changes of the scaffolds, we conducted DSC analysis. For each analysis, 0.1 mg samples were dried under vacuum and weighed. We examined the phase changes for each layer of the scaffold using a TA instruments, model Q200 instrument (New Castle, DE, USA) at a temperature range of 0 °C to 250 °C at scanning rate of 10 °C per minute.

2.8.7. Fluorescence Microscopy

Fluorescence microscopy was carried out to examine the interactions of FITC labeled laminin -2 antibody (B-4) with laminin bound assemblies and rhodamine labeled Type IV collagen with the biocomposites using a Phase Contrast Amscope Fluorescence Inverted Microscope (Irvine, CA, USA). To prepare samples for binding with laminin antibody, the laminin bound assemblies were washed and centrifuged with deionized water followed by the addition of BSA Blocker solution (1% BSA) in tris buffered saline to prevent non-specific binding. The sample was vortexed for two minutes and allowed to incubate at room temperature for 4 h. The sample was then centrifuged and washed once with TBS followed by washing with deionized water. To the sample, we then added FITC labeled laminin -2 antibody (B-4) (50 µg/mL). The sample was incubated overnight at 4 °C. Samples were then washed and centrifuged with deionized water and imaged on poly-L-lysine coated glass slides which was covered by a coverslip. Samples were then excited at 450 nm. A similar protocol was followed for examining the interactions of rhodamine labeled collagen IV antibody, where in the collagen IV-artemin-laminin-bound Ile-TMG-Ile assemblies were first washed and centrifuged, followed by the addition of 1% BSA blocking agent before incubation with the antibody. Finally samples were transferred to glass-slides covered with coverslips and imaged at 588 nm excitation.

2.8.8. UV-Vis Spectroscopy

To determine the binding efficiency and protein concentrations of the biocomposite assemblies after each layer of protein (laminin, artemin and Type IV collagen) was added, we carried out UV-Vis spectroscopy using a Nanodrop 2000 spectrophotometer (Waltham, MA, USA).

2.9. Statistical Analysis

We used two-tailed Student's t-test for carrying out statistical analysis. Studies were carried out in triplicate (n = 3). Data are presented as the mean value $\pm$ standard deviation (SD) of each sample group.

3. Results and Discussion

3.1. Self-Assembly of Ile-TMG-Ile

Molecular self-assembly of biomolecules transpires through weak, non-covalent interactions that include electrostatic, hydrophobic interactions, hydrogen bonds, van der Waals interactions and π–π stacking forces that result in the formation of stable and functional supramolecular structures [47]. Self-assembling peptides, in particular form unique supramolecular assemblies and offer several advantages. Thus, such materials pose a plethora of applications in tissue engineering as they are highly biocompatible and modifiable [48].

In this work, we designed a new bolaamphiphile by conjugating the amino acid Ile with the dicarboxylic acid 3,3 tetramethylene glutaric acid (TMG) resulting the in the formation of (2S,2'S,3S,3'S)-2,2'-((2,2'-(cyclopentane-1,1-diyl)bis(acetyl))bis(azanediyl))bis(3-methylpentanoic cid), abbreviated as Ile-TMG-Ile (Figure 1). In previous work, peptide based bolaamphiphiles containing amino acid moieties such as glycine, conjugated with dicarboxylic acids such as azelaic acid have been shown to self-assemble into nano and microtubes with closed ends or single layered sheets [49]. In an earlier study, we have shown that when phenylalanine was conjugated with dicarboxylic acids of different chain lengths, supramolecular assemblies of a variety of morphologies were formed depending upon the growth conditions used [50]. It has also been shown that peptide amphiphiles containing N-terminus palmitoylated groups such as CH_3-$(CH_2)_{14}$CO-NH-X-Ala_3-Glu_4-CO-NH_2, where in the amino acid X was varied between Ile, Phe or Val, self-assembled into micelles, nanoribbons or nanofibers depending upon the pH of the growth conditions [51].

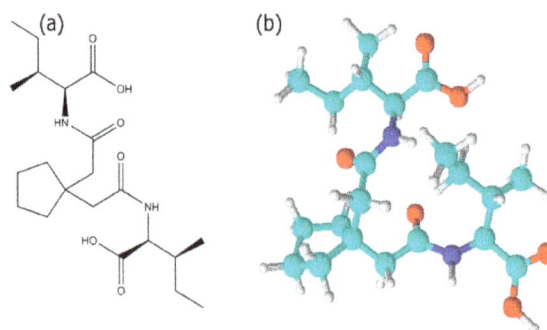

Figure 1. (**a**) Chemical structure of Ile-TMG-Ile; (**b**) ball and stick model.

Herein, we examined the self-assembly of Ile-TMG-Ile at pH values of 4 through 9 for a period of four weeks. To examine the morphologies of the formed assemblies, we conducted SEM and TEM microscopy. Figure 2 shows the SEM and TEM images of assemblies formed at varying pH. SEM analysis at pH 4 (Figure 2a) indicates the formation of short, thick nanofibers in the diameter range of 500 nm to 1 µm, while at pH 7 (Figure 2b) we observed the formation of long, multilayered nanoribbons several micrometers in length, with an average diameter of 500 nm to 1 µm. Figure 2c shows the structures of the assemblies formed under basic conditions (pH 9). Results indicated that under basic conditions, structures of a variety of shapes and sizes (spherical micelles, microtubes and fibers) in the range of 2 µm to 5 µm in diameter were formed. Corresponding TEM images, also indicated similar morphologies as shown in Figure 2d–f which correspond to assemblies formed at pH 4, 7 and 9 respectively. The sizes obtained by the TEM analysis, indicate that the average diameter of the nanofibers formed at pH 4 was found to be 20 nm, while those grown at pH 7 were in the size range of 500 nm to 1µm. The assemblies formed at pH 9 were found to be in the range of 200 nm to 500 nm. The size differences between TEM and SEM are attributed to the fact that the assemblies are intrinsically multiscale in nature and, vary in sizes. The TEM images display higher resolution and smaller sample sizes are most likely revealed by TEM analysis. Overall, self-assembly of Ile-TMG-Ile was found to be pH dependent.

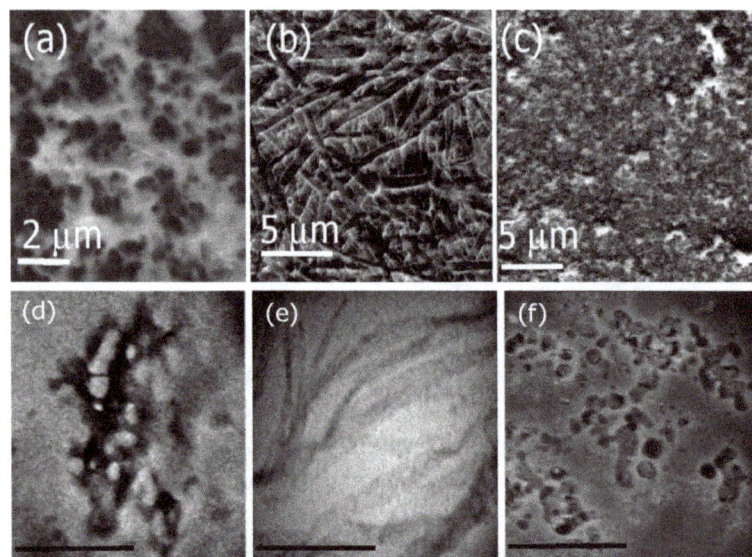

Figure 2. SEM images of Ile-TMG-Ile assemblies formed at (**a**) pH 4; (**b**) pH 7 and at (**c**) pH 9. TEM images of the assemblies are shown at (**d**) pH 4 (scale bar = 500 nm); (**e**) pH 7 (scale bar = 2 µm) and (**f**) pH 9 (scale bar = 1 µm nm).

The formation of short, thick nanofibers under acidic conditions is attributed to higher H-bonding interactions under acidic conditions as the carboxylic groups of the side chain isoleucines are likely to be protonated under those conditions. Additionally, assembly formation is promoted due to intermolecular H-bonding interactions between the –NH and O=C groups of the amide groups of the bolaamphiphile. Studies conducted previously with peptide amphiphiles such as bis(N-α-amido-glycylglycine)-1,7-heptane dicarboxylate have shown that at a pH range of 4 to 5, the formation of nanotubes is promoted due to higher H-bonding interactions and beta-sheet formation [52]. In the case of Ile-TMG-Ile, it is likely that beta-sheet formation is promoted, particularly due to the hydrophobicity of the Ile moieties which have been known to induce nanofiber formation [53]. Under neutral conditions, there appears to be a transition between beta-sheet structures to random coil due to changes in H-bonding interactions as the carboxyl groups are progressively deprotonated, resulting in the formation of uniform nanoribbons. Similar phenomena have been observed in the case of bola-glycolipids where changes in morphologies of supramolecular structures were observed, resulting in nanoribbon formation under neutral conditions. This was primarily attributed to a combination of hydrophobic interactions, chirality as well as well as changes in pH [54]. Under basic conditions, we observed a mixture of nanostructures as under those conditions the Ile-TMG-Ile bolaamphiphle is completely deprotonated and H-bonding is significantly diminished. Although C=O—NH amide H-bonding still exists under basic conditions, the carboxylate groups are negatively charged under those conditions and may result in repulsion between the negatively charged carboxylate groups. Thus, a variety of structures including micelles and few fibrillar structures are formed due to a combination of hydrophobic interactions, as well as amide-amide H-bonding and uniform assemblies are not formed.

We also monitored the growth of assemblies periodically using dynamic light scattering in all cases and over time. Results obtained after two weeks of growth are shown in Figure 3. As seen in the figure, the assemblies obtained were polydisperse. This is most likely because the assemblies are not uniform. They are mostly fibrillar, or ribbon shaped (in the case of assemblies grown at pH 4 and pH 7) or display a variety of morphologies as seen in the case of assemblies grown at pH 9. It is likely that aggregates of the assemblies at different stages of growth are observed. Overall, due to the formation of ribbon like structures under neutral conditions, we selected those assemblies for preparation of the scaffold.

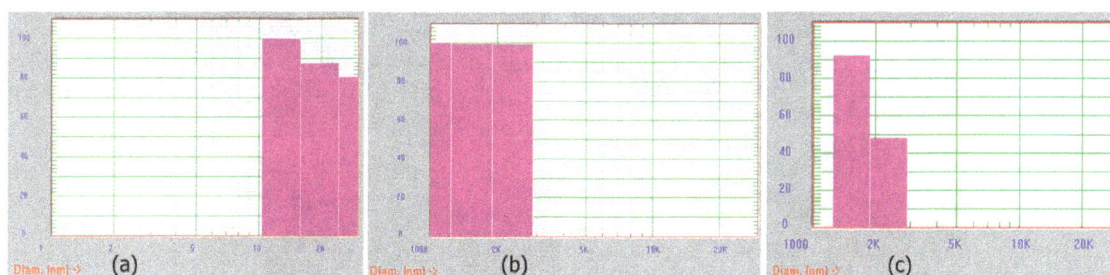

Figure 3. Dynamic light scattering analyses of assemblies formed at (**a**) pH 4; (**b**) pH 7 and (**c**) pH 9.

3.2. Functionalization of Ile-TMG-Ile and Preparation of Scaffold

To prepare the scaffolds that can be tailored for potential neural TE applications, we incorporated protein constituents that may aid in neural tissue regeneration due to their specific functional properties. The conjugation of each component was examined by SEM and TEM imaging as shown in Figure 4. We first conjugated the washed nanoribbons with laminin, a major component of the ECM of neural tissue. Upon incorporation of laminin, changes in morphology were observed (Figure 4a). The SEM image of laminin bound assemblies showed a relatively rough, gelatinous coating on the nanoribbons compared to the smooth surfaces in the absence of laminin as seen in Figure 2b. The corresponding TEM image (Figure 4e) showed a fibrous mesh like network upon conjugation with laminin. In general, laminin consists of both globular and rod-shaped domains and forms alpha-helical coiled coil structures [55]. Upon conjugation with the nanoribbons, laminin binds to nanoribbons, forming a gelatinous network intertwined with the nanoribbons. In general, laminin has been known to polymerize and form laminin networks, in cells due to interactions between the α-short arms of the amino terminal domain and β and γ-short arms of laminin [56]. It is likely that it wraps around Ile-TMG-Ile nanoribbon assemblies upon conjugation resulting in the mesh like networks.

Upon conjugation with artemin, further morphology changes were observed. In the SEM image (Figure 4b), we observed the incorporation of globular, rosette like structures throughout the gelatinous matrix. Similar structures were observed in the in the corresponding TEM image (Figure 4f). Previous studies have revealed that artemin monomers tend to self-assemble into rosette-like oligomers [57] and it is also known to be an exceptionally stable neurotrophic factor that can induce changes in the folding process of proteins as it functions as a molecular chaperone [58]. Thus, the morphology changes observed on the surfaces of the laminin bound nanoribbons further confirm the successful conjugation of artemin. We then conjugated the composite nanoribbons with Type IV collagen (Figure 4c,g). SEM and TEM images confirmed morphological changes after incorporation of Type IV collagen as the formation of large fibrillar mesh like structures were observed, integrated with rosette structures of artemin. In several studies, it has been shown that collagens tend to form long fibrillar structures, due to the formation of triple-helices and impart structural integrity to scaffolds [59]. Thus, our results confirm the formation of the biocomposite nanoribbons integrated with laminin, artemin and Type IV collagen. To impart electrical properties, essential for developing scaffolds for neural TE, we then incubated the conductive polymer polyaniline (PANI) with the biocomposite scaffolds. Previous studies have shown that PANI in the presence of dopants can self-assemble into nanostructures [60]. The presence of the amine groups of PANI, allow for electrostatic and H-bonding interactions between the carbonyl groups of the protein bound nanoribbons and PANI. Additionally, stacking interactions with the aromatic ring systems of PANI are also promoted between the proline and hydroxyproline moieties of Type IV Collagen. Distinct changes in morphology were observed upon incorporation of PANI into the biocomposite (Figure 4d,h) showing the formation of aggregates of PANI on the scaffold indicating its successful assimilation.

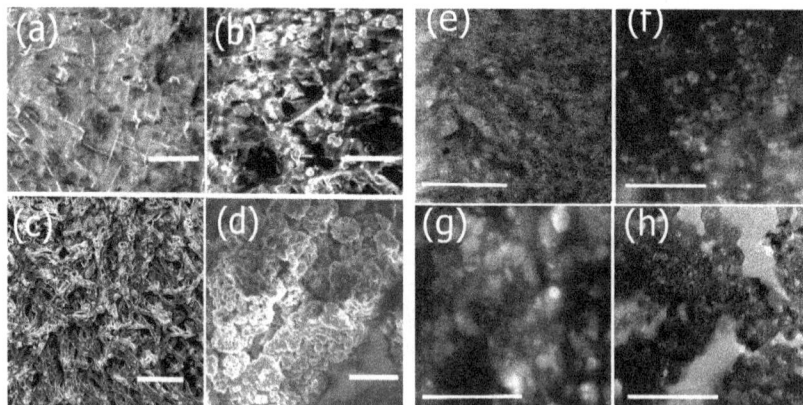

Figure 4. SEM and TEM images showing morphology changes after functionalization with each component. (**a**) SEM image of nanoribbons functionalized with laminin (scale bar = 2 μm); (**b**) SEM image showing subsequent conjugation with Artemin (scale bar = 2 μm); (**c**) SEM image after incorporation of Type IV collagen to laminin-artemin-bound nanoribbons (scale bar = 5 μm); (**d**) SEM image after incorporation of PANI to the functionalized biocomposite (scale bar = 5 μm). (**e**) TEM image of nanoribbons functionalized with laminin (scale bar = 2 μm); (**f**) TEM image showing subsequent conjugation with Artemin (scale bar = 2 μm); (**g**) TEM image after incorporation of Type IV collagen to laminin-artemin-bound nanoribbons (scale bar = 3 μm); (**h**) TEM image after incorporation of PANI to the functionalized biocomposite (scale bar = 3 μm).

3.3. FTIR Spectroscopy

To further confirm the formation of the scaffold, we conducted FTIR spectroscopy (Figure 5). As shown in Figure 5a, the self-assembled template nanoribbons showed characteristic peaks in the amide I region at 1658 cm^{-1} and at 1650 cm^{-1} with a shoulder at 1634 cm^{-1} along with peaks at 1450 cm^{-1} and 1413 cm^{-1} in the amide II region. These peaks are indicative of formation of a mix of random coil, alpha helical and beta-sheet structures [61] that resulted in the formation of nanoribbons. Previous studies using protein analogs with C terminus isoleucine are consistent with these findings of mostly alpha helical and random coil structures [62]. Additionally, a strong peak was observed at 1286 cm^{-1} and at 1075 cm^{-1}, attributed to C-O and C-H stretching respectively. Upon conjugation with laminin (Figure 5b), the amide I peaks were observed at 1656 cm^{-1} and at 1628 cm^{-1} indicating increased beta-sheet formation along with the presence of alpha helices. The amide II peaks were observed at 1550 cm^{-1} and at 1519 cm^{-1} while the C-O and C-H stretching peaks were seen at 1295 cm^{-1} and at 1059 cm^{-1} respectively [63]. It has been reported that laminins generally tend to polymerize into sheet like structures [64], which is consistent with our results. Furthermore, similar shifts were also seen after incorporation of laminin onto a poly(l-lactide-co- glycolide) scaffold, indicating successful conjugation of laminin with the nanoribbons [65]. After conjugation with artemin (Figure 5c), further shifts were observed. The amide I band was shifted to 1632 cm^{-1} with a shoulder at 1662 cm^{-1} while the amide II peaks shifted to 1535 cm^{-1} and 1514 cm^{-1}. The C-O and C-H stretching peaks were observed at 1299 cm^{-1} and at 1064 cm^{-1} with a shoulder at 1054 cm^{-1} respectively. These peaks are indicative of increase in beta-sheets along with the appearance of beta-turn structure.

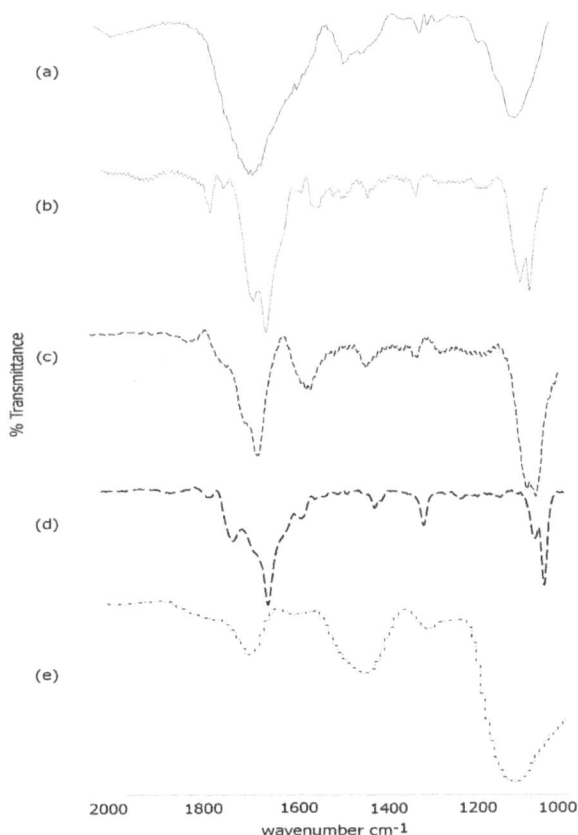

Figure 5. FTIR spectra of (**a**) self-assembled Ile-TMG-Ile nanoribbons; (**b**) Nanoribbons bound to laminin; (**c**) Artemin conjugated with laminin bound nanoribbons; (**d**) Type IV collagen conjugated with artemin-laminin bound nanoribbons; (**e**) PANI bound to Type IV collagen-artemin-laminin bound nanoribbons.

Upon incorporation of Type IV collagen (Figure 5d), the FTIR spectra showed peaks in the amide I region at 1629 cm^{-1} with a shoulder at 1652 cm^{-1} and a peak at 1699 cm^{-1}. The amide II peak was found to be at 1558 cm^{-1} while the C-O and C-H stretching peaks were observed at 1293 cm^{-1} and at 1035 cm^{-1} respectively. These changes confirm the incorporation of Type IV collagen. Furthermore, the secondary structure reveals the presence of alpha-helical content along with beta-strands and β and γ-turns [66] due to blending of Type IV collagen with artemin-laminin bound nanoribbons. Distinct changes were observed upon incorporation of PANI. As seen in Figure 4e, peaks were found at 1680 cm^{-1}, 1560 cm^{-1} and at 1430cm^{-1}, 1280 cm^{-1} and at 1112 cm^{-1}. The peaks at 1630 cm^{-1} at 1450 cm^{-1} are indicative of vibrations from quinoid rings and benzenoid ring systems as seen in PANI bound polystyrene nanocomposites [67]. These results further confirm the formation of the composite scaffold.

3.4. Fluorescence Microscopy

In order to confirm that the proteins retained biological activity after conjugation with the assemblies, we examined the binding affinity of the laminin bound assemblies as well as the Type IV collagen bound biocomposites with corresponding antibodies. We used FITC labeled laminin -2 antibody (B-4) for laminin bound assemblies and rhodamine labeled anti-collagen (Type IV) antibody for Type IV collagen bound biocomposites respectively and examined the interactions using fluorescence microscopy. In previous work, it has been shown that conjugation of different biological moieties including specific peptide sequences such as RGD, TAT peptides, or proteins such as transferrin, antibodies with nanomaterials not only increases the stability of the proteins but also enhances the applications of the nanomaterials themselves [68]. Furthermore, such conjugations allow for interactions with cell receptors or biological membranes, thereby promoting the use of such materials for a variety of biomedical applications. As shown in Figure 6, FITC conjugated anti-laminin bound antibodies efficiently bound to the laminin bound assemblies, (Figure 6a) and rhodamine conjugated anti-collagen IV antibodies bound to the biocomposite (Figure 6b). These results confirmed that laminin and Collagen IV retained their biological activity upon binding with the assemblies.

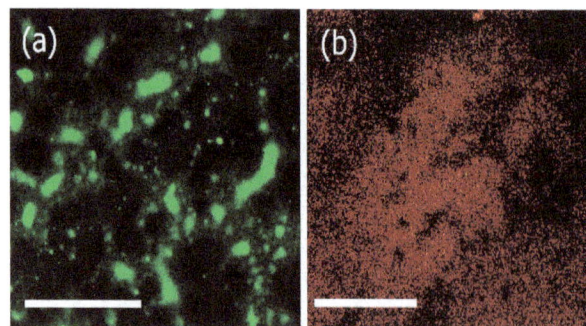

Figure 6. Fluorescence microscopy images of (**a**) FITC labeled laminin α-2 antibody bound to laminin bound assemblies; (**b**) Rhodamine labeled collagen IV antibody bound to biocomposite assemblies (collagen IV-artemin-laminin bound Ile-TMG-Ile). Scale bars = 20 μm.

3.5. Differential Scanning Calorimetry

In order to explore the thermal properties, we examined phase changes of the assemblies before and after conjugation with each component utilized in the formation of the scaffold (Figure 7). Short endothermic peaks were observed in the case of the Ile-TMG-Ile assemblies (Figure 7a) at 14.9 °C and at 36.7 °C followed by another endothermic peak at 99.6 °C, due to loss of loosely bound water. After functionalization with laminin, (Figure 7b) a large, broad endothermic peak was observed in the temperature range of 50 °C to 100 °C, followed by shallow endothermic peaks at 172.4 °C and at 232.1 °C. The significantly broad endothermic peak at the lower temperature is indicative loss of free water, due to the presence of hydrophilic amino acids in the protein. This is primarily due to the fact that hydrophilic groups can cause significant hydration and upon heating, changes in H-bonding interactions and consequently conformation changes in the laminin bound nanoribbons occur. This is a common occurrence in self-assembled peptides and proteins due to the rearrangement of components because of changes in inter and intra-molecular interactions [69]. The short peak at 171.4 °C is likely due to thermal melting, followed by crystallization at a higher temperature (232.1 °C). After further functionalization with artemin (Figure 7c) a similar broad endothermic peak was observed in the temperature range of 50 °C to 100 °C, (though the intensity was lower). Slight shifts were observed in the higher temperature range, with the thermal melting appearing at 172.6 °C and the subsequent short crystallization peak was observed at 242.1 °C. Upon binding with Type IV collagen, (Figure 7d) the composite once again showed a broad endothermic peak, between 50 °C to 100 °C, though the intensity of the peak was lesser than the previous layer, due to higher cross-linking in the presence of collagen and notably, peaks at higher temperature were diminished.

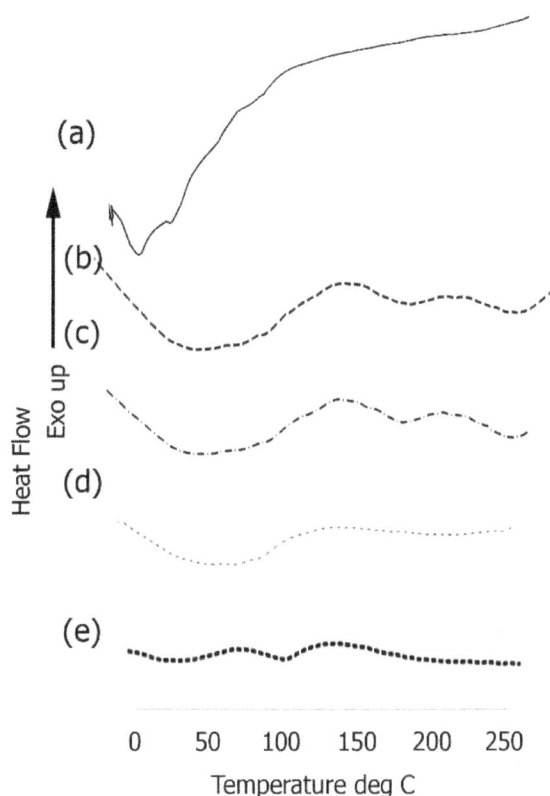

Figure 7. DSC thermograms of (**a**) self-assembled Ile-TMG-Ile nanoribbons; (**b**) Nanoribbons bound to laminin; (**c**) Artemin conjugated with laminin bound nanoribbons; (**d**) Type IV collagen conjugated with artemin-laminin bound nanoribbons; (**e**) PANI bound to Type IV collagen-artemin-laminin bound nanoribbons.

However, a significant change was observed upon incorporation of PANI (Figure 7e), where in the intensity of the broad endothermic peaks seen in the protein bound scaffold was significantly reduced. Relatively short endothermic peaks are observed at 52 °C and at 102 °C due loss of loosely bound water and no other significant peaks are observed. PANI is significantly hydrophobic compared to the proteinaceous components, resulting in shallow peaks due to loss of free unbound water. These results further confirm the integration of PANI.

3.6. Mechanical and Surface Properties

It is paramount that the designed scaffolds should be able to bear force loads in order to adequately support seeded cells and potentially boost the formation of new tissue. We utilized peak force microscopy to determine the mechanical properties of the scaffold. In general, nanoindentation was carried out using AFM to examine the changes in attractive and repulsive forces and the depth of indentation as the tip of the cantilever contacts the sample and is deformed. At least three to five points were selected for each sample to obtain force vs. separation curves and the data obtained was fit into Hertzian model to obtain the Young's modulus. The results obtained for average Young's Modulus values after incorporation of each of the components of the scaffold are shown in Table 1.

Table 1. Youngs Modulus values of constructs after consecutive incorporation of each layer.

Construct	Young's Modulus (GPa)
Self-Assembled nanoribbons	0.757
Nanoribbons bound to Laminin	1.805
Nanoribbons bound to Laminin-Artemin-Collagen	2.985
Nanoribbons bound to Laminin-Artemin-Collagen-PANI	5.522

These results indicate that as each protein layer was conjugated with the nanoribbons, the Young's Modulus (YM) was found to increase, demonstrating that each protein component consecutively increased the stiffness and mechanical strength of the scaffold. The YM values obtained were for dried scaffolds alone and are most likely higher than those one would expect for wet scaffolds, under in vivo or in vitro conditions. Previous nanoindentation studies have shown that native single collagen fibrils have an YM value in the range of 1 to 2 GPa [70]. It has also been reported that protein structures and self-assembled collagen based constructs designed to have properties to mimic the extracellular matrix display Young's Moduli averaging 1.2 GPa [71]. We also determined the Young's Modulus of control Type IV collagen by nanoindentation and the YM was found to be 532.2 ± 3 MPa, while control PANI films were found to have a YM of 1.52 ± 5 GPa. Thus, our results indicate that for the protein functionalized nanoribbon biocomposite, the overall mechanical strength of the scaffold increases due to the multi-component nature of the scaffold. Furthermore, it was found that the highest YM value was obtained after incorporation of PANI. A comparison of AFM images of scaffolds before and after incorporation of PANI and the corresponding force curves are shown in Figure 8.

AFM topography images indicate that the surface of the bicomposite before binding to PANI (Figure 8a) shows a more fibrillar structure, due to top layer of the scaffold being Type IV collagen, while after binding to PANI, more globular structures were observed to be deposited on the scaffold (Figure 8b). These results indicate changes in surface roughness and morphology of the scaffold after incorporation of PANI. Additionally, the force-curves also showed significant changes upon binding with PANI (Figure 8c,d). These results are consistent with previous nanoindentation studies, where it has been demonstrated that when PANI was deposited on vertical arrays of carbon nanotubes (CNTs), the Young's Modulus value dramatically increased due to strong electrostatic and π–π stacking interactions with CNTs [72]. It is expected that similar interactions occur between the biocomposite and PANI that leads to a higher YM. In general, the Young's Modulus values of pyrrolidinone and polyaniline films have been reported to be in the range of 200 MPa–5 GPa [73]. Thus, our results after incorporation of PANI with the biocomposite scaffold are within the values reported previously in the literature.

We also probed the changes in the surface roughness of the scaffold before and after incorporation of PANI. In general, it is known that surface roughness plays a key role in cell adhesion of scaffolds [74] For instance, it has been shown that silk fibroin bound PLA fibers with higher surface roughness promoted increased growth and adhesion of osteoblast cells [75]. For the biocomposite scaffold before incorporation of PANI, the average surface roughness (Ra) was found to be 154 nm, while the maximum roughness (Rmax) was determined to be 209 nm. We found that incorporation of PANI resulted in a significant increase in the surface roughness. The Ra was found to be 377 nm and the Rmax was found to be 2065 nm, further confirming the formation of the composite.

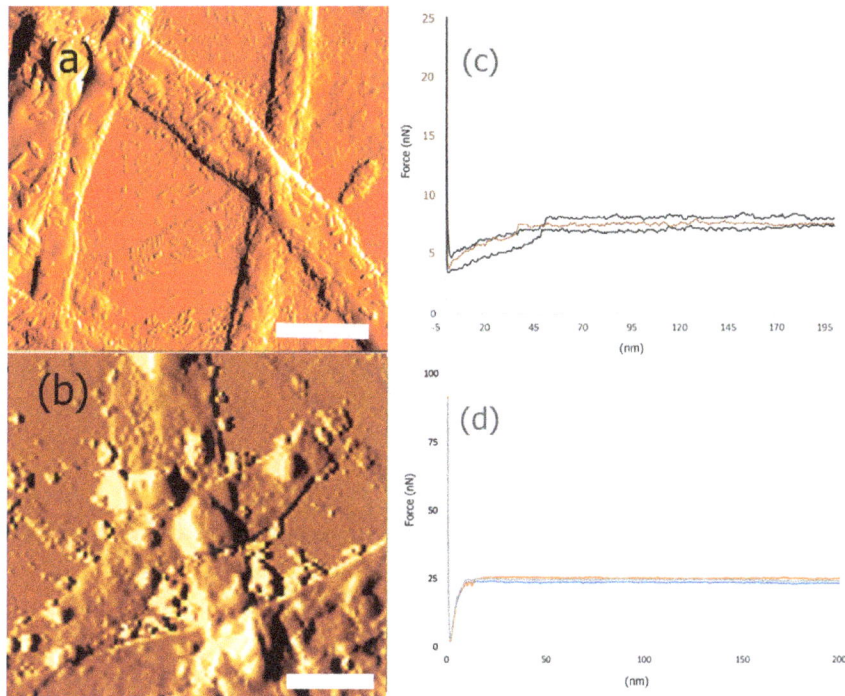

Figure 8. AFM amplitude image of (**a**) biocomposite nanoribbons bound to laminin, artemin and collagen. Scale Bar = 10 μm; (**b**) nanoribbons bound to laminin, artemin and collagen and PANI. Scale Bar = 5 μm. (**c**) Force curves obtained for biocomposite nanoribbons bound to laminin, artemin and collagen; (**d**) Force curves obtained for nanoribbons bound to laminin, artemin and collagen and PANI.

3.7. In Vitro Biodegradability Studies

When developing a scaffold for tissue engineering, the biodegradability of a scaffold plays a vital role. It has been reported that biodegradability promotes growth and proliferation of cells, as well as production of native ECM [76]. In comparison to non-biodegradable scaffolds, biodegradable composites are able to aid in the growth of new tissues at a highly expedited rate due to the increased proliferation and lack of hindrance for cell growth [77]. We examined the biodegradability of the formed scaffold in simulated body fluid buffer in order to mimic in vivo conditions. Figure 9 shows the results obtained over a period of three weeks in simulated body fluid. The scaffold shows a mass loss of 48.7% after 22 days. Overall, the scaffold showed degradation at a moderate rate, which is preferential to highly vascular nervous tissue [78]. This result is consistent with other polymer and basement membrane mimicking scaffolds for tissue engineering; such as, poly(3-hydroxybutyric acid) with chitin and chitosan which demonstrated biodegradability from 45–70% and collagen, hyaluronic acid and gelatin scaffolds which demonstrated 45% biodegradability [79].

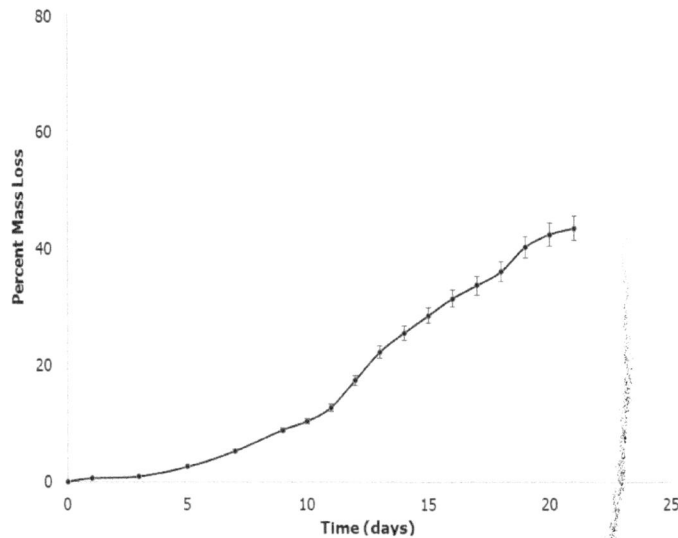

Figure 9. Biodegradation studies of the scaffolds showing mass loss over a period of 20 days.

3.8. Cell Studies

To examine if the formed scaffolds would be suitable for applications in neural TE, we conducted cell viability studies with rat neural cortical cells. We examined the effects of scaffolds before and after incorporation of PANI as shown in Figure 10. The data shown are representative of results obtained in the presence of 10 μM scaffolds in comparison with control untreated cells. Our results indicate that the cells continued to proliferate over a period of 72 h in the presence of scaffolds. Although cell proliferation continued over time, the rate of cell growth for the PANI bound construct was lower compared to the controls due to the known cytotoxicity of PANI [80]. This is most likely that proliferation continued due to the synergistic effects of the protein components of the scaffolds, which play a role in enhancing biocompatibility of the PANI bound scaffold. Previous studies have shown that laminin bound nanofibers significantly increased the attachment and neurite extension of cells in vitro [81] while artemin promotes axonal growth in damaged neural tissue, as well as regeneration of sensory neurons [82]. Other research has shown that after damage, artemin increases survivability of injured neurons [83]. The incorporation of Type IV collagen increases the influx of nutrients to growing and damaged neural tissue, increasing new growth and proliferation. We also studied the growth and proliferation in the presence of 6 μM and 13 μM scaffolds which showed similar trends of cell viability (data not shown). In general, no significant differences were observed in the proliferation at varying concentrations of the scaffold. These results indicated growth and proliferation of cortical cells continued and were comparative to the control throughout various time periods and concentrations of the construct.

To further investigate the cell proliferation of neural cortical cells and morphologies in the presence and absence of PANI bound scaffolds, we conducted phase contrast optical microscopy studies. The results obtained are shown in Figure 11. As seen in the figure, there was a major difference in the morphologies of the cells grown in the presence of PANI bound scaffolds over a period of seven days in comparison to those after 48 h. As shown in Figure 11a,b, in the absence of scaffold, cells continued to proliferate over time. However, we did not observe cluster formation or axonal outgrowths. Upon incubation with PANI bound scaffolds. After 48 h (Figure 11c), the cells appeared to relatively more elongated compared to controls However after seven days, we observed cluster formation and axonal outgrowth (Figure 11d). These results indicated that the cells efficiently continued to proliferate in the presence of the PANI bound scaffolds and enhanced neural cell growth as well as cell-cell adhesion, further confirming that the scaffolds were conducive to forming cell-scaffold matrices and provide an environment for the growth and support of neural cortical cells.

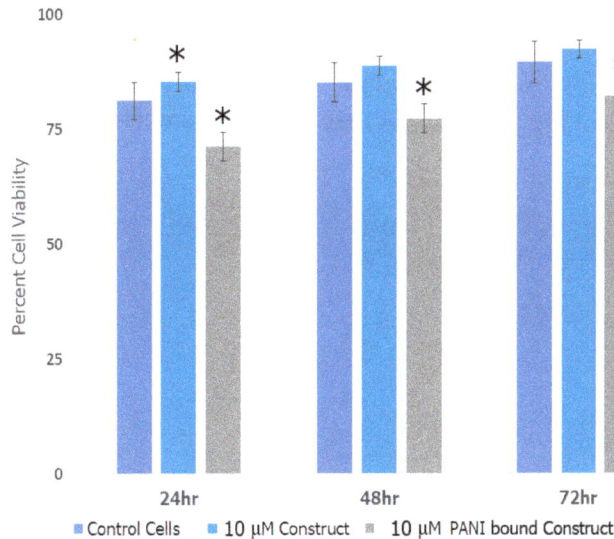

Figure 10. Cortical cell viability in the presence of scaffold constructs before and after incorporation of PANI. (* = p < 0.05 was determined to be statistically significant).

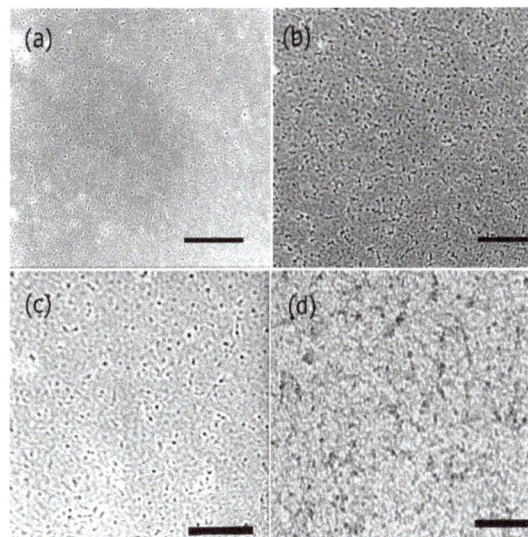

Figure 11. Phase contrast microscopy images showing the growth of neural cortical cells (**a**) control cells after 24 h; (**b**) control cells after 72 h; (**c**) cells with PANI bound scaffolds after 48 h and (**d**) cells with PANI bound scaffolds after 7 days. Scale bars: (**a**) 50 μm; (**b**) 50 μm; (**c**) 30 μm; (**d**) 50 μm.

3.9. Cyclic Voltammetry

To examine the electrochemical properties of the PANI bound scaffold, cyclic voltammetry was conducted. Prior to performing the experiment, the control PANI or the scaffold bound PANI were dried onto the platinum electrodes in vacuum overnight. After connecting the electrodes to the potentiostat, nitrogen was bubbled into the HCl cell solution to remove oxygen from the solution. Potential between -0.2 V to 0.9 V was applied at 10 mV/s to obtain I-V curves. At a voltage of 0.1 mV, an anodic oxidation peak was observed of approximately 2.6×10^{-7} Amps current, with a corresponding cathodic reduction peak at 0.27 mV of approximately -2.85×10^{-7} Amps current in the case of the control PANI. Cyclic voltammetry was then conducted with scaffold bound PANI. Our results indicated that oxidation peaks were observed at 0.213 mV at 1.13×10^{-5} amps current and at 0.56 mV at 1.12×10^{-5} Amps which correspond to the lecuoemeraldine-emeraldine and emeraldine-pernigraniline oxidation processes [84]. The reduction peaks were observed at cathodic currents of -1.39×10^{-5} Amps and at -1.54×10^{-5} Amps at 0.67 mV and at 0.32 mV respectively. These results are shown in Figure 12. Similar cyclic voltagrams have been observed for electrospun polycaprolactone and polyaniline fibers used in skeletal tissue engineering [85].

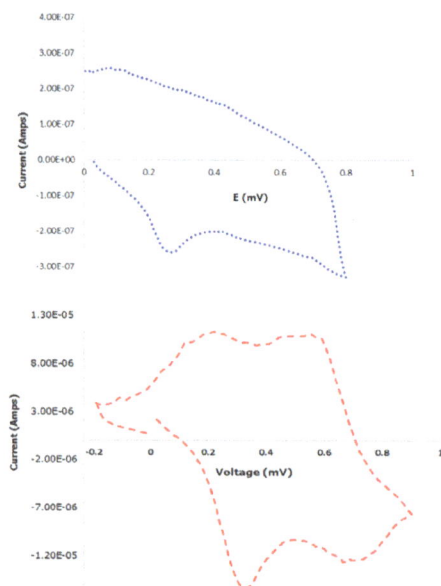

Figure 12. Comparison of cyclic voltagrams of control PANI (top) and PANI bound scaffold (bottom).

Cyclic voltammetry of polyaniline-carbon nanotubes also yielded similar voltammograms [86]. We also compared the cyclic voltagrams with scaffolds formed before incorporation of PANI which did not show any peaks (data not shown). To further examine the electrical properties of the scaffolds, we measured the resistance of the scaffolds to determine the conductivity. Our results showed that the PANI bound scaffolds displayed a conductance of 1.5×10^{-3} S/cm; while the scaffolds in the absence of PANI displayed zero conductivity. Control PANI displayed a conductivity of 2.0×10^{-3} S/cm, which was higher than the PANI bound scaffolds. Overall, these results are indicative that upon binding to PANI the scaffolds display electrical properties, compared to absence of PANI, which is essential for scaffolds for neural tissue regeneration.

4. Conclusions

In this work, 3,3 tetramethyleneglutaric acid and isoleucine were coupled and allowed to self-assemble into a nanoribbon matrix. The nanoribbons were then functionalized with laminin, a primary component in the neural cell basement membrane, along with artemin a glial cell line derived neurotropic factor and Type IV collagen another key component of the ECM of neural tissues. To impart electrical properties, we then incorporated polyaniline a conductive polymer into the scaffold matrix. Peak force microscopy studies revealed that the scaffolds had high mechanical strength and the Young's Modulus increased with conjugation with each protein layer. The scaffolds also displayed biodegradability. Furthermore, the scaffolds were found to promote cell proliferation and encouraged neurite outgrowths. Although cell proliferation was relatively lower for the PANI bound scaffolds, compared to scaffolds without PANI, our results indicated that the biological components of the scaffold overall aided in cell growth. Additionally, cyclic voltammetry conducted showed that PANI bound scaffolds displayed electrochemical properties. Thus, we have created supramolecular composite scaffolds that may be have applications in neural tissue engineering.

Acknowledgments: Andrew M. Smith and Harrison T. Pajovich would like to thank Fordham University Research Grants for financial support of this work. Ipsita A. Banerjee thanks NSF MRI Grant No. 1626378 for support of this work. The authors would also like to thank McMahon for his suggestions and help with the conductivity studies and Fath at the Queens College core facility for Imaging, Cellular and Molecular Biology for the use of the transmission electron microscope

Author Contributions: Andrew M. Smith synthesized the scaffolds, performed experiments and worked on data acquisition for the various studies conducted. Andrew M. Smith was also involved in writing the initial drafts. Harrison T. Pajovich performed some of the analytical experiments. Ipsita A. Banerjee conceived and designed the experiments, analyzed the data and wrote the paper.

Conflicts of Interest: The authors declare no conflict of interest. The founding sponsors had no role in the design of the study; in the collection, analyses, or interpretation of data; in the writing of the manuscript and in the decision to publish the results.

References

1. Saaty, T. Neurons the decision makers, Part I: The firing function of a single neuron. *Neural Netw.* **2017**, *86*, 102–114. [CrossRef] [PubMed]

2. Hammond, C.; Cayre, M.; Panatier, A.; Avignone, E. Neuron-glial cell cooperation. In *Cellular and Molecular Neurophysiology*, 4th ed.; Academic Press: Cambridge, MA, USA, 2015; pp. 25–37. ISBN 9780123970329.

3. Katsu-Jiménez, Y.; Alves, R.M.P.; Giménez-Cassina, A. Food for thought: Impact of metabolism on neuronal excitability. *Exp. Cell Res.* **2017**, *36*, 41–46. [CrossRef] [PubMed]

4. Gupta, M.K.; Jayaram, S.; Madugundu, A.K.; Chavan, S.; Advani, J.; Pandey, A.; Thongboonkerd, V.; Sirdeshmukh, R. Chromosome-centric human proteome project: Deciphering proteins associated with glioma and neurodegenerative disorders on chromosome 12. *J. Proteome Res.* **2014**, *13*, 3178–3190. [CrossRef] [PubMed]

5. Yu, X.; Bellamkonda, R.V. Tissue-Engineered scaffolds are Effective alternatives to autografts for bridging peripheral nerve gaps. *Tissue Eng.* **2003**, *9*, 421–430. [CrossRef] [PubMed]

6. Otto, R.; Penzis, R.; Gaube, F.; Adolph, O.; Föhr, K.J.; Warncke, P.; Robaa, D.; Appenroth, D.; Fleck, C.; Enzensperger, C.; et al. Evaluation of homobivalent carbolines as designed multiple ligands for the treatment of neurodegenerative disorders. *J. Med. Chem.* **2015**, *58*, 6710–6715. [CrossRef] [PubMed]

7. Carriel, V.; Alaminos, M.; Garzón, I.; Campos, A.; Cornelissen, M. Tissue engineering of the peripheral nervous system. *Expert Rev. Neurother.* **2014**, *14*, 301–318. [CrossRef] [PubMed]

8. Williams, D. Benefit and risk in tissue engineering. *Mater. Today* **2004**, *7*, 24–29. [CrossRef]

9. Evans, N.D.; Gentleman, E.; Polak, J.M. Scaffolds for stem cells. *Mater. Today* **2006**, *9*, 26–33. [CrossRef]

10. Gu, X.; Ding, F.; Williams, D.F. Neural tissue engineering options for peripheral nerve regeneration. *Biomaterials* **2014**, *35*, 6143–6156. [CrossRef] [PubMed]

11. Koss, K.M.; Unsworth, L.D. Neural tissue engineering: Bioresponsive nanoscaffolds using engineered self-assembling peptides. *Acta Biomater.* **2016**, *44*, 2–15. [CrossRef] [PubMed]

12. Skop, N.B.; Calderon, F.; Cho, C.H.; Gandhi, C.D.; Levison, S.W. Improvements in biomaterial matrices for neural precursor cell transplantation. *Mol. Cell. Ther.* **2014**, *2*, 19. [CrossRef] [PubMed]

13. Tanaka, M.; Sato, Y.; Haniu, H.; Nomura, H.; Kobayashi, S.; Takanashi, S.; Okamoto, M.; Takizawa, T.; Aoki, K.; Usui, Y.; et al. A three-dimensional block structure consisting exclusively of carbon nanotubes serving as bone regeneration scaffold and as bone defect filler. *PLoS ONE* **2017**, *12*, e0172601. [CrossRef] [PubMed]

14. Perkins, B.L.; Naderi, N. Carbon nanostructures in bone tissue engineering. *Open Orthop. J.* **2016**, *10*, 877–899. [CrossRef] [PubMed]

15. Palejwala, A.H.; Fridley, J.S.; Mata, J.A.; Samuel, E.L.G.; Luerssen, T.G.; Perlaky, L.; Kent, T.A.; Tour, J.M.; Jea, A. Biocompatibility of reduced graphene oxide nanoscaffolds following acute spinal cord injury in rats. *Surg. Neurol. Int.* **2016**, *7*, 75. [CrossRef] [PubMed]

16. Kohane, D.S.; Langer, R. Polymeric biomaterials in tissue engineering. *Pediatr. Res.* **2008**, *63*, 487–491. [CrossRef] [PubMed]

17. Chen, G.; Ushida, T.; Tateishi, T. Scaffold design for tissue engineering. *Macromol. Biosci.* **2002**, *2*, 67–77. [CrossRef]

18. Fukunishi, T.; Shoji, T.; Shinoka, T. Nanofiber composites in vascular tissue engineering. In *Nanofiber Composites for Biomedical Applications*; Woodhead Publishing: Cambridge, UK, 2017; pp. 455–481. ISBN 9780081001738.

19. Alvarez-Perez, M.A.; Guarino, V.; Cirillo, V.; Ambrosio, L. Influence of gelatin cues in PCL electrospun membranes on nerve outgrowth. *Biomacromolecules* **2010**, *11*, 2238–2246. [CrossRef] [PubMed]

20. Zhang, Z.; Xu, R.; Wang, Z.; Dong, M.; Cui, B.; Chen, M. Visible-Light neural stimulation on graphitic-carbon nitride/graphene photocatalytic fibers. *ACS Appl. Mater. Interfaces* **2017**, *9*, 34736–34743. [CrossRef] [PubMed]

21. Hemshekhar, M.; Thushara, R.M.; Chandranayaka, S.; Sherman, L.S.; Kemparaju, K.; Girish, K.S. Emerging roles of hyaluronic acid bioscaffolds in tissue engineering and regenerative medicine. *Int. J. Biol. Macromol.* **2016**, *86*, 917–928. [CrossRef] [PubMed]

22. Butterfield, K.C.; Conovaloff, A.W.; Panitch, A. Development of affinity-based delivery of NGF from a chondroitin sulfate biomaterial. *Biomatter* **2011**, *1*, 174–181. [CrossRef] [PubMed]

23. Weyers, A.; Linhardt, R.J. Neoproteoglycans in tissue engineering. *FEBS J.* **2013**, *280*, 2511–2522. [CrossRef] [PubMed]

24. Lei, J.; Yuan, Y.; Lyu, Z.; Wang, M.; Liu, Q.; Wang, H.; Yuan, L.; Chen, H. Deciphering the role of sulfonated unit in heparin-mimicking polymer to promote neural differentiation of embryonic stem cells. *ACS Appl. Mater. Interfaces* **2017**, *9*, 28209–28221. [CrossRef] [PubMed]

25. Zhou, K.; Thouas, G.A.; Bernard, C.C.; Nisbet, D.R.; Finkelstein, D.I.; Li, D.; Forsythe, J.S. Method to impart electro- and biofunctionality to neural scaffolds using graphene-polyelectrolyte multilayers. *ACS Appl. Mater. Interfaces* **2012**, *4*, 4524–4531. [CrossRef] [PubMed]

26. Cao, Z.; Gilbert, R.J.; He, W. Simple agarose-chitosan gel composite system for enhanced neuronal growth in three dimensions. *Biomacromolecules* **2009**, *10*, 2954–2959. [CrossRef] [PubMed]

27. Entekhabi, E.; Haghbin Nazarpak, M.; Moztarzadeh, F.; Sadeghi, A. Design and manufacture of neural tissue engineering scaffolds using hyaluronic acid and polycaprolactone nanofibers with controlled porosity. *Mater. Sci. Eng. C* **2016**, *69*, 380–387. [CrossRef] [PubMed]

28. Rad-Malekshahi, M.; Lempsink, L.; Amidi, M.; Hennink, W.E.; Mastrobattista, E. Biomedical Applications of Self-Assembling Peptides. *Bioconjug. Chem.* **2016**, *27*, 3–18. [CrossRef] [PubMed]

29. Mazza, M.; Hadjidemetriou, M.; De Lázaro, I.; Bussy, C.; Kostarelos, K. Peptide nanofiber complexes with siRNA for deep brain gene silencing by stereotactic neurosurgery. *ACS Nano* **2015**, *9*, 1137–1149. [CrossRef] [PubMed]

30. Stephanopoulos, N.; Freeman, R.; North, H.A.; Sur, S.; Jeong, S.J.; Tantakitti, F.; Kessler, J.A.; Stupp, S.I. Bioactive DNA-peptide nanotubes enhance the differentiation of neural stem cells into neurons. *Nano Lett.* **2015**, *15*, 603–609. [CrossRef] [PubMed]

31. Ellis-Behnke, R.G.; Liang, Y.-X.; You, S.-W.; Tay, D.K.C.; Zhang, S.; So, K.-F.; Schneider, G.E. Nano neuro knitting: peptide nanofiber scaffold for brain repair and axon regeneration with functional return of vision. *Proc. Natl. Acad. Sci. USA* **2006**, *103*, 5054–5059. [CrossRef] [PubMed]

32. Sur, S.; Pashuck, E.T.; Guler, M.O.; Ito, M.; Stupp, S.I.; Launey, T. A hybrid nanofiber matrix to control the survival and maturation of brain neurons. *Biomaterials* **2012**, *33*, 545–555. [CrossRef] [PubMed]

33. Hutton, J.C.; Schofield, P.J.; Williams, J.F.; Hollows, F.C. The failure of aldose reductase inhibitor 3,3'-tetramethylene glutaric acid to inhibit in vivo sorbitol accumulation in lens and retina in diabetes. *Biochem. Pharmacol.* **1974**, *23*, 2991–2998. [CrossRef]

34. Rosenberyg, D.; Artoul, S.; Segal, A.C.; Kolodney, G.; Radzishevsky, I.; Dikpoltsev, E.; Foltyn, V.N.; Inoue, R.; Mori, H.; Billard, J.-M.; et al. Neuronal D-Serine and Glycine Release Via the Asc-1 Transporter Regulates NMDA Receptor-Dependent Synaptic Activity. *J. Neurosci.* **2013**, *33*, 3533–3544. [CrossRef] [PubMed]

35. Arulmoli, J.; Pathak, M.M.; McDonnell, L.P.; Nourse, J.L.; Tombola, F.; Earthman, J.C.; Flanagan, L.A. Static stretch affects neural stem cell differentiation in an extracellular matrix-dependent manner. *Sci. Rep.* **2015**, *5*, 8499. [CrossRef] [PubMed]

36. Joo, S.; Yeon Kim, J.; Lee, E.; Hong, N.; Sun, W.; Nam, Y. Effects of ECM protein micropatterns on the migration and differentiation of adult neural stem cells. *Sci. Rep.* **2015**, *5*, 13043. [CrossRef] [PubMed]

37. Junka, R.; Valmikinathan, C.M.; Kalyon, D.M.; Yu, X. Laminin functionalized biomimetic nanofibers for nerve tissue engineering. *J. Biomater. Tissue Eng.* **2013**, *3*, 494–502. [CrossRef] [PubMed]

38. Tate, C.C.; Shear, D.A.; Tate, M.C.; Archer, D.R.; Stein, D.G.; LaPlaca, M.C. Laminin and fibronectin scaffolds enhance neural stem cell transplantation into the injured brain. *J. Tissue Eng. Regen. Med.* **2009**, *3*, 208–217. [CrossRef] [PubMed]

39. Baloh, R.H.; Tansey, M.G.; Lampe, P.A.; Fahrner, T.J.; Enomoto, H.; Simburger, K.S.; Leitner, M.L.; Araki, T.; Johnson, E.M.; Milbrandt, J. Artemin, a novel member of the GDNF ligand family, supports peripheral and central neurons and signals through the GFRα3-RET receptor complex. *Neuron* **1998**, *21*, 1291–1302. [CrossRef]

40. Detloff, M.R.; Smith, E.J.; Quiros Molina, D.; Ganzer, P.D.; Houlé, J.D. Acute exercise prevents the development of neuropathic pain and the sprouting of non-peptidergic (GDNF- and artemin-responsive) c-fibers after spinal cord injury. *Exp. Neurol.* **2014**, *255*, 38–48. [CrossRef] [PubMed]

41. Allen, S.J.; Watson, J.J.; Shoemark, D.K.; Barua, N.U.; Patel, N.K. GDNF, NGF and BDNF as therapeutic options for neurodegeneration. *Pharmacol. Ther.* **2013**, *138*, 155–175. [CrossRef] [PubMed]

42. Timpl, R.; Wiedmann, H.; van Delden, V.; Furthmayr, H.; Kühn, K. A network model for the organization of type IV collagen molecules in basement membranes. *Eur. J. Biochem.* **1981**, *120*, 203–211. [CrossRef] [PubMed]

43. Boeva, Z.A.; Sergeyev, V.G. Polyaniline: Synthesis, properties and application. *Polym. Sci. Ser. C* **2014**, *56*, 144–153. [CrossRef]

44. Magnuson, M.; Guo, J.-H.; Butorin, S.M.; Agui, A.; Sathe, C.; Nordgren, J. The electronic structure of polyaniline and doped phases studied by soft X-ray absorption and emission spectroscopies. *J. Chem. Phys.* **1999**, *111*, 4756–4763. [CrossRef]

45. Li, M.; Guo, Y.; Wei, Y.; MacDiarmid, A.G.; Lelkes, P.I. Electrospinning polyaniline-contained gelatin nanofibers for tissue engineering applications. *Biomaterials* **2006**, *27*, 2705–2715. [CrossRef] [PubMed]

46. He, Q.; Shi, J.; Zhu, M.; Chen, Y.; Chen, F. The three-stage in vitro degradation behavior of mesoporous silica in simulated body fluid. *Microporous Mesoporous Mater.* **2010**, *131*, 314–320. [CrossRef]

47. Sun, L.; Zheng, C.; Webster, T.J. Self-assembled peptide nanomaterials for biomedical applications: Promises and pitfalls. *Int. J. Nanomed.* **2017**, 73–86. [CrossRef] [PubMed]

48. Wan, S.; Borland, S.; Richardson, S.M.; Merry, C.L.R.; Saiani, A.; Gough, J.E. Self-assembling peptide hydrogel for intervertebral disc tissue engineering. *Acta Biomater.* **2016**, *46*, 29–40. [CrossRef] [PubMed]

49. Kogiso, M.; Ohnishi, S.; Yase, K.; Masuda, M.; Shimizu, T. Dicarboxylic oligopeptide bolaamphiphiles: Proton-triggered self-assembly of microtubes with loose solid surfaces. *Langmuir* **1998**, *14*, 4978–4986. [CrossRef]

50. Menzenski, M.Z.; Banerjee, I.A. Self-assembly of supramolecular nanostructures from phenylalanine derived bolaamphiphiles. *New J. Chem.* **2007**, *31*, 1674. [CrossRef]

51. Cote, Y.; Fu, I.W.; Dobson, E.T.; Goldberger, J.E.; Nguyen, H.D.; Shen, J.K. Mechanism of the pH-controlled self-assembly of nanofibers from peptide amphiphiles. *J. Phys. Chem. C* **2014**, *118*, 16272–16278. [CrossRef] [PubMed]

52. Matsui, H.; Gologan, B. Crystalline glycylglycine bolaamphiphile tubules and their pH-sensitive structural transformation. *J. Phys. Chem. B* **2000**, *104*, 3383–3386. [CrossRef]

53. Loo, Y.; Zhang, S.; Hauser, C.A.E. From short peptides to nanofibers to macromolecular assemblies in biomedicine. *Biotechnol. Adv.* **2012**, *30*, 593–603. [CrossRef] [PubMed]

54. Cuvier, A.-S.; Berton, J.; Stevens, C.V.; Fadda, G.C.; Babonneau, F.; Van Bogaert, I.N.A.; Soetaert, W.; Pehau-Arnaudet, G.; Baccile, N. pH-triggered formation of nanoribbons from yeast-derived glycolipid biosurfactants. *Soft Matter* **2014**, *10*, 3950–3959. [CrossRef] [PubMed]

55. Macdonald, P.R.; Lustig, A.; Steinmetz, M.O.; Kammerer, R.A. Laminin chain assembly is regulated by specific coiled-coil interactions. *J. Struct. Biol.* **2010**, *170*, 398–405. [CrossRef] [PubMed]

56. Hussain, S.-A.; Carafoli, F.; Hohenester, E. Determinants of laminin polymerization revealed by the structure of the α5 chain amino-terminal region. *EMBO Rep.* **2011**, *12*, 276–282. [CrossRef] [PubMed]

57. De Graaf, J.; Amons, R.; Möller, W. The primary structure of artemin from Artemia cysts. *Eur. J. Biochem.* **1990**, *193*, 737–741. [CrossRef] [PubMed]

58. Hassani, L.; Sajedi, R.H. Effect of artemin on structural transition of lactoglobulin. *Spectrochim. Acta A Mol. Biomol. Spectrosc.* **2013**, *105*, 24–28. [CrossRef] [PubMed]

59. Liu, G.Y.; Agarwal, R.; Ko, K.R.; Ruthven, M.; Sarhan, H.T.; Frampton, J.P. Templated assembly of collagen fibers directs cell growth in 2D and 3D. *Sci. Rep.* **2017**, *7*, 9628. [CrossRef] [PubMed]

60. Huang, K.; Wan, M. Self-assembled polyaniline nanostructures with photoisomerization function. *Chem. Mater.* **2002**, *14*, 3486–3492. [CrossRef]

61. Kong, J.; Yu, S. Fourier transform infrared spectroscopic analysis of protein secondary structures. *Acta Biochim. Biophys. Sin. (Shanghai)* **2007**, *39*, 549–559. [CrossRef] [PubMed]

62. Almlén, A.; Vandenbussche, G.; Linderholm, B.; Haegerstrand-Björkman, M.; Johansson, J.; Curstedt, T. Alterations of the C-terminal end do not affect in vitro or in vivo activity of surfactant protein C analogs. *Biochim. Biophys. Acta. Biomembr.* **2012**, *1818*, 27–32. [CrossRef] [PubMed]

63. Langham, A.A.; Waring, A.J.; Kaznessis, Y.N. Comparison of interactions between beta-hairpin decapeptides and SDS/DPC micelles from experimental and simulation data. *BMC Biochem.* **2007**, *8*, 11. [CrossRef] [PubMed]

64. Baker, B.R.; Garrell, R.L. g-Factor analysis of protein secondary structure in solutions and thin films. *Faraday Discuss.* **2004**, *126*, 209. [CrossRef] [PubMed]

65. Wen, X.; Wang, Y.; Guo, Z.; Meng, H.; Huang, J.; Zhang, L.; Zhao, B.; Zhao, Q.; Zheng, Y.; Peng, J. Cauda equina-derived extracellular matrix for fabrication of nanostructured hybrid scaffolds applied to neural tissue engineering. *Tissue Eng. Part A* **2015**, *21*, 1095–1105. [CrossRef] [PubMed]

66. Vass, E.; Hollósi, M.; Besson, F.; Buchet, R. Vibrational spectroscopic detection of beta- and gamma-turns in synthetic and natural peptides and proteins. *Chem. Rev.* **2003**, *103*, 1917–1954. [CrossRef] [PubMed]

67. Pud, A.A.; Nikolayeva, O.A.; Vretik, L.O.; Noskov, Y.V.; Ogurtsov, N.A.; Kruglyak, O.S.; Fedorenko, E.A. New nanocomposites of polystyrene with polyaniline doped with lauryl sulfuric acid. *Nanoscale Res. Lett.* **2017**, *12*, 493. [CrossRef] [PubMed]

68. Arruebo, M.; Valladares, M.; Gonzalez-Fernandez, A. Antibody conjugated nanoparticles for biomedical applications. *J. Nanomaterials* **2009**, *2009*, 37. [CrossRef]

69. Gröschel, A.H.; Müller, A.H.E. Self-assembly concepts for multicompartment nanostructures. *Nanoscale* **2015**, *7*, 11841–11876. [CrossRef] [PubMed]

70. Heim, A.J.; Matthews, W.G.; Koob, T.J. Determination of the elastic modulus of native collagen fibrils via radial indentation. *Appl. Phys. Lett.* **2006**, *89*. [CrossRef]

71. Muiznieks, L.D.; Keeley, F.W. Molecular assembly and mechanical properties of the extracellular matrix: A fibrous protein perspective. *Biochim. Biophys. Acta* **2012**, *1832*, 866–875. [CrossRef] [PubMed]

72. Ding, J.; Li, X.; Wang, X.; Zhang, J.; Yu, D.; Qiu, B. Fabrication of vertical array CNTs/polyaniline composite membranes by microwave-assisted in situ polymerization. *Nanoscale Res. Lett.* **2015**, *10*, 493. [CrossRef] [PubMed]

73. Wei, Y.; Jang, G.W.; Hsueh, K.F.; Scherr, E.M.; MacDiarmid, A.G.; Epstein, A.J. Thermal transitions and mechanical properties of films of chemically prepared polyaniline. *Polymer (Guildf)* **1992**, *33*, 314–322. [CrossRef]

74. Xu, C.; Yang, F.; Wang, S.; Ramakrishna, S. In vitro study of human vascular endothelial cell function on materials with various surface roughness. *J. Biomed. Mater. Res. A* **2004**, *71*, 154–161. [CrossRef] [PubMed]

75. Chen, B.-Q.; Kankala, R.K.; Chen, A.-Z.; Yang, D.-Z.; Cheng, X.-X.; Jiang, N.-N.; Zhu, K.; Wang, S.-B. Investigation of silk fibroin nanoparticle-decorated poly(l-lactic acid) composite scaffolds for osteoblast growth and differentiation. *Int. J. Nanomed.* **2017**, *12*, 1877–1890. [CrossRef] [PubMed]

76. Nguyen, K.T.; West, J.L. Photopolymerizable hydrogels for tissue engineering applications. *Biomaterials* **2002**, *23*, 4307–4314. [CrossRef]

77. Sheikh, Z.; Najeeb, S.; Khurshid, Z.; Verma, V.; Rashid, H.; Glogauer, M. Biodegradable materials for bone repair and tissue engineering applications. *Materials (Basel)* **2015**, *8*, 5744–5794. [CrossRef] [PubMed]

78. Serbo, J.V.; Gerecht, S. Vascular tissue engineering: biodegradable scaffold platforms to promote angiogenesis. *Stem Cell Res. Ther.* **2013**, *4*, 8. [CrossRef] [PubMed]

79. Ikejima, T.; Inoue, Y. Crystallization behavior and environmental biodegradability of the blend films of poly(3-hydroxybutyric acid) with chitin and chitosan. *Carbohydr. Polym.* **2000**, *41*, 351–356. [CrossRef]

80. Bober, P.; Humpolíček, P.; Pacherník, J.; Stejskal, J.; Lindfors, T. Conducting polyaniline based cell culture substrate for embryonic stem cells and embryoid bodies. *RSC Adv.* **2015**, *5*, 50328–50335. [CrossRef]

81. Neal, R.A.; Lenz, S.M.; Wang, T.; Abebayehu, D.; Brooks, B.P.C.; Ogle, R.C.; Botchwey, E.A. Laminin- and basement membranepolycaprolactone blend nanofibers as a scaffold for regenerative medicine. *Nanomater. Environ.* **2014**, *2*, 1–12. [CrossRef] [PubMed]

82. Wong, L.E.; Gibson, M.E.; Arnold, H.M.; Pepinsky, B.; Frank, E. Artemin promotes functional long-distance axonal regeneration to the brainstem after dorsal root crush. *Proc. Natl. Acad. Sci. USA* **2015**, *112*, 6170–6175. [CrossRef] [PubMed]

83. Önger, M.E.; Delibaş, B.; Türkmen, A.P.; Erener, E.; Altunkaynak, B.Z.; Kaplan, S. The role of growth factors in nerve regeneration. *Drug Discov. Ther.* **2016**, *3*, 285–291. [CrossRef] [PubMed]

84. Shin, Y.J.; Kim, S.H.; Yang, D.H.; Kwon, H.; Shin, J.S. Amperometric glucose biosensor by means of electrostatic layer-by-layer adsorption onto polyaniline-coated polyester films. *J. Ind. Eng. Chem.* **2010**, *16*, 380–384. [CrossRef]

85. Ku, S.H.; Lee, S.H.; Park, C.B. Synergic effects of nanofiber alignment and electroactivity on myoblast differentiation. *Biomaterials* **2012**, *33*, 6098–6104. [CrossRef] [PubMed]

86. Gajendran, P.; Saraswathi, R. Polyaniline-carbon nanotube composites. *Pure Appl. Chem.* **2008**, *80*, 2377–2395. [CrossRef]

PERMISSIONS

LIST OF CONTRIBUTORS

Ioana Chiulan, Adriana Nicoleta Frone and Denis Mihaela Panaitescu
Polymer Department, National Institute for R&D in Chemistry and Petrochemistry ICECHIM, 202 Splaiul Independentei, 060021 Bucharest, Romania

Calin Brandabur
Symme3D and LTHD Corporation SRL, 300425 Timisoara, Romania

Marcelie Priscila de Oliveira Rosso and Karina Torres Pomini
Department of Biological Sciences (Anatomy), Bauru School of Dentistry, University of São Paulo (USP), Alameda Dr. Octávio Pinheiro Brisola 9-75, Vila Nova Cidade Universitária, Bauru, São Paulo CEP 17012-901, Brazil

Rogério Leone Buchaim
Department of Biological Sciences (Anatomy), Bauru School of Dentistry, University of São Paulo (USP), Alameda Dr. Octávio Pinheiro Brisola 9-75, Vila Nova Cidade Universitária, Bauru, São Paulo CEP 17012-901, Brazil
Medical School, Discipline of Human Morphophysiology, University of Marilia (UNIMAR), Av. Higino Muzi Filho, 1001 Campus Universitário, Jardim Araxa, Marília, São Paulo CEP 17525-902, Brazil

Natália Kawano and Gabriela Furlanette
Medical School, Discipline of Human Morphophysiology, University of Marilia (UNIMAR), Av. Higino Muzi Filho, 1001 Campus Universitário, Jardim Araxa, Marília, São Paulo CEP 17525-902, Brazil

Daniela Vieira Buchaim
Medical School, Discipline of Human Morphophysiology, University of Marilia (UNIMAR), Av. Higino Muzi Filho, 1001 Campus Universitário, Jardim Araxa, Marília, São Paulo CEP 17525-902, Brazil
Medical School, Discipline of Neuroanatomy, University Center of Adamantina (UNIFAI), Rua Nove de Julho, 730, Centro, Adamantina, São Paulo CEP 17800-000, Brazil

Mulugeta Gizaw, Jeffrey Thompson, Addison Faglie and Shih-Feng Chou
Department of Mechanical Engineering, College of Engineering, The University of Texas at Tyler, Tyler, TX 75799, USA

Shih-Yu Lee
School of Nursing, College of Nursing and Health Sciences, The University of Texas at Tyler, Tyler, TX 75799, USA

Pierre Neuenschwander
Department of Cellular and Molecular Biology, The University of Texas Health Science Center at Tyler, Tyler, TX 75708, USA

Krzysztof Wrzesinski and Stephen J. Fey
Tissue Culture Engineering Laboratory, Department of Biochemistry and Molecular Biology, University of Southern Denmark, 5230 Odense, Denmark
CelVivo IVS, 5491 Blommenslyst, Denmark

Nora Freyer, Selina Greuel, Fanny Knöspel, Florian Gerstmann, Lisa Storch and Katrin Zeilinger
Berlin-Brandenburg Center for Regenerative Therapies (BCRT), Charité–Universitätsmedizin Berlin, 13353 Berlin, Germany

Georg Damm and Daniel Seehofer
Department of Hepatobiliary Surgery and Visceral Transplantation, University of Leipzig, 04103 Leipzig, Germany

Jennifer Foster Harris and Rashi Iyer
Los Alamos National Laboratory, Los Alamos, NM 87545, USA

Frank Schubert
StemCell Systems GmbH, Berlin 12101, Germany

Catherine Jauregui and Parthasarathy Madurantakam
Philips Institute, School of Dentistry, Virginia Commonwealth University, Richmond, VA 23298, USA
Department of General Practice, School of Dentistry, Virginia Commonwealth University, Richmond, VA 23298, USA

Suyog Yoganarasimha
Department of Biomedical Engineering, School of Engineering, Virginia Commonwealth University, Richmond, VA 23284, USA

Benjamin A. Minden-Birkenmaier and Gary L. Bowlin
Department of Biomedical Engineering, University of Memphis, 3806 Norriswood Ave., Memphis, TN 38152, USA

Zhenzhen Xu, Andrew Koo, Sifon Ndon, Henry Hsia and Biraja C. Dash
Department of Surgery (Plastic), Yale School of Medicine, New Haven, CT 06510, USA

Lawrence Lin
Department of Public Health Studies, Johns Hopkins University, Baltimore, MD 21218, USA

Francois Berthiaume
Department of Biomedical Engineering, Rutgers University, The State University New Jersey, Piscataway, NJ 08901, USA

Alan Dardik
Department of Surgery (Vascular), Yale School of Medicine, New Haven, CT 06510, USA

Dominik Egger and Cornelia Kasper
Department of Biotechnology, University of Natural Resources and Life Sciences, Muthgasse 18, 1190 Vienna, Austria

Carla Tripisciano and Viktoria Weber
Christian Doppler Laboratory for Innovative Therapy Approaches in Sepsis, Danube University Krems, Dr.-Karl-Dorrek-Straße 30, 3500 Krems, Austria

Massimo Dominici
Division of Oncology, Department of Medical and Surgical Sciences for Children & Adults, University-Hospital of Modena and Reggio Emilia, Via Università 4, 41121 Modena, Italy
Technopole of Mirandola TPM, 41037 Mirandola, Modena, Italy

Ioanna Christou, Panagiotis Mallis, Efstathios Michalopoulos, Theofanis Chatzistamatiou and Catherine Stavropoulos-Giokas
Hellenic Cord Blood Bank, Biomedical Research Foundation Academy of Athens, 4 Soranou Ephessiou Street, 115 27 Athens, Greece

George Mermelekas, Jerome Zoidakis and Antonia Vlahou
Biotechnology division, Biomedical Research Foundation Academy of Athens, 4 Soranou Ephessiou Street, 115 27 Athens, Greece

Andrew M. Smith, Harrison T. Pajovich and Ipsita A. Banerjee
Department of Chemistry, Fordham University, 441 East Fordham Road, Bronx, New York, NY 10458, USA

Index